Plasmid Biology

Edited by

Patricia Marques

Delve Publishing
www.delvepublishing.com

Plasmid Biology

Patricia Marques

Delve Publishing
2010 Winston Park Drive,
2nd Floor
Oakville, ON L6H 5R7
Canada
www.delvepublishing.com
Tel: 001-289-291-7705
001-905-616-2116
Fax: 001-289-291-7601
Email: orders@arclereducation.com

ISBN: 978-1-77361-172-3 (Hardcover)

DECLARATION

Some content or chapters in this book are open access copyright free published research work, which is published under Creative Commons License and are indicated with the citation. We are thankful to the publishers and authors of the content and chapters as without them this book wouldn't have been possible.

ABOUT THE EDITOR

Patricia Marques

Patricia obtained her PhD form University College Dublin in 2010. Her interests are on Microbiology and Parasitology. She is currently working as a Postdoc at University of Maryland Baltimore, USA on Chlamydia infections

TABLE OF CONTENTS

SECTION IV: SPECIFIC PLASMID SYSTEMS

SECTION V: VIRULENCE AND ANTIBIOTIC RESISTANCE

SECTION VI: PLASMID ECOLOGY AND EVOLUTION

SECTION VII: PLASMIDS AS GENETIC TOOLS

SECTION VIII: PLASMID CURING SYSTEMS

CONTRIBUTORS

Patricia Marques

Carsten Kuenne
Institute of Medical Microbiology, Justus-Liebig University, Giessen, Germany,

Sonja Voget
Goettingen Genomics Laboratory, Institute for Microbiology and Genetics, Georg-August

Jordan Pischimarov
Institute of Medical Microbiology, Justus-Liebig University, Giessen, Germany,

Sebastian Oehm
[3]Bioinformatics Resource Facility, Center for Biotechnology, Bielefeld University, Bielefeld, Germany

Alexander Goesmann[3]
Bioinformatics Resource Facility, Center for Biotechnology, Bielefeld University, Bielefeld, Germany

Rolf Daniel
Goettingen Genomics Laboratory, Institute for Microbiology and Genetics, Georg-August

Torsten Hain
Institute of Medical Microbiology, Justus-Liebig University, Giessen, Germany,

Trinad Chakraborty
Institute of Medical Microbiology, Justus-Liebig University, Giessen, Germany, University Goettingen, Goettingen, Germany,

Marco Constante
EMBL/CRG Systems Biology Research Unit, Centre for Genomic Regulation (CRG) and UPF, Barcelona, Spain

Raik Gru nberg,
EMBL/CRG Systems Biology Research Unit, Centre for Genomic Regulation (CRG) and UPF, Barcelona, Spain

Mark Isalan
EMBL/CRG Systems Biology Research Unit, Centre for Genomic Regulation (CRG) and UPF, Barcelona, Spain

Mukta M. Watve
Indian Institute of Science Education and Research, Pune, Maharashtra, India

Neelesh Dahanukar
Indian Institute of Science Education and Research, Pune, Maharashtra, India

Milind G. Watve
Indian Institute of Science Education and Research, Pune, Maharashtra, India

Chen Chris Gong
Department of Theory and Bio-Systems, Max Planck Institute of Colloids and Interfaces, Potsdam, Germany,
Institute of Physics and Astronomy, University of Potsdam, Potsdam, Germany,

Stefan Klumpp
Department of Theory and Bio-Systems, Max Planck Institute of Colloids and Interfaces, Potsdam, Germany,
Institute for Nonlinear Dynamics, Georg August University of Go¨ttingen, Go¨ttingen, Germany

Uli Klümper
Department of Environmental Engineering, Technical University of Denmark, Kongens Lyngby, Denmark

Ariadni Droumpali
Department of Environmental Engineering, Technical University of Denmark, Kongens Lyngby, Denmark

Arnaud Dechesne
Department of Environmental Engineering, Technical University of Denmark, Kongens Lyngby, Denmark

Barth F. Smets
Department of Environmental Engineering, Technical University of Denmark, Kongens Lyngby, Denmark

Barbara E. Funnell
Department of Molecular Genetics, University of Toronto, Toronto, ON, Canada

David O. Wood
Department of Microbiology and Immunology, Laboratory of Molecular Biology, University of South Alabama College of Medicine, Mobile, Alabama, United States of America,

Andria Hines
[1]Department of Microbiology and Immunology, Laboratory of Molecular Biology, University of South Alabama College of Medicine, Mobile, Alabama, United States of America,

Aimee M. Tucker
Department of Microbiology and Immunology, Laboratory of Molecular Biology, University of South Alabama College of Medicine, Mobile, Alabama, United States of America,

Andrew Woodard
Department of Microbiology and Immunology, Laboratory of Molecular Biology, University of South Alabama College of Medicine, Mobile, Alabama, United States of America,

Lonnie O. Driskell
Department of Microbiology and Immunology, Laboratory of Molecular Biology, University of South Alabama College of Medicine, Mobile, Alabama, United States of America,

Nicole Y. Burkhardt
Department of Entomology, University of Minnesota, St. Paul, Minnesota, United States of America

Timothy J. Kurtti
Department of Entomology, University of Minnesota, St. Paul, Minnesota, United States of America

Gerald D. Baldridge
Department of Entomology, University of Minnesota, St. Paul, Minnesota, United States of America

Ulrike G. Munderloh
Department of Entomology, University of Minnesota, St. Paul, Minnesota, United States of America

Shin Okazaki
Graduate School of Agriculture, Tokyo University of Agriculture and Technology, Tokyo, Japan,

Rujirek Noisangiam
School of Biotechnology, Institute of Agricultural Technology, Suranaree University of Technology, Nakhon Ratchasima, Thailand,

Takashi Okubo
Graduate School of Life Science, Tohoku University, Sendai, Japan,

Takakazu Kaneko
Faculty of Life Sciences, Kyoto Sangyo University, Motoyama, Kamigamo, Kita-Ku, Kyoto 603-8555, Japan,

Kenshiro Oshima
Center of Omics and Bioinformatics, Graduate School of Frontier Sciences, University of Tokyo, Tokyo, Japan,

Masahira Hattori
Center of Omics and Bioinformatics, Graduate School of Frontier Sciences, University of Tokyo, Tokyo, Japan,

Kamonluck Teamtisong
School of Biotechnology, Institute of Agricultural Technology, Suranaree University of Technology, Nakhon Ratchasima, Thailand,

Pongpan Songwattana
School of Biotechnology, Institute of Agricultural Technology, Suranaree University of Technology, Nakhon Ratchasima, Thailand,

Panlada Tittabutr
School of Biotechnology, Institute of Agricultural Technology, Suranaree University of Technology, Nakhon Ratchasima, Thailand,

Nantakorn Boonkerd
School of Biotechnology, Institute of Agricultural Technology, Suranaree University of Technology, Nakhon Ratchasima, Thailand,

Kazuhiko Saeki
Department of Biological Sciences, Faculty of Science, Kyousei Science Center for Life and Nature, Nara Women's University, Kitauoya Nishimachi, Nara 630-8506, Japan,

Shusei Sato
Graduate School of Life Science, Tohoku University, Sendai, Japan,

Toshiki Uchiumi
Graduate School of Science and Engineering, Kagoshima University, Kagoshima, Japan

Kiwamu Minamisawa
Graduate School of Life Science, Tohoku University, Sendai, Japan,

Neung Teaumroong
School of Biotechnology, Institute of Agricultural Technology, Suranaree University of Technology, Nakhon Ratchasima, Thailand,

Bryon A. Nicholson1
Department of Veterinary Microbiology and Preventive Medicine, College of Veterinary Medicine, Iowa State University, Ames, Iowa, United States of America,

Aaron C. West
Department of Chemistry, College of Liberal Arts and Sciences, Iowa State University, Ames, Iowa, United States of America,

Paul Mangiamele1
Department of Veterinary Microbiology and Preventive Medicine, College of Veterinary Medicine, Iowa State University, Ames, Iowa, United States of America,

Nicolle Barbieri
Department of Veterinary Microbiology and Preventive Medicine, College of Veterinary Medicine, Iowa State University, Ames, Iowa, United States of America,

Yvonne Wannemuehler
Department of Veterinary Microbiology and Preventive Medicine, College of Veterinary Medicine, Iowa State University, Ames, Iowa, United States of America,

Lisa K. Nolan
Department of Veterinary Microbiology and Preventive Medicine, College of Veterinary Medicine, Iowa State University, Ames, Iowa, United States of America,

Catherine M. Logue
Department of Veterinary Microbiology and Preventive Medicine, College of Veterinary Medicine, Iowa State University, Ames, Iowa, United States of America,

Ganwu Li
Department of Veterinary Diagnostic and Production Animal Medicine, College of Veterinary Medicine, 1802 University Blvd, Iowa State University, Ames, Iowa, 50011, United States of America,
Department of Veterinary Preventive Medicine, College of Veterinary Medicine, Nanjing Agricultural University, Nanjing, Jiangsu, 210095, P. R. China

Masaki Shintani
Department of Applied Chemistry and Biochemical Engineering, Graduate School of Engineering, Shizuoka University, Shizuoka, Japan
Department of Bioscience, Graduate School of Science and Technology, Shizuoka University, Shizuoka, Japan

Zoe K. Sanchez
Department of Applied Chemistry and Biochemical Engineering, Graduate School of Engineering, Shizuoka University, Shizuoka, Japan

Kazuhide Kimbara
Department of Applied Chemistry and Biochemical Engineering, Graduate School of Engineering, Shizuoka University, Shizuoka, Japan

Ajamete Kaykas
Howard Hughes Medical Institute, Department of Pharmacology, and Center for Developmental Biology, University of Washington School of Medicine, Seattle, WA 98195, USA

Randall T Moon
Howard Hughes Medical Institute, Department of Pharmacology, and Center for Developmental Biology, University of Washington School of Medicine, Seattle, WA 98195, USA

Xiankai Liu
State Key Laboratory of Pathogen and Biosecurity, Beijing Institute of Biotechnology, Beijing, China

Dongshu Wang
State Key Laboratory of Pathogen and Biosecurity, Beijing Institute of Biotechnology, Beijing, China

Huagui Wang
State Key Laboratory of Pathogen and Biosecurity, Beijing Institute of Biotechnology, Beijing, China

Erling Feng
State Key Laboratory of Pathogen and Biosecurity, Beijing Institute of Biotechnology, Beijing, China

Li Zhu
State Key Laboratory of Pathogen and Biosecurity, Beijing Institute of Biotechnology, Beijing, China

Hengliang Wang
State Key Laboratory of Pathogen and Biosecurity, Beijing Institute of Biotechnology, Beijing, China

PREFACE

Plasmids are DNA molecules present in almost all bacteria, together with the Chromosomal DNA. They can play an important roles in the adaptation of the bacteria to different environments.

Plasmids, which vary widely in size from a few thousand to hundreds of thousands of base pairs, are most often circular molecules of double-stranded DNA. However, some bacteria have linear plasmids, and some plasmids, can be single-stranded DNA. The number of plasmid copies varies among plasmids, and some bacteria can even harbor more than one type of plasmid. Thus, a cell can harbor two or more different types of plasmids, with hundreds of copies of some plasmid types and only one or a few copies of other types.

This book starts with the Introduction of concepts in Plasmid biology, replication, function, regulation and incompatibility. Followed by examples of Plasmid (Section 1) in different bacteria (Listeria, Lactoccocus, Borrelia and Rickettsia) and in eukaryotes.

The next chapters deal with the different processes in plasmid biology, replication of copy number (Section 2), maintenance and inheritance (Section 3), systems specific for specific plasmids (symbiosis, flagella, and linear plasmids) (Section 4), virulence and antibiotic resistance in Salmonella and E.coli (Section 5), and ecology and evolution (Section 6).

Section 7 gives examples of Plasmids has genetic tools for interfering RNA, centromere plasmid, plasmid has a shuttle for Chlamydia genetic transformation and seamless insert-plasmid.

Section 8, shows examples of plasmid curing systems using plasmid incompatibility and interference.

This book is a general of overview of Plasmids and their Biology.

Editor

INTRODUCTION

Patricia Marques

In addition to the chromosome, bacterial cells often contain plasmids. These DNA molecules are found in essentially all types of bacteria and play an important role in bacterial adaptation and evolution. They also are important tools in molecular biology.

Plasmids, which vary widely in size from a few thousand to hundreds of thousands of base pairs (a size comparable to that of the bacterial chromosome), are most often circular molecules of double-stranded DNA. However, some bacteria have linear plasmids, and some plasmids, most often those from gram-positive bacteria, can accumulate single-stranded DNA. The number of plasmid copies varies among plasmids, and some bacteria can even harbor more than one type of plasmid. Thus, a cell can harbor two or more different types of plasmids, with hundreds of copies of some plasmid types and only one or a few copies of other types.

Like chromosomes, plasmids encode proteins and RNA molecules and replicate as the cell grows, and the replicated copies are usually distributed into each daughter cell when the cell divides. However, unlike chromosomes, plasmids generally do not encode functions essential to bacterial growth. Instead, they provide gene products that can benefit the bacterium under certain circumstances but are not always essential. In cases where a plasmid is almost as big and carries essential genes, which one is the chromosome and which is a plasmid? Probably a better criterion for whether a DNA is a plasmid or the chromosome is the nature of it s origin of replication. In all known cases, one of the large DNAs has a typical bacterial origin of replication with an *oriC* site and closely linked *dnaA, dnaN,* and *gyrA* genes, among others, while the other DNA has a typical plasmid origin with *repABC-like* genes more characteristic of plasmids.

NAMING PLASMIDS

Before methods for physical detection of plasmids became available, plasmids made their presence known by conferring phenotypes on the cells harboring them. Consequently, many plasmids were named after the genes they carry. For example, R-factor plasmids contain genes for resistance to several antibiotics (hence the name R for resistance). These were the first plasmids discovered, when *Shigella* and *Escherichia coli* strains resistant to a number of antibiotics were isolated from the fecal flora of patients in Japan in the late 1950s. The ColEl plasmid, from which many of the cloning vectors were derived, carries a gene for the protein colicin El, a bacteriocin that kills bacteria that do not carry this plasmid. The Toi plasmid contains genes for the degradation of toluene, and the Ti plasmid carries genes for tumor initiation in plants. This system of nomenclature has led to some confusion, because plasmids carry various genes besides the ones for which they were originally named. Also, many of these plasmids have been altered beyond recognition to make plasmid cloning vectors and for other purposes.

To avoid further confusion, the naming of plasmids is now standarized. Plasmids are given a number and letter names much like bacterial strains. A small "p" for plasmid, precedes capital letters that describe the plasmid or sometimes give the initials of the person or persons who isolated or constructed it. These letters are often followed by numbers to identify the particular construct. When the plasmid is further altered, a different number is assigned to indicate the change. For example, plasmid pBR322 was constructed by Bolivar and Rodriguez from the ColE1 plasmid and is derivative number 322 of the plasmids they constructed. pBR325 is pBR322 with a chloramphenicol resistance gene inserted.

FUNCTIONS ENCODED BY PLASMIDS

Depending on their size, plasmids can encode a few or hundreds of different proteins. Plasmid genes usually give bacteria a selective advantage under only some conditions. Gene products encoded by plasmids include enzymes for the utilization of unusual carbon sources such as toluene, resistance to substances such as heavy metals and antibiotics, synthesis of antibiotics, and synthesis of toxins and proteins that allow the successful infection of higher organisms. Table 1. lists a few naturally occurring plasmids and some traits they encode, as well as the host in which they were originally found.

Table 1: Some naturally occurring plasmids and the traits they carry

Plasmid	Trait	Original source
ColE1	Bacteriocin which kills *E.coli*	*E. coli*
Toi	Degradation of toluene and benzoic acid	*Pseudomonas putida*
TI	Tumor initiation in plants	*Agrobacterium tumefaciens*
pJP4	2,4-D(dichlorophenoxyacetic acid) degradation	*Alcaligenes eutrophus*
pSym	Nodulation on roots of legume plants	*Rhizobium meliloti*
SCP1	Antibiotic methylenomycinbiosynthesis	*Streptomyces coelicolor*
RK2	Resistance to ampicillin, tetracycline, and kanamycin	*Klebsiella aerogenes*

PLASMID STRUCTURE

Most plasmids are circular with no free ends, although a few known plasmids are linear (Casjens et al., 2012; Casjens et al., 2017; Chaconas

& Kobryn, 2010). In a circular plasmid, all of the nucleotides in each strand are joined to another nucleotide on each side by covalent bonds to form continuous strands that are wrapped around each other. Such DNAs are said to be covalently closed circular. This structure prevents the strands from separating, and there are no ends to rotate, so that the plasmid can be supercoiled. In a DNA that is supercoiled, the two strands are wrapped around each other more or less often than once in about 10.5 bp, as predicted from the Watson-Crick double-helical structure of DNA. If they are wrapped around each other more often than once every 10.5 bp, the DNA is positively supercoiled; if they are wrapped around each other less often, the DNA is negatively supercoiled. Like the chromosome, covalently closed circular plasmid DNAs are usually negatively supercoiled. Because DNA is stiff, the negative supercoiling introduces stress, and this stress is partially relieved by the plasmid wrapping up on itself. This makes the plasmid more compact, so that it runs more quickly in an agarose gel. In the cell, the DNA wraps around proteins, which relieves some of the stress. The remaining stress facilitates some reactions involving the plasmid, such as separation of the two DNA strands for replication or transcription.

PURIFYING PLASMIDS (Engebrecht, Heilig, & Brent, 2001)

The structure of plasmids can be used to purify them away from the chromosomal and other DNA in the cell. Many purification procedures are based on the relatively small size of most plasmids. The plasmids often do not precipitate at the salt concentrations at which the chromosome precipitates. Therefore, if extracts of cells are treated at high salt concentrations, the chromosome often precipitates and can be removed by centrifugation while the much smaller plasmids stay in the supernatant.

Other purification steps take advantage of the fact that most plasmids are covalently closed circular and supercoiled. This procedure is based on the fact that covale ntly closed circular DNAs bind less EtBr than do linear or nicked circular DNAs. EtBr intercalates (inserts itself) between DNA bases, pushing the bases apart and rotating the two strands of the DNA around each other. If the two strands of the DNA are not free to rotate, as in a covalently closed circular plasmid, the binding of EtBr eventually introduces positive supercoils and increases the stress on the DNA until no more EtBr can bind. EtBr bound to DNA makes it *less* dense in salt solutions made with heavy atoms such as cesium chloride (CsCI). As a consequence, if the DNA is mixed with a solution of CsCI and EtBr and centrifuged to establish a gradient of CsCI concentration, the covalently closed circular plasmid DNAs will band lower, at a position where the solution is more dense.

Large, low-copy-number plasmids are much more difficult to detect. Most methods for detecting large plasmids involve separating them from the chromosome directly by electrophoresis on agarose gels. The cells are often broken open directly on the agarose gel to avoid breaking the large plasmid DNA. The plasmid, because of its unique size, makes a sharp band on the gel, distinct from that due to chromosomal DNA, which is usually broken and so gives a more diffuse band. Also, methods such as pulsed-field gel electrophoresis have been devised to allow the separation of long pieces of DNA based on size. These methods depend on periodic changes in the direction of the electric field. The molecules attempt to reorient themselves each time the field shifts, and the longer molecules will take longer to reorient than the shorter ones and so move more slowly on the gel. Such methods have allowed the separation of DNA molecules hundreds of thousands of base pairs long and the detection of very large plasmids.

PROPERTIES OF PLASMIDS

To exist free of the chromosome, plasmids must have the ability to replicate independently. DNA molecules that can replicate autonomously in the cell are called replicons. Plasmids, phage DNA, and the chromosomes are all replicons, at least in some types of cells. To be a replicon in a particular type of cell, a DNA molecule must have at least one origin of replication, or *ori* site, where replication begins. In addition, the cell must contain the proteins that enable replication to initiate at this site. Plasmids encode only a few of the proteins required for their own replication. In fact, many encode only one of the proteins needed for initiation at the *ori* site. All of the other required proteins, including DNA polymerases, ligases, primases,helicases, and so on, are borrowed from the host.

Each type of plasmid replicates by one of two general mechanisms, which is determined along with other properties by its *ori* region. The plasmid replication origin is often named *oriV* for *ori* vegetative, to distinguish it from *oriT,* which is the site at which DNA transfer initiates in plasmid conjugation. Most of the evidence for the mechanisms described below came from observations of replicating plasmid DNA with the electron microscope.

Theta Replication (Lilly & Camps, 2015)

Some plasmids begin replication by opening the two strands of DNA at the *ori* region, creating a structure that looks like the Greek letter q, theta replication. In this process, an RNA primer begins replication, which can proceed in one or both directions around the plasmid. In the first case, a single replication fork moves around the molecule until it returns to the origin and then the two daughter DNA separate. In the other case (bidirectional replication), two replication forks move out from the *ori* region, one in either direction, and replication is complete (and the two daughter DNAs separate) when the two forks meet somewhere on the other side of the molecule.

The theta mechanism is the most common form of DNA replication, especially in gram-negative bacteria. It is used not only by most plasmids, including ColEI, RK2, F, and Pl, but also by the chromosome in most bacteria.

Rolling-Circle Replication (Khan, 2000)

Another type of replication is called rolling-circle replication because it was first discovered in a type of phage where the template circle seems to roll like a signet ring which has been dipped in ink and rolled on paper, making a copy of its design on the paper. Plasmids which replicate by this mechanism are called RC plasmids. This type of plasmid is widespread, being found in both gramnegative and gram-positive eubacteria as well as archaea. In an RC plasmid, the replication occurs in two stages. In the first stage, the double-stranded circular plasmid DNA replicates to form another double-stranded circular DNA and a single-stranded circular DNA. This stage is analogous to the replication of the DNA of some single-stranded DNA phages and to DNA transfer during plasmid conjugation. In the second stage, the complementary strand is synthesized on the single-stranded DNA to make another double-stranded DNA. First the Rep protein recognizes and binds to a palindromic sequence which contains the double-strand origin (DSO) on the DNA. Binding of the Rep protein to this sequence might allow the formation of a cruciform structure by base pairing between the inverted repeated sequences in the cruciform. Once the cruciform forms, the Rep protein can make a nick in the sequence. It is important for the models that the Rep protein is also known to function as a dimer, at least in some plasmids. After the Rep protein has made a break in the DSO sequence, it remains covalently attached to the phosphate at the 5' end of the DNA at the nick through a tyrosine in one copy of the Rep protein in the dimer. The DNA polymerase III (the replicative polymerase) uses the free 3' hydroxyl end at the break as a primer to replicate around the circle, displacing one of the strands. It may use a host helicase to help separate the strands, or the Rep protein itself may have the helicase activity, depending on the plasmid. Once the circle is complete, the 5' phosphate is transferred from the tyrosine on the Rep protein to the 3' hydroxyl on the other end of the displaced strand, producing a single-stranded circular DNA. This process is called a phosphotransferase reaction and requires little energy. The same reaction is used to reform a circular plasmid after conjugational transfer.

It is less certain what happens to the newly formed double-stranded DNA when the DNA polymerase III has made its way all around the circle and gets back to the site of the DSO. Why does it not just keep going, making a longer molecule with individual genomes linked head to tail, in a structure called a concatemer? Such strutures are created when some phage DNAs replicate by a rolling-circle mechanism. One idea is that the DNA polymerase III does keep going past the DSO for a short distance, creating another doublestranded DSO. The other copy of Rep protein in the dimer may then nick the newly created DSO, transferring the 5' end to itself as above. This might inactivate the Rep protein, releasing it with a short oligonucleotide attached. Other reactions, probably involving host DNA ligase, then cause the nick to be resealed, resulting in a circular double-stranded DNA molecule.

The displaced circular single-stranded DNA now replicates by a completely different mechanism using only host-encoded proteins. The RNA polymerase first makes a primer at a different origin, the single-strand origin (SSO), and this RNA then primes replication around the circle by DNA polymerase III. However, the RNA polymerase does not make this primer until the singlestranded DNA is completely displaced during the first stage of replication. This delay is accomplished by locating the SSO immediately counterclockwise of the DSO, so that the SSO does not appear in the displaced DNA until the displacement of the single-stranded DNA is almost complete. After the entire complementary strand has been synthesized, the 5' exonuclease activity of DNA polymerase I removes the RNA primer, replacing it with DNA, and host DNA ligase joins the ends to make another double-stranded plasmid. The net result is two new double-stranded plasmids synthesized from the original double-stranded plasmid.

In order for the complementary strand of the displaced single-stranded DNA to be synthesized, the RNA polymerase of the host cell must recognize the SSO on the DNA. In some hosts the SSO is not well recognized and single-stranded DNA accumulates. For this reason, some RC plasmids were originally called single-stranded DNA plasmids, although we now know that this is nor their normal state. Broad-host-range RC plasmids presumably have an SSO that is recognized by the RNA polymerases of a wide variety of hosts, which allows them to make the complementary strand of the displaced singlestranded DNA in a variety of hosts.

The Rep protein is used only once for every round of plasmid DNA replication and is destroyed after the round is completed. This allows the replication of the plasmid to be controlled by the amount of Rep protein in the cell and keeps the total number of plasmid molecules in the cell within narrow limits dictated by the copy number.

Replication of Linear plasmids (Hinnebusch & Tilly, 1993)

Linear DNAs face a problem with replicating the lagging strand, the strand that ends with a 5' phosphate, all the way to the end of the DNA. This has been called the "primer problem" because DNA polymerases cannot initiate the synthesis of a new strand of DNA. They can only add nucleotides to a preexisting primer, and, in a linear DNA, there is no upstream primer on this strand from which to grow. Different linear DNAs solve the primer problem in different ways. Some linear plasmids have hairpin ends, with the 5' and 3' ends joined to each other. These plasmids replicate from an internal origin of replication to form dimeric circles, composed of two plasmids joined head to tail to form a circle. These dimeric circles are then resolved into individual linear plasmid DNAs by enzymes called prototelomerases, which cut the dimeric circles into the two linear plasmid DNAs and join the ends to each other to form the closed hairpins. Other linear plasmids seem to replicate their ends by a completely different mechanism. They have extensive inverted repeated sequences at their ends and a terminal protein attached to their 5' ends. It is not clear how these plasmids replicate, but they might use some sort of slippage mechanism, using the terminal protein as either a recombinase or a primer or both. It is interesting that bacteria with linear plasmids also often have linear chromosomes and the two DNAs replicate by similar mechanisms.

FUNCTIONS OF THE *ORI* REGION

In most plasmids, the genes for proteins required for replication are located very close to the *ori* sequences at which they act. Thus, only a very small region surrounding the plasmid *ori* site is required for replication. As a consequence, the plasmid still replicates if most of its DNA is removed, provided that the *ori* region remains and the plasmid DNA is still circular. Smaller plasmids are easier to use as cloning vectors, so often the only part of the orginal plasmid that remains in a cloning vector is the *ori* region.

In addition, the genes in the *ori* region often determine many other properties of the plasmid. Therefore, any DNA molecule with the *ori* region of a particular plasmid will have most of the characteristics of that plasmid.

PLASMIDS HOST RANGE

The host range of a plasmid includes all the types of bacteria in which the plasmid can replicate; it is usually determined by the *ori* region. Some plasmids, such as those with *ori* regions of the ColE1 plasmid type, including p8R322, pET, and pUC, have a narrow host range. These plasmids replicate only in *E. coli* and some other closely related bacteria such as *Salmonella* and *Klebsiella* species. In contrast, plasmids with a broad host range include the RK2 and RSFl010 plasmids, as well as the rolling-circle plasmids from gram-positive bacteria. Plasmids with the *ori* region of RK2 can replicate in most types of gram-negative bacteria, and RSFl010-derived plasmids even replicate in some types of gram-positive bacteria. Many of the plasmids isolated from gram-positive bacteria also have quite broad host ranges. For example, pUBl10, which was first isolated from the gram-positive *Staphylococcusaureus,* also replicates in many other gram-positive bacteria, including *Bacillus subtilis.* However, most plasmids isolated from gram-negative bacteria do not replicate in gram-positive bacteria and vice versa, which reflects the evolutionary divergence of these groups. Broad-host-range plasmids must encode all of their own proteins required for initiation of replication, and so they do not have to depend on the host cell for any of these functions. They also must be able to express these genes in many types of bacteria. Apparently, the promoters and ribosome initiation sites for the replication genes of broad-host-range plasmids have evolved so that they can be recognized in a wide variety of bacteria.

REGULATION OF COPY NUMBER (del Solar & Espinosa, 2000; Friehs, 2004)

Another characteristic of plasmids that is determined mostly by their *ori* region is their copy number, or the average number of that particular plasmid per cell. More precisely, we define the copy number as the number of copies of the plasmid in a newborn cell immediately after cell division. All plasmids must regulate their replication; otherwise they would fill up the cell and become too great a burden for the host, or their replication would not keep up with the cell replication and they would be progressively lost during cell division. Some plasmids, such as pIJ101 of *Streptomyces coelicolor,* replicate enough times to populate the cell with hundreds of copies. However, others, such as the F plasmid of *E. coli,* replicate only once or a few times during the cell cycle. Table 2. lists the copy numbers of these and other plasmids.

Table 2. Copy number of some plasmids	
Plasmid	Aproximate copy number
F	1
P1 prophage	1
RK2	4-7 (E. coli)
pBR322	16
pUC18	~30-50
pIJ101	40-300

The regulation mechanisms used by plasmids with higher copy numbers often differ greatly from those used by plasmids with lower copy numbers. Plasmids that have high copy numbers, such as ColE1 plasmids, need only have a mechanism that inhibits the initiation of plasmid replication when the number of plasmids in the cell reaches a certain level. Consequently, these molecules are called relaxed plasmids. By contrast, low-copy-number plasmids such as F must replicate only once or very few times during each cell cycle and so must have a righter mechanism for regulating their replication. Hence, these are called stringent plasmids. Much more is understood about the regulation of replication of relaxed plasmids than about the regulation of replication of stringent plasmids.

The regulation of relaxed plasmids falls into three general categories. Some plasmids are regulated by an antisense RNA, sometimes called a counter transcribed (ctRNA) because it is transcribed from the same region of the plasmid but from the opposite strand of an RNA essential for plasmid replication. Because the ctRNA is transcribed from the opposite strand from the essential RNA, it is able to hybridize to the essential RNA and inhibit its function. The ctRNA of these plasmids is often assisted in its inhibitory role by a protein. Other plasmids are regulated by a ctRNA alone, which inhibits the translation of a protein essential for replication. Yet others are regulated by a protein alone, which binds to repeated sequences in the plasmid DNA called iterons, there by inhibiting plasmid replication.

Incompatibility (Novick & Hoppensteadt, 1978)

Another function of plasmids that is controlled by the *ori* region is incompatibility. Incompatibility refers to the ability of two plasmids to coexist stably in the same cell. Many bacteria, as they are isolated from nature, contain more than one type of plasmid. These plasmid types coexist stably in the bacterial cell and remain there even after many cell generations. In fact, bacterial cells containing multiple types of plasmids are not cured of each plasmid any more frequently than if the other plasmids were not there.

However, sometimes two different types of plasmid cannot coexist stably in the same cell. In this case, one or the other plasmid is lost as the cells multiply; this loss is more frequent than would occur if the plasmids were not occupying the cells along with the other plasmid. If two plasmids cannot coexist stably, they are said to be members of the same incompatibility (Inc) group. If two plasmids can coexist stably, they belong to different Inc groups. There are a number of ways in which plasmids can be incompatible. One way is if they can each regulate the other's replication. Another way is if they share the same partitioning *(par)* functions, which are often closely associated with replication control in the *ori* region. There may be hundreds of different Inc groups, and plasmids are usually classified by the Inc group to which they belong. For example, RP4, also called RK2, is an IncP (incompatibility group P) plasmid. In contrast, RSF101 is an lncQ plasmid: it can therefore be stably maintained with RP4 because it belongs to a different Inc group but cannot be stably maintained with another lncQ plasmid.

MECHANISMS TO PREVENT CURING OF PLASMIDS

Cells that have lost a plasmid during cell division are said to be cured of the plasmid. Several mechanisms prevent curing, including plasmid addiction systems, site-specific recombinases that resolve multimers, and partitioning systems. The last two are reviewed below.

Resolution of multimeric plasmids (Crozat, Fournes, Cornet, Hallet, & Rousseau, 2014)

The possibility that a cell will lose a plasmid during cell division is increased if the plasmids form dimers or higher multimers during replication. A dimer consists of two individual copies of the plasmid molecules linked head to tail to form a larger circle, and a multimer links more than two such monomers. Such dimers and multimers probably occur as a result of recombination between monomers. Recombination between two monomers forms a dimer, and subsequent recombination can form higher and higher multimers. Also, rolling-circle replication of RC plasmids can form multimers if termination after each round of replication is not efficient. Multimers may replicate more efficiently than monomers, perhaps because they have more than one origin of replication, so they tend to accumulate if the plasmid has no effective way to remove them. The formation of multimers creates a particular problem when the plasmid attempts to segregate into the daughter cells on cell division. One reason is that multimers lower the effective copy number. Each multimer segregates into the daughter cells as a single plasmid, and if all of the plasmid is taken up in one large multimer, it can segregate into only one daughter cell. Therefore, multimers greatly increase the chance of a plasmid being lost during cell division.

To avoid this problem, many plasmids have site-specific recombination systems that resolve multimers as soon as they form. These systems can be either chromosomally encoded or encoded by the plasmid itself. A site-specific recombination system promotes recombination between specific sites on the plasmid if the same site occurs more than once in the molecule, as it would in a dimer or multimer. This recombination has the effect of resolving multimers into separate monomeric plasmid molecules.

Partitioning (Gerdes, Moller-Jensen, & Bugge Jensen, 2000; Mann, 1985; Williams & Thomas, 1992)

The most effective mechanism that plasmids have to avoid being lost from dividing cells is their set of partitioning systems. These systems ensure that at least one copy of the plasmid segregates into each daughter cell during cell division. The functions involved in these systems are called Par functions, and in many ways they are analogous to the Par functions involved in chromosome segregation.

CONSTRUCTING A PLASMID CLONING VECTOR

A cloning vector is an autonomously replicating DNA (replicon) into which other DNAs can be inserted. Any DNA inserted into the cloning vector replicates passively with the vector so that many copies (clones) of the original piece of DNA can be obtained.

Plasmids offer many advantages as cloning vectors, and many plasmids have been engineered to serve as plasmid cloning vectors. They generally do not kill the host cell and are relatively easy to purify to obtain the cloned DNA. They also can be made relatively small because few plasmid-encoded functions are required for their replication. In fact, in one of the first cloning experiments, a frog gene was cloned into plasmid pSC101 (Cohen, Chang, Boyer, & Helling, 1973).

Finding the Plasmid *ori* Region

The first step in making a plasmid cloning vector from a wild-type plasmid we have isolated is to find the *ori* region of the plasmid. The *ori* region is usually responsible for most of the properties of the plasmid, including replication, copy number control, and partitioning.

Recombinant DNA techniques are particularly suited for locating and studying these regions. The plasmid is cut into several pieces with a restriction endonuclease and the pieces are ligated (joined) to another piece of DNA that has a selectable marker, such as resistance to ampicillin. For the experiment to work, the second piece of DNA cannot have a functional origin of replication. The ligated mixture is then used to transform bacteria, and the antibiotic-resistant transformants are selected by plating the mixture on agar plates containing growth medium and the antibiotic. The only DNA molecules able to replicate and also confer antibiotic resistance on the cells are hybrids with both the *ori* region of the plasmid and the piece of DNA with the antibiotic resistance gene. Therefore, only cells harboring these hybrid molecules can grow on the antibiotic-containing medium. We can further localize the *ori* region by cutting the fragment known to contain it into even smaller pieces and repeating the process above until the smallest piece from the plasmid that can function as an origin has been identified.

INTRODUCING SELECTABLE GENES

The next step is to introduce genes into the cloning vector that confer selectable properties on cells containing the plasmid. This allows us to select cells which contain the plasmid vector and to move it from one host cell to another. Genes that confer resistance to an antibiotic make convenient selectable genes. Cells containing the plasmid can then be selected by plating on medium containing the antibiotic, and only cells containing the plasmid will multiply to form a colony. Antibiotic resistance genes are often taken from transposons and other plasmids. Some antib iotic resistance genes which have been introduced into cloning vectors arc the chloramphenicol resistance gene, Cam^r, of transposon Tn9; the tetracycline resistance gene, Tet^r, of plasmid pSCl0l; the ampicillin resistance gene, Amp^r, of transposon Tn3; and the kanamycin resistance gene, Kan^r, of transposon Tn5. The antibiotic resistance gene that is chosen depends on the uses to which the cloning vector will be put. Some antibiotic resistance genes such as the Tet^r gene from pSCl01 are expressed only in some types of bacteria closely related to *E. coli,* while others such as Kan^r from Tn5 are expressed in most gram-negative bacteria.

Introducing Unique Restriction Sites

Since many applications of plasmid cloning vectors require that clones be introduced into restriction sites, it is necessary that our cloning vector have some restriction sites that are unique. If a site is unique, the cognate restriction endonuclease cuts the vector at only that one site when it is used to cut the plasmid. We can then clone pieces of foreign DNA into the unique site, and the cloning vector will remain intact. There are many tricks for removing the extra sites. We can change the sequence at some of the sites by site-specific mutagenesis so that they are no longer recognized by the restriction endonuclease. Alternatively, we can make a partial digest of the plasmid so that it is cut at only one of the sites. The site at which it was cut can then be removed by filling in the overhang with DNA polymerase, if the restriction endonuclease leaves a 5' overhang, or by removing the overhang with a single-strand-specific DNase before religating.

Not only do we want to have unique sites for some restriction nucleases in our cloning vector, but also these sites should be located in the plasmid in a way which makes them most useful for cloning. In many plasmid cloning vectors, the unique sites are located in a selectable gene, so that insertion of a foreign piece of DNA in the site inactivates the selectable gene. Normally, during a cloning operation, only a small percentage of the cloning vectors pick up a foreign DNA insert. Ifthose that have picked up an insert no longer confer the selectable trait, for example resistance to an antibiotic, they can be more easily identified. Many cloning vectors also have the unique restriction sites for many different restriction endonucleases all grouped in one small region on the plasmid called a polyclonal or multiple-restriction site. This offers the convenience of choosing among a variety of restriction cndonucleases for cloning, and the cloned DNA is always inserted at the same site, independent of the restriction endonuclease used. Polyclonalsites can also be used for directional cloning. If the cloning vector is cut by two different incompatible restriction endonucleases with unique sites within the polyclonal site, the resulting overhangs cannot pair to recyclize the plasmid. The plasmid can recyclize only if it picks up a piece of foreign DNA. If the piece of DNA to be cloned has overhangs for the two different sites at its ends, it is usually cloned in only one orientation into the polyclonal site. The unique restriction sites can also be placed so that genes cloned into them will be expressed from promoters and translation initiation regions (TIRs) on the plasmid. These are called expression vectors and can be used to express foreign genes in *E. coli* and ocher convenient hosts. Such vectors can also be used to attach affinity tags to proteins to aid in their purification.

Examples of Plasmid Cloning Vectors

A number of plasmid cloning vectors have been engineered for special purposes. Almost all of these plasmids have at least some of the features mentioned above for a desirable cloning vector.

To reiterate:

- They are small, so that the plasmid can easily isolated and introduced into various bacteria.
- They have a relatively high copy number, so that the plasmid can be easily purified in sufficient quantities.
- They carry easily selectable traits, such as a gene con ferring resistance to an antibiotic, which can be used to select cells that contain the plasmid.
- They have one or a few sites for restriction endonucleases, which cut DNA and allow the insertion of foreign DNAs. Also, these sites usually occur in selectable genes to facilitate the detection (plasmids with foreign DNA inserts by insertional inactivation).

Many plasmid cloning vectors have other special properties that aid in particular experiments. For example, some contain the sequences recognized by phage packaging systems, so that they can be packaged into phage heads. Expression vectors can be used to make foreign proteins in bacteria. Mobilizable plasmids have mobilization sites and so can be transferred by conjugation to other cells. Some broad-host-range vectors have *ori* regions that allow them to replicate in many types of bacteria or even in organisms from different kingdoms. Shuttle vectors contain more than one type of replication origin and so can replicate in unrelated organisms.

Shuttle Vectors

Sometimes, an experiment requires that a plasmid cloning vector to be transferred from one organism into another. If the two organisms are not related, the same plasmid *ori* region is not likely to function in both organisms. Such applications require the use of shuttle vectors, so named because they can be used to "shuttle" genes between the two organisms. A shuttle vector has two origins of replication, one that functions in each organism. Shuttle vectors also must contain selectable genes that can be expressed in both organisms.

In most cases, one of the organisms in which the shuttle vector can replicate is *E. coli.* The genetic tests can be performed with the other organism, but the plasmid can be purified and otherwise manipulated by the refined methods developed for *E. coli.* Some shuttle vectors can replicate in gram-positive bacteria and *E. coli,* whereas others can *be* used in lower or *even* higher eukaryotes.

REFERENCES

1. Casjens, S. R., Gilcrease, E. B., Vujadinovic, M., Mongodin, E. F., Luft, B. J., Schutzer, S. E., Et Al. (2017). Plasmid Diversity And Phylogenetic Consistency In The Lyme Disease Agent Borrelia Burgdorferi. BMC Genomics, 18(1), 165-017-3553-5. Doi:10.1186/S12864-017-3553-5
2. Casjens, S. R., Mongodin, E. F., Qiu, W. G., Luft, B. J., Schutzer, S. E., Gilcrease, E. B., Et Al. (2012). Genome Stability Of Lyme Disease Spirochetes: Comparative Genomics Of Borrelia Burgdorferi Plasmids. Plos One, 7(3), E33280. Doi:10.1371/Journal.Pone.0033280
3. Chaconas, G., & Kobryn, K. (2010). Structure, Function, And Evolution Of Linear Replicons In Borrelia. Annual Review Of Microbiology, 64, 185-202. Doi:10.1146/Annurev.Micro.112408.134037
4. Cohen, S. N., Chang, A. C., Boyer, H. W., & Helling, R. B. (1973). Construction Of Biologically Functional Bacterial Plasmids In Vitro. Proceedings Of The National Academy Of Sciences Of The United States Of America, 70(11), 3240-3244.
5. Crozat, E., Fournes, F., Cornet, F., Hallet, B., & Rousseau, P. (2014). Resolution Of Multimeric Forms Of Circular Plasmids And Chromosomes. Microbiology Spectrum, 2(5), 10.1128/Microbiolspec.PLAS-0025-2014. Doi:10.1128/Microbiolspec.PLAS-0025-2014
6. Del Solar, G., & Espinosa, M. (2000). Plasmid Copy Number Control: An Ever-Growing Story. Molecular Microbiology, 37(3), 492-500. Doi:Mmi2005

7. Engebrecht, J., Heilig, J. S., & Brent, R. (2001). Preparation Of Bacterial Plasmid DNA. Current Protocols In Immunology, Chapter 10, Unit 10.3. Doi:10.1002/0471142735.Im1003s27
8. Friehs, K. (2004). Plasmid Copy Number And Plasmid Stability. Advances In Biochemical Engineering/Biotechnology, 86, 47-82.
9. Gerdes, K., Moller-Jensen, J., & Bugge Jensen, R. (2000). Plasmid And Chromosome Partitioning: Surprises From Phylogeny. Molecular Microbiology, 37(3), 455-466. Doi:Mmi1975
10. Hinnebusch, J., & Tilly, K. (1993). Linear Plasmids And Chromosomes In Bacteria. Molecular Microbiology, 10(5), 917-922.
11. Khan, S. A. (2000). Plasmid Rolling-Circle Replication: Recent Developments. Molecular Microbiology, 37(3), 477-484. Doi:Mmi2001
12. Lilly, J., & Camps, M. (2015). Mechanisms Of Theta Plasmid Replication. Microbiology Spectrum, 3(1), PLAS-0029-2014. Doi:10.1128/Microbiolspec.PLAS-0029-2014
13. Mann, N. H. (1985). Plasmid Partitioning In Escherichia Coli. Microbiological Sciences, 2(10), 299-302.
14. Novick, R. P., & Hoppensteadt, F. C. (1978). On Plasmid Incompatibility. Plasmid, 1(4), 421-434.
15. Williams, D. R., & Thomas, C. M. (1992). Active Partitioning Of Bacterial Plasmids. Journal Of General Microbiology, 138(1), 1-16. Doi:10.1099/00221287-138-1-1

SECTION I:
EXAMPLES OF PLASMIDS

CHAPTER

1

COMPARATIVE ANALYSIS OF PLASMIDS IN THE GENUS LISTERIA

Carsten Kuenne[1], Sonja Voget[2], Jordan Pischimarov[1], Sebastian Oehm[3], Alexander Goesmann[3], Rolf Daniel[2], Torsten Hain[1], Trinad Chakraborty[1]

[1]Institute of Medical Microbiology, Justus-Liebig University, Giessen, Germany,

[2]Goettingen Genomics Laboratory, Institute for Microbiology and Genetics, Georg-August University Goettingen, Goettingen, Germany,

[3]Bioinformatics Resource Facility, Center for Biotechnology, Bielefeld University, Bielefeld, Germany

ABSTRACT

Background

We sequenced four plasmids of the genus *Listeria*, including two novel plasmids from *L. monocytogenes* serotype 1/2c and 7 strains as well as one from the species *L. grayi*. A comparative analysis in conjunction with 10 published *Listeria* plasmids revealed a common evolutionary background.

Principal Findings

All analysed plasmids share a common replicon-type related to theta-replicating plasmid pAMbeta1. Nonetheless plasmids could be broadly divided into two distinct groups based on replicon diversity and the genetic content of the respective plasmid groups. *Listeria* plasmids are characterized by the presence of a large number of diverse mobile genetic elements and a commonly occurring translesion DNA polymerase both of which have probably contributed to the evolution of these plasmids. We detected small non-coding RNAs on some plasmids that were homologous to those present on the chromosome of *L. monocytogenes* EGD-e. Multiple genes involved in heavy metal resistance (cadmium, copper, arsenite) as well as multidrug efflux (MDR, SMR, MATE) were detected on all listerial plasmids. These factors promote bacterial growth and survival in the environment and may have been acquired as a result of selective pressure due to the use of disinfectants in food processing environments. MDR efflux pumps have also recently been shown to promote transport of cyclic diadenosine monophosphate (c-di-AMP) as a secreted molecule able to trigger a cytosolic host immune response following infection.

Conclusions

The comparative analysis of 14 plasmids of genus *Listeria* implied the existence of a common ancestor. Ubiquitously-occurring MDR genes on plasmids and their role in listerial infection now deserve further attention.

INTRODUCTION

The genus *Listeria* comprises six non-pathogenic species *L. marthii, L. innocua, L. welshimeri, L. seeligeri, L. grayi,* and *L. rocourtiae*, and two species with pathogenic potential viz. *L. monocytogenes* and *L. ivanovii*, which can cause both human and animal infections [1]–[7]. Since *L. monocytogenes* exhibits resistance towards heat and cold stress it can proliferate in food processing environments [8] and thus colonize dairy and meat products which have caused several outbreaks as well as sporadic cases of listeriosis [9]. Three serotypes of the species *L. monocytogenes* viz. 1/2a, 1/2b and 4b are responsible for 95% of all human clinical infections [10].

Extrachromosomal DNA was previously detected in many *L. monocytogenes* wildtype strains with rates of isolation ranging from 0–79% with an overall average of 30% [11]–[15]. Two studies which examined 173 [14] and 322 [16] isolates of *L. monocytogenes* respectively found an overrepresentation of plasmids in strains from food and the environment in comparison to those obtained from clinical cases. It was shown that plasmids were found more frequently (75%) in recurrent *L. monocytogenes* strains sampled from food/processing environments than in those from sporadic strains (35%) [17]. Plasmids were also more frequently associated with serogroup 1 strains compared to those from serogroup 4. It was determined that 95% of the *L. monocytogenes* plasmid-positive strains were resistant towards cadmium versus only 13% of the plasmid-negative strains [14] and that the *cadAC* genes were similar to those previously detected in *Staphylococcus aureus* [18]. Only in two cases antibiotic resistance of *L. monocytogenes* could be traced to a plasmid [19], [20]. Plasmids were also previously described for *L. innocua* [2] and *L. grayi* [12]. Furthermore plasmids pAMbeta1 and pIP501 of *Streptococcus* could be transferred to *L. monocytogenes* where they stably replicated underlining the broad-host range of these replicons and their potential for horizontal transfer between strains of these genera [12], [21]. The contribution of plasmids to the infectious process has not been examined and their evolutionary history is not yet well understood apart from homologies to other gram-positive plasmids such as with plasmid pXO_2 from *Bacillus anthracis* which is required for the pathogenic properties of this species [22]–[24].

RESULTS AND DISCUSSION

Listeria plasmids overview

We determined the entire sequences of plasmids from *L. monocytogenes* 7 UG1 SLCC2482, *L. monocytogenes* 1/2c UG1 SLCC2372, *L. monocytogenes* 1/2b UG1 SLCC2755 and *L. grayi*subspecies *grayi* UG1 DSM20601. For comparative analysis we included sequences of the plasmids pLM33 of *L. monocytogenes* Lm1, pCT100 of *L. monocytogenes* DRDC8, pLM80 of *L. monocytogenes* H7858, pLM5578 of *L. monocytogenes* 08-5578 and pLI100 of *L. innocua*Clip11262 which were downloaded from the NCBI website as well as further five gapped

L. monocytogenes plasmid sequences from strains FSL J1.194, FSL R2-503, FSL N1-017, FSL F2-515 and J0161 which were retrieved from the Broad Institute (http://www.broad.mit.edu) database. All plasmid contigs were remapped and reannotated.

It should be noted that plasmids sequenced by the Broad Institute were found to contain a large number of SNPs leading to truncated genes. A recent study assumed that higher selective pressure was responsible for this phenomenon [23], but other studies with this data have also indicated truncations in many essential housekeeping genes on the chromosomes of these strains [25] indicating an alternative explanation i.e. sequencing errors. Indeed the average sizes of coding sequences from *L. monocytogenes* plasmids sequenced in this study vary between 260 and 264 while those obtained from the Broad study range from 131 to 245 amino acids, respectively (Table 1).

Table 1: General features of 14 plasmids of genus *Listeria*

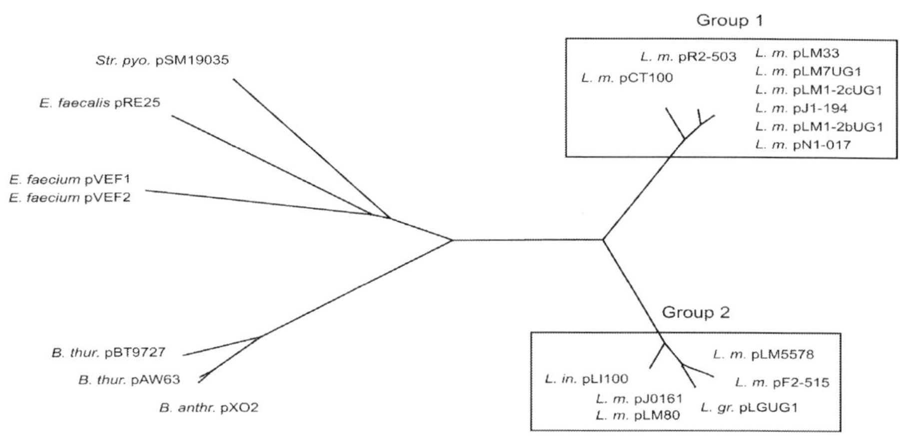

Since it is not feasible to include locus tags for up to 14 homologs of a gene we decided to only include a gene name or annotation in the text which can be used in conjunction with a homology matrix to identify the respective loci. Furthermore a public Geco server [26] including all plasmids of this study as well as their reference annotations was set up

Phylogenetic Clustering Based on Replication Protein

All plasmids contained a similar minimal replicon consisting of three genes necessary for replication (*repA*) and partitioning (*repB, repC*) as well as the origin of replication [27]–[32] and a gene encoding a DNA polymerase IV. The replicon is a member of the pAMbeta1 family of theta-replicating plasmids and its proteins are most closely related to plasmids from the genera *Bacillus* (pXO_2, pAW63, pBT9727), *Streptococcus* (pSM19035) and *Enterococcus* (pRE25, pVEF1, pVEF2) with protein identities ranging from between 36–56%. An exception to this homology was found to be RepC which shows no sequence similarity but a similar location, size and orientation as its putative functional homologs in the plasmids of the aforementioned genera. The genes encoding *repB*/C are overlapping indicating an operon. Interestingly, the translesion DNA polymerase has previously been suggested to stimulate spontaneous deletions during DNA repair [33], [34] and could thus contribute to variation and adaptation of both plasmid and host genes when present.

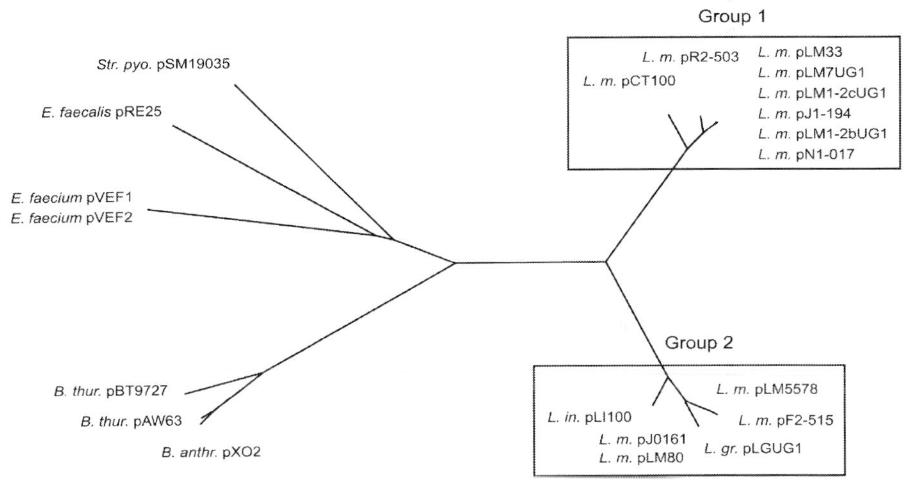

Figure 1: Phylogenetic tree of the replication initiation protein.

To examine the relationship of the plasmid backbones we joined all fragments of the RepA proteins and used CLUSTALW [35] to create a phylogenetic tree (Figure 1). This methodology clearly confirms the relatedness of these plasmids to those present in other gram-positive strains and separated the plasmids of genus *Listeria* into two distinct phylogenetic groups consisting of *L. monocytogenes* serotypes 7, 1/2c, 1/2b, 4b FSL N1-017 and serogroup 4 DRDC8 in one cluster and *L. monocytogenes* serotype 1/2a, 4b H7858, *L. innocua* and *L. grayi*in the other. The plasmid of *L. monocytogenes* 1/2b strain F2-515 was an exception to this since it clustered with group 2 confirming observations from a previous study [23]. Plasmid sizes generally correlate with the clustering based on the replication initiation protein ranging from 32–57 kb in group 1 and 77–83 kb in group 2 (Table 1), again with the exception of F2-515 which belongs to group 2 but has a size similar to plasmids of group 1.

Phylogenetic tree based on the replication initiation protein RepA of plasmids of genus *Listeria* and related genera. In the case of pF2-515 and pN1-017 multiple proteins had to be merged due to premature stopcodons. Clustalw [35] was used to create the multiple sequence alignment which was visualized using Dendroscope [60]. The clustering of the replication initiation proteins shows a clear separation into two phylogenetic groups.

Comparative genetic analysis

The replicon-based distinction is mirrored by the gene content to some extent, which indicates a highly similar set of genes for most plasmids of group 1, with a more heterogenous distribution for group 2 (Figure 2). Genes were considered homologs if BlastP found a sequence identity of at least 30% covering more than 80% of both proteins. Apart from the replicon no other feature is conserved overall, but all plasmids contain a cadmium resistance operon (*cadA/C*) [18] with the exception of pLGUG1 which lacks *cadC* and harbors a transposase at that relative position.

Whole-sequence alignment of 14 plasmids of genus *Listeria* as computed by Mauve [59]. Relevant loci are marked specifically. These include genes involved in heavy metal detoxification (copper, cadmium, arsenite), multidrug resistance (MDR: SMR, MATE), phage defense (R/M systems, Abi), oxidative stress response, an incomplete type IV secretion system (T4SS), a PemIK stable inheritance module, a Kdp-type potassium transport system and a sequence duplication. It should be noted that Mauve was not able to identify all homologies due to algorithmic limitations.

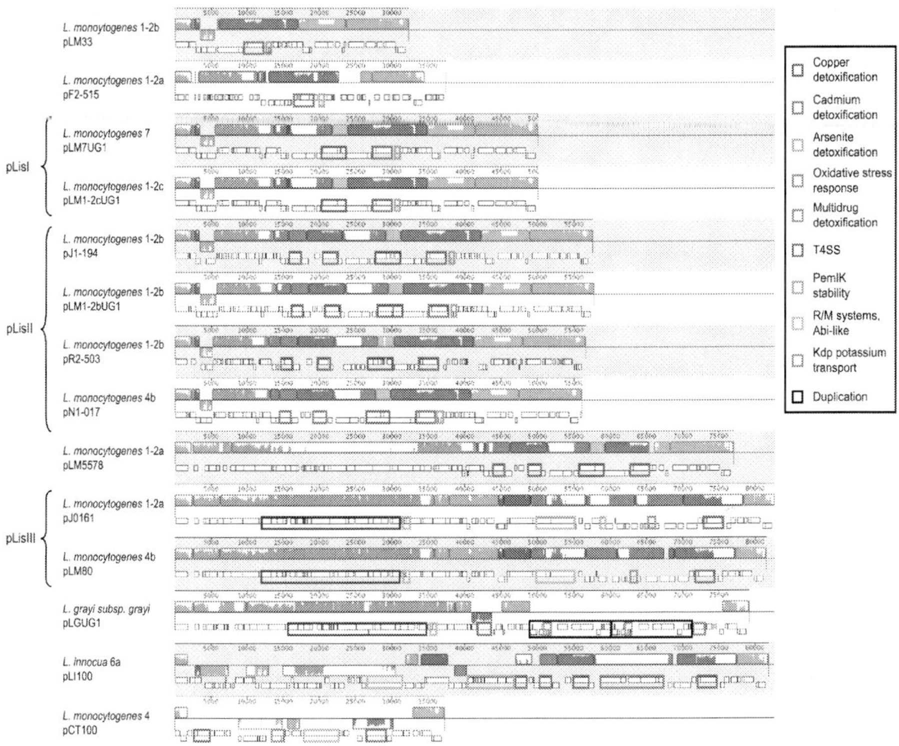

Figure 2: Complete alignment of plasmid sequences.

Apart from pCT100, plasmids of group 1 are closely related to each other and mainly differ by two putative indel events. Most of them can be grouped into two types which will be referred to as pLisI (pLM7UG1, pLM1-2cUG1) and pLisII (pJ1-194, pR2-503, pLM1-2bUG1, pN1-017). The smallest plasmid pLM33 was previously described to contain multiple transposases, remnants of *Listeria* phage A006, CRISPR associated protein Cas5 implied in phage defense and a Clp protease related to *Lactobacillus* which is involved in environmental stress response [23]. Despite clustering with group 2 according to its rep-protein, the sequence of plasmid pF2-515 shows a much higher homology to plasmids of group 1 which is not reflected by its gene-content due to a large number of false stop codons leading to a computed average protein length of only 131 amino acids. The indel between pLM33 and pLisI is 18 kb in size and contains multiple transposases, a copper-transporting P-type ATPase and a multicopper oxidase (MCO) implied in copper detoxification [36].

Either one or both copper resistance genes show putatively premature stopcodons in pLM7UG1, pJ1-194, pR2-503 and pN1-017. All plasmids excepting the smallest plasmids, pLM33, pF2-515 and pCT100 contained a protein with a FIC domain (filamentation induced by cyclic adenosine monophosphate) which is implicated in the disruption of cellular functions following transfer to the host cell cytoplasm during infection [37]. The plasmids of type pLisI and pLisII differ by 6 kb. This region encodes a transposase, a cadmium-transporting ATPase as well as an NADH peroxidase and a periplasmic component of an ABC-type glycine/betaine transport system closely related to *Aerococcus viridans* ATCC 11563 with an identity of 97% and 55% respectively. NADH peroxidases are described as being necessary for decomposition of hydrogen peroxide accumulated during aerobic growth [38]. This could play a role in intracellular survival against hydrogen peroxide stress [39] as well as in defense against disinfectants which are also known to induce general and oxidative stress responses [40]. All plasmids of group 1 apart from pCT100, and pLGUG1 of group 2, carry the sequence of the PemIK toxin/antitoxin stable maintenance system described for *Lactobacillus salivarius* UCC118 plasmid pSF118–20 [41]. In most cases one of the genes (pLM7UG1, pLM1-2cUG1, pJ1-194, pLM1-2bUG1, pR2-503) could not be identified by the gene prediction, which implies a decay of this functionality in those strains.

Only half of the sequence of pCT100 is shared with any other plasmid of the genus *Listeria*, including the replicon, a cadmium resistance system, a copper transporter and an insertion of 6 kb shared with pLI100 related to 12 kb plasmid pEW104 of *Lactococcus lactis subsp. cremoris* W10. Sequences present in this 6 kb fragment include two genes probably involved in replication and a single-gene type I restriction and modification (R/M) system called *Lla*GI which was shown to confer decreased bacteriophage sensitivity to its host [42]. The other half of pCT100 consists of multiple genes indicated in copper detoxification and a Na^{+}-driven multidrug efflux pump which belongs to the MATE family (multidrug and toxic compound extrusion) [43].

All plasmids of group 2 excepting pLGUG1 showed extensive homology to plasmids of group 1 further indicating a common ancestor. Plasmids pLM5578 and pLI100 share 76% and 45% nucleotide identity with pLisII respectively, with only 18% identity for the plasmids pJ0161 and pLM80. The latter two plasmids are closely related to each other and will be referred to as type pLisIII. Both contain a 40 kb region that is similar to plasmid pXO2 from *B. anthracis* which can also be found in pLGUG1 and to some extent in pLM5578 which is thought to include an incomplete type IV secretion system [22]. This system was shown to be insufficient for conjugation in pXO2 [31]. They also share a locus of 20 kb containing 24 genes including multiple transposases and two restriction modifications systems, one of them related to type III R/M system LlaFl of *Lactococcus lactis* [44]. This locus also harbors a gene encoding a triphenylmethane reductase described for the degradation of toxic synthetic dyes in *Citrobacter*[45] and a system of two genes *ebrAB*, which create a heterodimer channel involved in multidrug efflux in *Bacillus subtilis* [46]. The latter system belongs to the small multidrug resistance family (SMR) and is implied in resistance towards ethidium bromide and quarternary ammonium compounds [47] commonly found in disinfectants and could support persistence in food processing environments. Plasmid pLMJ0161 contains a specific insertion of a gene encoding an Abi-like protein implied in phage resistance [48]. The second half of pLGUG1 harbors a specific insertion spanning 24 kb which consists of a duplicated sequence comprising 17 genes most of them hypothetical. In addition this locus includes a Tn552-family transposase and a PemI/PemK post segregational killing system. Other pLGUG1 specific genes encode a MATE family multidrug efflux pump distantly related to the one found in pCT100. In pLI100 regions homologous to pLisII are interrupted by multiple pLI100-specific insertions. One of these consists of six genes related to potassium transport which span a region of 10 kb and are located on the same strand implying an operon. This Kdp-ATPase system [49] consists of a two-component signal transduction system (*kdpD/E*) and a potassium-transporting ATPase (*kdpA/B/C*). It is widely distributed among bacteria and archeae and plays a vital role in osmotic adaptation and pH regulation [50]. A homologue of this system exists in the chromosome of all fully sequenced strains of genus *Listeria* (data not shown) and contributes to growth during osmotic stress and low temperature in *L. monocytogenes* [51]. Another pLI100-specific insertion is an arsenite resistance operon related to integrative conjugative element ICESde3396 of *Streptococcus dysgalactiae subsp. equisimilis* strain NS3396 (EU142041) spanning 12 kb and consisting of seven genes which may contribute to the survival of *L. innocua* in the environment. Adjacent to this region a coenzyme A disulfide reductase gene was identified which is implied in oxidative stress response in *Borrelia burgdorferi* bb0728 [52].

In general, the plasmids of genus *Listeria* harbor a large and diverse number of mobile genetic elements. Between 6 and 24 genes per plasmid were annotated as transposase, resolvase, integrase, recombinase or invertase. This suggests that plasmids may act as an evolutionary sink for mobile genetic elements which may have shaped the diversity and evolution of plasmids in the genus *Listeria*.

Small non-coding RNA

Recently small non-coding RNAs (sRNAs) have become a focus of research because of their roles in bacterial regulatory mechanisms [53]. Interestingly we could identify multiple putative sRNAs on listerial plasmids. One class of RNA was already described for plasmids where they are predominantly implied in replication control, segregation and conjugation [54]. To identify additional sRNAs a software called sRNAdb (J. Pischimarov, unpublished), which employs BlastN, was used to find sequence similarities previously described in *L. monocytogenes*. Using a cutoff of 60% identity and 80% coverage, four putative sRNAs [55] were identified on plasmids of genus *Listeria*. These comprise two pairs of homologues being *rli28/rli50*and *rli44/rli46*. Homologous sequences to *rli28/rli50* could be identified in pLM80-like, pLI100 and pCT100 while *rli44/rli46* could be found in pLM33, pF2-515, pLisII, pLisIII and pLGUG1.

Conclusion

Here we report on the completion of four new plasmid sequences, including two novel plasmids from *L. monocytogenes* serotype 1/2c and 7 strains as well as one from the species *L. grayi*. In the comparative analysis presented here we compared sequences of 14 plasmids from three species using additional sequences either previously published or deposited in databases. We found that all plasmids share a common replicon-type related to theta-replicating plasmid pAMbeta1 [56] implying a common ancestor. Nonetheless a phylogenetic division must have occurred when considering the replication initiation protein. This division was mostly mirrored by the genetic content which showed clear distinctions between those groups apart from two atypical plasmids (pF2-515, pCT100). Based on regions of synteny, we are able to trace diversification and evolution driven by indels that account for the range of plasmid sizes detected. The presence of a commonly occurring translesion repair DNA polymerase on all plasmids suggests a mechanism by which genetic deletions are generated. Since plasmids of genus *Listeria* are related to *Bacillus*, *Enterococcus* and *Streptococcus* and were described to be transferable between some of these genera [12], [21], it is likely that exchange amongst these bacteria takes place in many different environmental niches e.g. gut and soil. Also, the unexpected detection of a large number of mobile genetics elements present on these plasmids imply that these could be involved in increasing genetic diversity or even altering gene expression both at chromosomal and episomal sites. Furthermore we found multiple independent systems involved in defense against phages (type I and III restriction systems, Abi-like) in group 2 implying a role for plasmids in the dissemination of these genes to ward off bacteriophage infection. The detection of small non-coding RNAs on a number of plasmids that were homologous to those present on the chromosome of *L. monocytogenes* EGD-e suggests that sRNAs might be transfered via plasmid conjugation.

The overrepresentation of plasmids in studies examining strains from food and the environment [14], [16] and in recurrent *L. monocytogenes* strains sampled from food/processing facilities [17]is an intriguing observation. However, the presence of multiple genes involved in heavy metal resistance (cadmium, copper, arsenite) as well as multidrug efflux (MDR, SMR, MATE) and oxidative stress response (peroxidase, reductase) on listerial plasmids could assist survival and their presence may have resulted from selective pressure due to the use of disinfectants in food processing environments. Finally, we note that MDR efflux pumps have recently been shown to promote cyclic diadenosine monophosphate (c-di-AMP) as a secreted molecule able to trigger the cytosolic host response following infection [57]. The implication for the presence of MDR genes on plasmids and their role in listerial infection now deserves further scrutiny.

MATERIALS AND METHODS

Public data sources

Contigs of five gapped *L. monocytogenes* plasmids were downloaded from the homepage of the Broad Institute (http://www.broad.mit.edu/annotation/genome/listeria_group) originating from strains FSL J1.194 (2.44), FSL R2-503 (2.52, 2.53, 2.54), FSL N1-017 (2.75, 2.76, 2.77), FSL F2-515 (2.1405, 2.1406, 2.1407, 2.1408, 2.1409, 2.1410, 2.1411, 2.1412, 2.1413, 2.1414, 2.1415) and J0161 (1.50, 1.51). The plasmids pLM33 of *L. monocytogenes* Lm1 (GU244485), pCT100 of *L. monocytogenes* DRDC8 (U15554), pLM80 of *L. monocytogenes* H7858 (AADR01000010, AADR01000058), pLM5578 of *L. monocytogenes* 08-5578 (CP001603) and pLI100 of *L. innocua* Clip11262 (AL592102) were downloaded from the GenBank database (http://www.ncbi.nlm.nih.gov/genbank/index.html).

Isolation and Sequencing

The remaining strains from *L. monocytogenes* 7 UG1 SLCC2482, *L. monocytogenes* 1/2c UG1 SLCC2372, *L. monocytogenes* 1/2b UG1 SLCC2755 and *L. grayi subspecies grayi* UG1 DSM20601 were isolated using Epicentre's MasterPure gram-positive DNA purification kit as recommended by the manufacturer. The DNA was sequenced on a 454 GS-FLX System to coverages between 16–57x. The resulting reads were assembled *de novo* with the 454 Newbler assembler and mapped vs. published plasmids to identify homologous contigs. PCR-based techniques were used to close the remaining gaps which were sequenced with Sanger ABI Big Dye technology. The sequencing was performed by Roche (Germany), Goettingen Genomics Laboratory (Goettingen, Germany) and Agowa (Berlin, Germany).

Bioinformatics

All contigs of gapped plasmids were scaffolded according to finished plasmids and joined to a consecutive sequence using the spacer "nnnnnttaattaattaannnnn" to prevent the gene prediction from crossing contig borders. All sequences were then reordered to a putative origin adjacent to the replication initiation gene (*repA*) as described for the homolog replicon of *B. anthr*acis plasmid pXO2 [29] and automatically annotated using the GenDB system [58]. The annotation was corrected based on a comparative syntheny analysis as offered by Geco [26]. In order to compute a phylogenetic tree Clustalw [35] was applied on replication initiation proteins using standard parameters. A multiple sequence alignment of the complete plasmid sequences was created with the Mauve software [59] using a progressive alignment including seed families to increase sensitivity. Mauve was not able to identify all homologies correctly with any combination of parameters. The chosen alignment is the optimal result considering false positives/negatives (data not shown).

REFERENCES

1. 1. Graves LM, Helsel LO, Steigerwalt AG, Morey RE, Daneshvar MI, et al. (2010) *Listeria marthii* sp. nov., isolated from the natural environment, Finger Lakes National Forest. Int J Syst Evol Microbiol 60: 1280–1288.LM GravesLO HelselAG SteigerwaltRE MoreyMI Daneshvar2010*Listeria marthii* sp. nov., isolated from the natural environment, Finger Lakes National Forest.Int J Syst Evol Microbiol6012801288
2. 2. Glaser P, Frangeul L, Buchrieser C, Rusniok C, Amend A, et al. (2001) Comparative genomics of *Listeria* species. Science 294: 849–852.P. GlaserL. FrangeulC. BuchrieserC. RusniokA. Amend2001Comparative genomics of *Listeria*species. Science294849852
3. 3. Hain T, Steinweg C, Kuenne CT, Billion A, Ghai R, et al. (2006) Whole-genome sequence of *Listeria welshimeri* reveals common steps in genome reduction with *Listeria innocua* as compared to *Listeria monocytogenes*. J Bacteriol 188: 7405–7415.T. HainC. SteinwegCT KuenneA. BillionR. Ghai2006Whole-genome sequence of *Listeria welshimeri* reveals common steps in genome reduction with *Listeria innocua* as compared to *Listeria monocytogenes*.J Bacteriol18874057415
4. 4. Steinweg C, Kuenne CT, Billion A, Mraheil MA, Domann E, et al. (2010) Complete genome sequence of *Listeria seeligeri*, a nonpathogenic member of the genus *Listeria*. J Bacteriol 192:

1473–1474.C. SteinwegCT KuenneA. BillionMA MraheilE. Domann2010Complete genome sequence of *Listeria seeligeri*, a nonpathogenicmemberofthegenus*Listeria*.JBacteriol19214731474

5. 5. Leclercq A, Clermont D, Bizet C, Grimont PA, Le Fleche-Mateos A, et al. (2009) *Listeria rocourtiae* sp. nov. Int J Syst Evol Microbiol. A. LeclercqD. ClermontC. BizetPA GrimontA. Le Fleche-Mateos2009*Listeria rocourtiae* sp. nov.Int J Syst Evol Microbiol
6. 6. Chakraborty T, Hain T, Domann E (2000) Genome organization and the evolution of the virulence gene locus in *Listeria* species. Int J Med Microbiol 290: 167–174.T. ChakrabortyT. HainE. Domann2000Genome organization and the evolution of the virulence gene locus in *Listeria* species.Int J Med Microbiol290167174
7. 7. Guillet C, Join-Lambert O, Le MA, Leclercq A, Mechai F, et al. (2010) Human listeriosis caused by *Listeria ivanovii*. Emerg Infect Dis 16: 136–138.C. GuilletO. Join-LambertMA LeA. LeclercqF. Mechai2010Human listeriosis caused by *Listeria ivanovii*.Emerg Infect Dis16136138
8. 8. Azizoglu RO, Osborne J, Wilson S, Kathariou S (2009) Role of growth temperature in freeze-thaw tolerance of *Listeria* spp. Appl Environ Microbiol 75: 5315–5320.RO AzizogluJ. OsborneS. WilsonS. Kathariou2009Role of growth temperature in freeze-thaw tolerance of *Listeria* spp.Appl Environ Microbiol7553155320
9. 9. McLauchlin J, Mitchell RT, Smerdon WJ, Jewell K (2004) *Listeria monocytogenes* and listeriosis: a review of hazard characterisation for use in microbiological risk assessment of foods. Int J Food Microbiol 92: 15–33.J. McLauchlinRT MitchellWJ SmerdonK. Jewell2004*Listeria monocytogenes* and listeriosis: a review of hazard characterisation for use in microbiological risk assessment of foods.Int J Food Microbiol921533
10. 10.Kathariou S (2002) *Listeria monocytogenes* virulence and pathogenicity, a food safety perspective. J Food Prot 65: 1811–1829.S. Kathariou2002*Listeria monocytogenes*virulence and pathogenicity, a food safety perspective.J Food Prot6518111829
11. 11. Fistrovici E, Collins-Thompson DL (1990) Use of plasmid profiles and restriction endonuclease digest in environmental studies

of *Listeria* spp. from raw milk. Int J Food Microbiol 10: 43–50.E. FistroviciDL Collins-Thompson1990Use of plasmid profiles and restriction endonuclease digest in environmental studies of *Listeria* spp. from raw milk.Int J Food Microbiol104350

12. 12. Perez-Diaz JC, Vicente MF, Baquero F (1982) Plasmids in *Listeria*. Plasmid 8: 112–118.JC Perez-DiazMF VicenteF. Baquero1982Plasmids in *Listeria*.Plasmid8112118

13. 13. Peterkin PI, Gardiner MA, Malik N, Idziak ES (1992) Plasmids in *Listeria monocytogenes* and other *Listeria* species. Can J Microbiol 38: 161–164.PI PeterkinMA GardinerN. MalikES Idziak1992Plasmids in *Listeria monocytogenes* and other *Listeria*species.Can J Microbiol38161164

14. 14. Lebrun M, Loulergue J, Chaslus-Dancla E, Audurier A (1992) Plasmids in *Listeria monocytogenes* in relation to cadmium resistance. Appl Environ Microbiol 58: 3183–3186.M. LebrunJ. LoulergueE. Chaslus-DanclaA. Audurier1992Plasmids in *Listeria monocytogenes* in relation to cadmium resistance.Appl Environ Microbiol5831833186

15. 15. Kolstad J, Caugant DA, Rorvik LM (1992) Differentiation of *Listeria monocytogenes*isolates by using plasmid profiling and multilocus enzyme electrophoresis. Int J Food Microbiol 16: 247–260.J. KolstadDA CaugantLM Rorvik1992Differentiation of *Listeria monocytogenes* isolates by using plasmid profiling and multilocus enzyme electrophoresis.Int J Food Microbiol16247260

16. 16. McLauchlin J, Hampton MD, Shah S, Threlfall EJ, Wieneke AA, et al. (1997) Subtyping of *Listeria monocytogenes* on the basis of plasmid profiles and arsenic and cadmium susceptibility. J Appl Microbiol 83: 381–388.J. McLauchlinMD HamptonS. ShahEJ ThrelfallAA Wieneke1997Subtyping of *Listeria monocytogenes* on the basis of plasmid profiles and arsenic and cadmium susceptibility.J Appl Microbiol83381388

17. 17. Harvey J, Gilmour A (2001) Characterization of recurrent and sporadic *Listeria monocytogenes* isolates from raw milk and nondairy foods by pulsed-field gel electrophoresis, monocin typing, plasmid profiling, and cadmium and antibiotic resistance determination. Appl Environ Microbiol 67: 840–847.J. HarveyA.

Gilmour2001Characterization of recurrent and sporadic *Listeria monocytogenes*isolates from raw milk and nondairy foods by pulsed-field gel electrophoresis, monocin typing, plasmid profiling, and cadmium and antibiotic resistance determination.Appl Environ Microbiol67840847

18. 18. Lebrun M, Audurier A, Cossart P (1994) Plasmid-borne cadmium resistance genes in *Listeria monocytogenes* are similar to *cadA* and *cadC* of *Staphylococcus aureus* and are induced by cadmium. J Bacteriol 176: 3040–3048.M. LebrunA. AudurierP. Cossart1994Plasmid-borne cadmium resistance genes in *Listeria monocytogenes* are similar to *cadA* and *cadC* of *Staphylococcus aureus* and are induced by cadmium.J Bacteriol17630403048
19. 19. Poyart-Salmeron C, Carlier C, Trieu-Cuot P, Courtieu AL, Courvalin P (1990) Transferable plasmid-mediated antibiotic resistance in *Listeria monocytogenes.* Lancet 335: 1422–1426.C. Poyart-SalmeronC. CarlierP. Trieu-CuotAL CourtieuP. Courvalin1990Transferable plasmid-mediated antibiotic resistance in *Listeria monocytogenes*.Lancet33514221426
20. 20. Hadorn K, Hachler H, Schaffner A, Kayser FH (1993) Genetic characterization of plasmid-encoded multiple antibiotic resistance in a strain of *Listeria monocytogenes*causing endocarditis. Eur J Clin Microbiol Infect Dis 12: 928–937.K. HadornH. HachlerA. SchaffnerFH Kayser1993Genetic characterization of plasmid-encoded multiple antibiotic resistance in a strain of *Listeria monocytogenes* causing endocarditis.Eur J Clin Microbiol Infect Dis12928937
21. 21. Flamm RK, Hinrichs DJ, Thomashow MF (1984) Introduction of pAMbeta1 into*Listeriamonocytogenes*byconjugationandhomology between native *L. monocytogenes*plasmids. Infect Immun 44: 157–161.RK FlammDJ HinrichsMF Thomashow1984Introduction of pAM beta 1 into *Listeria monocytogenes* by conjugation and homology between native *L. monocytogenes* plasmids.Infect Immun44157161
22. 22. Nelson KE, Fouts DE, Mongodin EF, Ravel J, DeBoy RT, et al. (2004) Whole genome comparisons of serotype 4b and 1/2a strains of the food-borne pathogen *Listeria monocytogenes* reveal new insights into the core genome components of this species. Nucleic

Acids Res 32: 2386–2395.KE NelsonDE FoutsEF MongodinJ. RavelRT DeBoy2004Whole genome comparisons of serotype 4b and 1/2a strains of the food-borne pathogen *Listeria monocytogenes* reveal new insights into the core genome components of this species. Nucleic Acids Res3223862395

23. 23. Canchaya C, Giubellini V, Ventura M, de Los Reyes-Gavilan CG, Margolles A (2010) Mosaic-like sequences containing transposon, phage and plasmid elements among *Listeria monocytogenes* plasmids. Appl Environ Microbiol. C. CanchayaV. GiubelliniM. VenturaCG de Los Reyes-GavilanA. Margolles2010Mosaic-like sequences containing transposon, phage and plasmid elements among *Listeria monocytogenes*plasmids.Appl Environ Microbiol
24. 24. Gilmour MW, Graham M, Van DG, Tyler S, Kent H, et al. (2010) High-throughput genome sequencing of two *Listeria monocytogenes* clinical isolates during a large foodborne outbreak. BMC Genomics 11: 120.MW GilmourM. GrahamDG VanS. TylerH. Kent2010High-throughput genome sequencing of two *Listeria monocytogenes* clinical isolates during a large foodborne outbreak. BMC Genomics11120
25. 25. Orsi RH, Borowsky ML, Lauer P, Young SK, Nusbaum C, et al. (2008) Short-term genome evolution of *Listeria monocytogenes* in a non-controlled environment. BMC Genomics 9: 539.RH OrsiML BorowskyP. LauerSK YoungC. Nusbaum2008Short-term genome evolution of *Listeria monocytogenes* in a non-controlled environment.BMC Genomics9539
26. 26. Kuenne CT, Ghai R, Chakraborty T, Hain T (2007) GECO—linear visualization for comparative genomics. Bioinformatics 23: 125–126.CT KuenneR. GhaiT. ChakrabortyT. Hain2007GECO—linear visualization for comparative genomics.Bioinformatics23125126
27. 27. Weaver KE, Clewell DB, An F (1993) Identification, characterization, and nucleotide sequence of a region of *Enterococcus faecalis* pheromone-responsive plasmid pAD1 capable of autonomous replication. J Bacteriol 175: 1900–1909. KE WeaverDB ClewellF. An1993Identification, characterization, and nucleotide sequence of a region of *Enterococcus faecalis* pheromone-responsive plasmid pAD1 capable of autonomous replication.J Bacteriol17519001909

28. 28. Wilcks A, Smidt L, Okstad OA, Kolsto AB, Mahillon J, et al. (1999) Replication mechanism and sequence analysis of the replicon of pAW63, a conjugative plasmid from *Bacillus thuringiensis*. J Bacteriol 181: 3193–3200.A. WilcksL. SmidtOA OkstadAB KolstoJ. Mahillon1999Replication mechanism and sequence analysis of the replicon of pAW63, a conjugative plasmid from *Bacillus thuringiensis*.J Bacteriol18131933200

29. 29. Tinsley E, Naqvi A, Bourgogne A, Koehler TM, Khan SA (2004) Isolation of a minireplicon of the virulence plasmid pXO2 of *Bacillus anthracis* and characterization of the plasmid-encoded RepS replication protein. J Bacteriol 186: 2717–2723.E. TinsleyA. NaqviA. BourgogneTM KoehlerSA Khan2004Isolation of a minireplicon of the virulence plasmid pXO2 of *Bacillus anthracis* and characterization of the plasmid-encoded RepS replication protein.J Bacteriol18627172723

30. 30. Francia MV, Fujimoto S, Tille P, Weaver KE, Clewell DB (2004) Replication of *Enterococcus faecalis* pheromone-responding plasmid pAD1: location of the minimal replicon and oriV site and RepA involvement in initiation of replication. J Bacteriol 186: 5003–5016.MV FranciaS. FujimotoP. TilleKE WeaverDB Clewell2004Replication of *Enterococcus faecalis* pheromone-responding plasmid pAD1: location of the minimal replicon and oriV site and RepA involvement in initiation of replication.J Bacteriol18650035016

31. 31. Van der Auwera GA, Andrup L, Mahillon J (2005) Conjugative plasmid pAW63 brings new insights into the genesis of the *Bacillus anthracis* virulence plasmid pXO2 and of the *Bacillus thuringiensis* plasmid pBT9727. BMC Genomics 6: 103.GA Van der AuweraL. AndrupJ. Mahillon2005Conjugative plasmid pAW63 brings new insights into the genesis of the *Bacillus anthracis* virulence plasmid pXO2 and of the *Bacillus thuringiensis* plasmid pBT9727.BMC Genomics6103

32. 32. Francia MV, Weaver KE, Goicoechea P, Tille P, Clewell DB (2007) Characterization of an active partition system for the *Enterococcus faecalis* pheromone-responding plasmid pAD1. J Bacteriol 189: 8546–8555.MV FranciaKE WeaverP. GoicoecheaP. TilleDB Clewell2007Characterization of an active partition system

for the *Enterococcus faecalis*pheromone-responding plasmid pAD1.J Bacteriol18985468555

33. 33. Friedberg EC, Wagner R, Radman M (2002) Specialized DNA polymerases, cellular survival, and the genesis of mutations. Science 296: 1627–1630.EC FriedbergR. WagnerM. Radman2002Specialized DNA polymerases, cellular survival, and the genesis of mutations.Science29616271630
34. 34. Koskiniemi S, Andersson DI (2009) Translesion DNA polymerases are required for spontaneous deletion formation in *Salmonella typhimurium*. Proc Natl Acad Sci U S A 106: 10248–10253.S. KoskiniemiDI Andersson2009Translesion DNA polymerases are required for spontaneous deletion formation in *Salmonella typhimurium*.Proc Natl Acad Sci U S A1061024810253
35. 35. Thompson JD, Higgins DG, Gibson TJ (1994) CLUSTAL W: improving the sensitivity of progressive multiple sequence alignment through sequence weighting, position-specific gap penalties and weight matrix choice. Nucleic Acids Res 22: 4673–4680.JD ThompsonDG HigginsTJ Gibson1994CLUSTAL W: improving the sensitivity of progressive multiple sequence alignment through sequence weighting, position-specific gap penalties and weight matrix choice.Nucleic Acids Res2246734680
36. 36. Kosman DJ (2010) Multicopper oxidases: a workshop on copper coordination chemistry, electron transfer, and metallophysiology. J Biol Inorg Chem 15: 15–28.DJ Kosman2010Multicopper oxidases: a workshop on copper coordination chemistry, electron transfer, and metallophysiology.J Biol Inorg Chem151528
37. 37. Roy CR, Mukherjee S (2009) Bacterial FIC Proteins AMP Up Infection. Sci Signal 2: e14.CR RoyS. Mukherjee2009Bacterial FIC Proteins AMP Up Infection.Sci Signal2e14
38. 38. Gibson CM, Mallett TC, Claiborne A, Caparon MG (2000) Contribution of NADH oxidase to aerobic metabolism of *Streptococcus pyogenes*. J Bacteriol 182: 448–455.CM GibsonTC MallettA. ClaiborneMG Caparon2000Contribution of NADH oxidase to aerobic metabolism of *Streptococcus pyogenes*.J Bacteriol182448455

39. 39. La CS, Sauvageot N, Giard JC, Benachour A, Posteraro B, et al. (2007) Comparative study of the physiological roles of three peroxidases (NADH peroxidase, Alkyl hydroperoxide reductase and Thiol peroxidase) in oxidative stress response, survival inside macrophages and virulence of *Enterococcus faecalis*. Mol Microbiol 66: 1148–1163.CS LaN. SauvageotJC GiardA. BenachourB. Posteraro2007Comparative study of the physiological roles of three peroxidases (NADH peroxidase, Alkyl hydroperoxide reductase and Thiol peroxidase) in oxidative stress response, survival inside macrophages and virulence of *Enterococcus faecalis*.Mol Microbiol6611481163
40. 40. Ceragioli M, Mols M, Moezelaar R, Ghelardi E, Senesi S, et al. (2010) Comparative transcriptomic and phenotypic analysis of the responses of *Bacillus cereus* to various disinfectant treatments. Appl Environ Microbiol 76: 3352–3360.M. CeragioliM. MolsR. MoezelaarE. GhelardiS. Senesi2010Comparative transcriptomic and phenotypic analysis of the responses of *Bacillus cereus* to various disinfectant treatments.Appl Environ Microbiol7633523360
41. 41. Fang F, Flynn S, Li Y, Claesson MJ, van Pijkeren JP, et al. (2008) Characterization of endogenous plasmids from *Lactobacillus salivarius* UCC118. Appl Environ Microbiol 74: 3216–3228.F. FangS. FlynnY. LiMJ ClaessonJP van Pijkeren2008Characterization of endogenous plasmids from *Lactobacillus salivarius* UCC118. Appl Environ Microbiol7432163228
42. 42. Madsen A, Josephsen J (2001) The *Lla*GI restriction and modification system of *Lactococcus lactis* W10 consists of only one single polypeptide. FEMS Microbiol Lett 200: 91–96.A. MadsenJ. Josephsen2001The *Lla*GI restriction and modification system of *Lactococcus lactis* W10 consists of only one single polypeptide. FEMS Microbiol Lett2009196
43. 43. Kuroda T, Tsuchiya T (2009) Multidrug efflux transporters in the MATE family. Biochim Biophys Acta 1794: 763–768.T. KurodaT. Tsuchiya2009Multidrug efflux transporters in the MATE family. Biochim Biophys Acta1794763768
44. 44. Su P, Im H, Hsieh H, Kang'A S, Dunn NW (1999) *Lla*FI, a type III restriction and modification system in *Lactococcus lactis*. Appl

Environ Microbiol 65: 686–693.P. SuH. ImH. HsiehS. Kang'ANW Dunn1999*Lla*FI, a type III restriction and modification system in *Lactococcus lactis*.Appl Environ Microbiol65686693

45. 45. Kim MH, Kim Y, Park HJ, Lee JS, Kwak SN, et al. (2008) Structural insight into bioremediation of triphenylmethane dyes by Citrobacter sp. triphenylmethane reductase. J Biol Chem 283: 31981–31990.MH KimY. KimHJ ParkJS LeeSN Kwak2008Structural insight into bioremediation of triphenylmethane dyes by Citrobacter sp. triphenylmethane reductase.J Biol Chem2833198131990

46. 46. Zhang Z, Ma C, Pornillos O, Xiu X, Chang G, et al. (2007) Functional characterization of the heterooligomeric EbrAB multidrug efflux transporter of *Bacillus subtilis*. Biochemistry 46: 5218–5225.Z. ZhangC. MaO. PornillosX. XiuG. Chang2007Functional characterization of the heterooligomeric EbrAB multidrug efflux transporter of *Bacillus subtilis*.Biochemistry4652185225

47. 47. Masaoka Y, Ueno Y, Morita Y, Kuroda T, Mizushima T, et al. (2000) A two-component multidrug efflux pump, EbrAB, in *Bacillus subtilis*. J Bacteriol 182: 2307–2310.Y. MasaokaY. UenoY. MoritaT. KurodaT. Mizushima2000A two-component multidrug efflux pump, EbrAB, in *Bacillus subtilis*.J Bacteriol18223072310

48. 48. Chopin MC, Chopin A, Bidnenko E (2005) Phage abortive infection in lactococci: variations on a theme. Curr Opin Microbiol 8: 473–479.MC ChopinA. ChopinE. Bidnenko2005Phage abortive infection in lactococci: variations on a theme.Curr Opin Microbiol8473479

49. 49. Ballal A, Basu B, Apte SK (2007) The Kdp-ATPase system and its regulation. J Biosci 32: 559–568.A. BallalB. BasuSK Apte2007The Kdp-ATPase system and its regulation.J Biosci32559568

50. 50. Booth IR (1985) Regulation of cytoplasmic pH in bacteria. Microbiol Rev 49: 359–378.IR Booth1985Regulation of cytoplasmic pH in bacteria.Microbiol Rev49359378

51. 51. Brondsted L, Kallipolitis BH, Ingmer H, Knochel S (2003) *kdpE* and a putative RsbQ homologue contribute to growth of *Listeria monocytogenes* at high osmolarity and low temperature. FEMS Microbiol Lett 219: 233–239.L. BrondstedBH KallipolitisH. IngmerS. Knochel2003*kdpE* and a putative RsbQ homologue

contribute to growth of *Listeria monocytogenes* at high osmolarity and low temperature.FEMS Microbiol Lett219233239

52. 52. Boylan JA, Hummel CS, Benoit S, Garcia-Lara J, Treglown-Downey J, et al. (2006) Borrelia burgdorferi bb0728 encodes a coenzyme A disulphide reductase whose function suggests a role in intracellular redox and the oxidative stress response. Mol Microbiol 59: 475–486.JA BoylanCS HummelS. BenoitJ. Garcia-LaraJ. Treglown-Downey2006Borrelia burgdorferi bb0728 encodes a coenzyme A disulphide reductase whose function suggests a role in intracellular redox and the oxidative stress response.Mol Microbiol59475486
53. 53. Waters LS, Storz G (2009) Regulatory RNAs in bacteria. Cell 136: 615–628.LS WatersG. Storz2009Regulatory RNAs in bacteria. Cell136615628
54. 54. Brantl S (2007) Regulatory mechanisms employed by cis-encoded antisense RNAs. Curr Opin Microbiol 10: 102–109.S. Brantl2007Regulatory mechanisms employed by cis-encoded antisense RNAs.Curr Opin Microbiol10102109
55. 55. Toledo-Arana A, Dussurget O, Nikitas G, Sesto N, Guet-Revillet H, et al. (2009) The *Listeria* transcriptional landscape from saprophytism to virulence. Nature 459: 950–956.A. Toledo-AranaO. DussurgetG. NikitasN. SestoH. Guet-Revillet2009The *Listeria*transcriptional landscape from saprophytism to virulence. Nature459950956
56. 56. Bruand C, Ehrlich SD (1998) Transcription-driven DNA replication of plasmid pAMbeta1 in *Bacillus subtilis*. Mol Microbiol 30: 135–145.C. BruandSD Ehrlich1998Transcription-driven DNA replication of plasmid pAMbeta1 in *Bacillus subtilis*. Mol Microbiol30135145
57. 57. Woodward JJ, Iavarone AT, Portnoy DA (2010) c-di-AMP Secreted by Intracellular *Listeria monocytogenes* Activates a Host Type I Interferon Response. Science. JJ WoodwardAT IavaroneDA Portnoy2010c-di-AMP Secreted by Intracellular *Listeria monocytogenes* Activates a Host Type I Interferon Response.Science
58. 58.Meyer F, Goesmann A, McHardy AC, Bartels D, Bekel T, et al. (2003) GenDB—an open source genome annotation system for

prokaryote genomes. Nucleic Acids Res 31: 2187–2195.F. MeyerA. GoesmannAC McHardyD. BartelsT. Bekel2003GenDB—an open source genome annotation system for prokaryote genomes.Nucleic Acids Res3121872195

59. 59. Darling AC, Mau B, Blattner FR, Perna NT (2004) Mauve: multiple alignment of conserved genomic sequence with rearrangements. Genome Res 14: 1394–1403.AC DarlingB. MauFR BlattnerNT Perna2004Mauve: multiple alignment of conserved genomic sequence with rearrangements.Genome Res1413941403

60. 60. Huson DH, Richter DC, Rausch C, Dezulian T, Franz M, et al. (2007) Dendroscope: An interactive viewer for large phylogenetic trees. BMC Bioinformatics 8: 460.DH HusonDC RichterC. RauschT. DezulianM. Franz2007Dendroscope: An interactive viewer for large phylogenetic trees.BMC Bioinformatics8460

CHAPTER

2

A BIOBRICK LIBRARY FOR CLONING CUSTOM EUKARYOTIC PLASMIDS

Marco Constante, Raik Gru nberg, Mark Isalan

EMBL/CRG Systems Biology Research Unit, Centre for Genomic Regulation (CRG) and UPF, Barcelona, Spain

ABSTRACT

Researchers often require customised variations of plasmids that are not commercially available. Here we demonstrate the applicability and versatility of standard synthetic biological parts (biobricks) to build custom plasmids. For this purpose we have built a collection of 52 parts that include multiple cloning sites (MCS) and common protein tags, protein reporters and selection markers, amongst others. Importantly, most of the parts are designed in a format to allow fusions that maintain the reading frame. We illustrate the collection by building several model contructs, including concatemers of protein binding-site motifs, and a variety of plasmids for eukaryotic stable cloning and chromosomal insertion. For example, in 3 biobrick iterations, we make a cerulean-reporter plasmid for cloning fluorescent protein fusions. Furthermore, we use the collection to implement a recombinase-mediated DNA insertion (RMDI), allowing chromosomal site-directed exchange of genes. By making one recipient stable cell line, many standardised cell lines can subsequently be generated, by fluorescent fusion-gene exchange. We propose that this biobrick collection may be distributed peer-to-peer as a stand-alone library, in addition to its distribution through the Registry of Standard Biological Parts

INTRODUCTION

The construction of specific plasmid DNA sequences is a routine technique in molecular biology laboratories [1]. The most widely used DNA sequences (for instance, a promoter followed by a fluorescent protein, a multiple cloning site and a polyadenylation signal) are easily found in commercially-available plasmids [2], [3]. However, when requirements start to become more stringent, multiple plasmid modifications are required and, although many changes may be relatively simple to perform, multiple modifications may become time-consuming cloning challenges.

Complex, multi-factor plasmids have to be built frequently for the applications of synthetic biology [4], [5], [6], often with the combinatorial use of different DNA parts [7]. To simplify such types of constructions an idempotent cloning system has recently been developed by Tom Knight [8] (described in the BioBrick Foundation Request for Comments #10, BBF RFC 10; http://biobricks.org/). Briefly, this system uses a specific set of restriction enzyme sites at the 3′ and 5′ ends of each DNA cassette ('biobrick'), such that a biobrick 'A' may be fused with a biobrick 'B' to produce 'AB'. 'AB' contains an uncleavable 'scar' sequence between 'A' and 'B' and, importantly, the exact same set of restriction enzyme sites as the initial biobricks, at the 3′ and 5′ ends. In other words, every biobrick fusion product is itself a new biobrick and may even be used iteratively for the assembly of concatemers (See Figure 1). Given the physical idempotent characteristics of the system, biobricks may be fused together in any combination of parts, with few restrictions in the number of the biobricks, and no restrictions on the order ('BA' would be as simple to construct as 'AB').

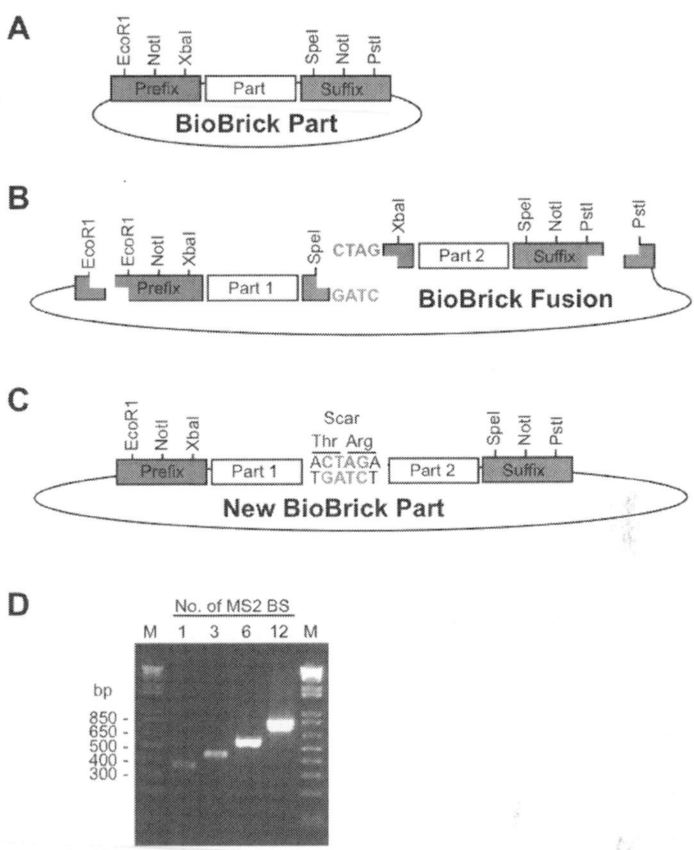

Figure 1. The biobrick assembly principle[8], [10].

(A) Each biobrick part has the same prefix and suffix, containing restriction enzyme sites. (B) Following restriction digests, a two-insert ligation into the biobrick vector results in a biobrick fusion. (C) The new biobrick part regenerates the original prefix and suffix, but contains an in-frame Thr-Arg scar in protein-coding fusions. (D) MS2 binding site concatemers (MS2 BS), built with iterative biobrick assembly, from 1 to 12-copies (4 steps). M=marker (1 kb ladder). The upstream and downstream sequences between the primer annealing sites and the biobricks contribute 312 bp, while each MS2 BS is 39 bp.

The RFC 10 Biobrick format [8] is itself very useful and has formed the core of engineering challenges such as the annual International Genetically Engineered Machine (iGEM) competitions [9], where students are asked to engineer systems using biobricks.

It is a well-documented system with a large and growing collection of parts that use the prefix GAATTCGCGGCCGCTTCTAGAG (or GAATTCGCGGCCGCTTCTAG for protein coding parts starting with ATG) and the suffix TACTAGTAGCGGCCGCTGCAG. However, the original scheme has certain limitations, such as the difficulty of fusing parts coding for protein domains, since the fusion 'scar' changes the codon reading frame from biobrick 'A' to biobrick 'B'. Phillips and Silver [10] resolved this issue by presenting a slight modification to the initial design (removing the last G of the prefix and the first T of the suffix thereby conserving the reading frame in the scar). Special attention is required when preparing new parts in this format (BBF RFC 23), since the removal of the last G of the prefix means that a Dam methylation site may be formed if the part starts with TC, which blocks restriction with Xba I. Since the scar resulting from a RFC 23 fusion (ACTAGA) codes for threonine-arginine, a chimeric protein 'AB' may be assembled from two RFC 23 biobricks 'A' and B'. When preparing such RFC 23 protein fusions, users should be aware that the arginine from the scar may be problematic since it is positively charged.

It is of interest to note that biobricks in the RFC 23 format may still be assembled with biobricks in the RFC 10 format without requiring new restriction enzymes. Although such a RFC 23-RFC 10 cross-fusion does not maintain the 6-bp scar required for the maintenance of a codon frame, N-terminal fusions of RFC 23 protein coding parts may still be performed as long as the frameshift is corrected by an adaptor part. In addition, the RFC 23-RFC 10 cross-fusion may be useful for assemblies that do not require the maintenance of a reading frame or a specific scar size, such as linking non-coding biobricks (e.g. transcription factor binding sites). We refer to this compatibility between RFC 23 and RFC 10 as 'assembly compatibility', that should not be confused with 'RFC 23 compatibility' that is reserved for fusion between biobricks that maintain the reading frame. Further biobrick formats (with only limited cross-compability) were later proposed and are documented as BBF RFCs.

Engineering with the Biobrick system is usually accompanied with an abstraction hierarchy perspective. With this perspective in mind, one uses 'DNA' to design basic 'parts' that may then be assembled into a single functional 'device' with a specific function. Finally, several of these 'devices' may be put together to design a synthetic 'system' with a high-level functionality. This hierarchy facilitates synthetic engineering since, ideally, a researcher working, for instance, at the 'system' level does not need to know the specific details on how to engineer and build the 'devices' used in that 'system'.

Synthetic biologists are now in the process of developing and characterizing basic 'parts' or part sets ('devices' [11]) that may be used to engineer biological machines [12], [13], [14] and to model them [15]. These parts are available from the Registry of Standard Biological Parts (http://partsregistry.org/) which is supported by the BioBricks Foundation (http://bbf.openwetware.org/).

Although the registry is available to any researcher, whether or not they work in the field of synthetic biology, we believe that the size of the registry, the existence of different biobrick formats (not always compatible with each other), and the limited communication of this resource to non-synthetic biologists, may hamper the evolution and distribution of biobricks. We therefore developed a cloning tool that should be of interest outside the field of synthetic biology; a biobrick toolkit for the construction of custom eukaryotic expression plasmids, using frequently-used components. We originally noted that many biobrick parts that one would require for eukaryotic projects were apparently missing from the registry. These parts included: several multiple cloning sites with different reading frames; an extensive range of reporter proteins; eukaryotic selection markers; eukaryotic internal ribosomal entry sites; and protein epitope tags. Importantly, most of these parts should be compatible with each other [10], [16]. In addition, the novel use of MCS biobricks facilitates the cloning of DNA fragments of interest (e.g. open reading frames, promoters), without having to first remove any internal cleavage sites of the biobrick enzymes in order to render them 'biobrick compatible'.

This work is intended to highlight the potential of the biobrick system to construct custom plasmids for eukaryotic cell lines, for both synthetic and non-synthetic biologists. Since many parts that we consider important did not exist in the registry, we have built a distribution of 9 previously existing and 43 new biobricks. We demonstrate the utility of the new biobricks with some examples of assemblies. In addition, we use the created library to establish a gene-switching system, from tdTomato to

EGFP expression, using recombinase-mediated DNA insertion (RMDI). Finally, we propose that this collection be distributed peer-to-peer as an independent library, as well as being available from the registry, and discuss some advantages in having this type of collection.

MATERIALS AND METHODS

Preparation of Parts

New parts were designed following the recommendations in BBF RFC 23. Briefly, parts were formatted to contain the prefix 5′-GAATTCGCGGCCGCTTCTAGA-3′ and the suffix 5′-ACTAGTAGCGGCCGCTGCAG-3′. Construction of the parts was done using 1) oligonucleotide inserts, 2) PCR with oligonucleotide-directed mutagenesis when required [1] or 3) synthesized directly by GenScript Corporation (Piscataway, NJ). Importantly, the part sequences do not start with a TC since this forms a Dam methylase site (GATC) and digestion with XbaI can be inhibited. The Biobrick sequences are available at the Registry of Standard Biological Parts website (http://partsregistry.org/), at GenBank (http://www.ncbi.nlm.nih.gov/genbank/) and in the Supporting Information S1.

Biobrick Assemblies

Assemblies were performed similarly to the method presented in the Biobrick assembly kit from New England Biolabs (NEB; Ref. E0546S) and the Gingko Bioworks manual (http://ginkgobioworks.com/support/). Alternatively, one may use the adapted streamlined protocol we have previously reported [14].

Briefly, 500 ng of the upstream part are digested with 20000 U of EcoRI (NEB) and 10000 U of SpeI (NEB) in NEB EcoRI buffer. At the same time, 500 ng of the downstream part are digested with 20000 U of XbaI (NEB) and 20000 U of PstI (NEB) in NEB Buffer 3. Finally and concomitantly, 500 ng of the destination plasmid are digested with 20000 U of EcoRI and 20000 U of PstI in NEB EcoRI Buffer. Digestions are performed at 37°C for 1 hour followed by heat inactivation at 80°C for 20 minutes. Note that the recipient pSB1A* plasmids in the library contain a *ccdB* death cassette that is removed in the digestion.

Furthermore, each assembly should contain an antibiotic resistance gene that is not present in either the upstream or the downstream part (e.g. if the upstream part is in the pSB1AK3 plasmid, containing ampicillin and kanamycin resistance, and the downstream part is in the pSB1AC3 plasmid, containing ampicillin and chloramphenicol resistance, the destination plasmid should be the pSB1AT3 plasmid since it contains the tetracycline resistance not present in the other plasmids).

Since the destination plasmid contains the *ccdB* death cassette, a two-way ligation of the fragments can be done using the Roche rapid DNA ligation kit, according to the manufacturer's instructions, without need for gel isolation or other purifications. Briefly, 4 μl of the upstream part digestion, 4 μl of the downstream part digestion and 2 μl of the destination plasmid digestion are used. 2 μl of the ligation product are transformed in one shot competent Top10 *E. coli*(Invitrogen), according to the manufacturer's instructions, and are subsequently plated onto LB agar plates. The plates contain antibiotic for the resistance provided by the destination plasmid (in the example above one should use LB agar plates containing tetracycline) and are incubated overnight.

PCRs of Biobricks (from plasmid DNA or bacterial colony) were performed using NEB Taq polymerase, using the manufacturer's instructions, and using the standard sequencing primers BBa_G00100 (5′-TGCCACCTGACGTCTAAGAA-3′) and BBa_G00101 (5′-ATTACCGCCTTTGAGTGAGC-3′) with 20 cycles at 55°C annealing temperature, with appropriate denaturation and elongation steps. PCR products were analyzed on agarose gels and positive colonies were grown overnight for DNA isolation using the Qiagen spin miniprep kit according to the manufacturer's instructions. All clones were verified by DNA sequencing.

Cell Culture and Transfection

HEK293 cells were purchased from ATCC. Cells were propagated in Dulbecco's Modified Eagle Medium (Gibco) supplemented with 10% fetal bovine serum (Gibco) and 1% penicillin-streptomycin (Gibco). Transfection of the cells was performed with lipofectamine 2000 (Invitrogen), according to the manufacturer's instructions, in 6-well plates: cells were incubated for 4 hours with 4 μg of DNA and 10 μl of lipofectamine reagent. Stable clones were prepared by selection with 0.5 mg/ml G418 (Sigma).

Cloning small Espin into the MCS of a custom-made plasmid

Rat small Espin (sEspin) [17] was amplified by PCR, from a plasmid kindly provided by the group of Hernán López-Schier, using the primers sEspinF (5′-AAGAGGGATCCATGAACTCCC-3′) and sEspinR (5′-CTTCTTACCGGTTTACTTAGGGATCTCCCCCTTC-3′) at 55°C annealing temperature. The PCR product was cloned into the Biobrick custom plasmid using the restriction enzymes BamHI and AgeI.

Microscopy

After transfection, cells were allowed to recover overnight and were dispensed into glass bottom culture dishes (MatTek). 24 h later, the medium was replaced with PBS for observation under the microscope.

Recombinase-mediated DNA inclusion plasmid preparation

The plasmid used for the creation of the stable clone expressing tdTomato was prepared using an assembly of the biobricks between the CMV promoter and polyadenylation (pA) signal. The assembled plasmid was digested with XbaI and PstI and cloned into pEGFP-C1 digested with NheI and PstI, thereby removing the EGFP sequence and providing the CMV upstream and pA downstream of the biobrick insert. Neomycin resistance used for selection is provided by the pEGFP-C1 plasmid backbone under the expression of the SV40 early promoter. The EGFP plasmid for inclusion was prepared using the biobrick collection.

Flow cytometry

Flow-assisted cell sorting (FACS) analysis was performed with FACSCanto (BD) and cell separation with FACSARIA II (BD).

Results

Library

To create a distribution for the construction of custom plasmids, we have put together a set of 9 existing and 43 new biobricks and plasmid backbones. We believe that these should be useful for the creation of custom-made plasmids for use in eukaryotic cell lines (see Table 1). Importantly, a set of multiple cloning site (MCS) biobricks, covering all three different frames has been built, essential for the combination of the biobrick assembly strategy with classical cloning strategies (Fig. 2).

```
MCS1a      GGGCCCAAGCTCGAGAGCGGATCCTCAAGCTT
MCS1b     AGGGCCCAAGCTCGAGAGCGGATCCTCAAGCTT
MCS1c    AAGGGCCCAAGCTCGAGAGCGGATCCTCAAGCTT
            ApaI       XhoI       BamHI    HindIII

MCS2a      AGATCTACGAGCTCATGGTACCATACCGGT
MCS2b     AAGATCTACGAGCTCATGGTACCATACCGGT
MCS2c    AAAGATCTACGAGCTCATGGTACCATACCGGT
           BglII      SacI       KpnI     AgeI

MCS3a     TGTACAGCAGGCCTTCGTCGACGAGCATGCCGATCGATA
MCS3b    ATGTACAGCAGGCCTTCGTCGACGAGCATGCCGATCGATA
MCS3c  AATGTACAGCAGGCCTTCGTCGACGAGCATGCCGATCGATA
         BsrGI     StuI      SalI      SphI      ClaI
```

Figure 2. Multiple Cloning Site (MCS) biobricks.

The uniqueness of each cloning site is dependent on whether the remaining biobricks and backbones used for the custom plasmid also contain the restriction site. Blunt end restriction enzymes are represented in italic. These biobricks link classical cloning to the biobrick system.

Table 1. List of biobrick parts for eukaryotic plasmids

Group	Nickname	Description	Registry	GenBank
Backbones	pSB1A3*	High copy number plasmid carrying ampicillin resistance.	pSB1A3	
	pSB1AK3*	High copy number plasmid carrying ampicillin and kanamycin resistance.	pSB1AK3	

	pSB1AC3*	High copy number plasmid carrying ampicillin and chloramphenicol resistance.	pSB1AC3	
	pSB1AT3*	High copy number plasmid carrying ampicillin and tetracyclin resistance.	pSB1AT3	
Construction	Kozak	Simple Kozak sequence protein head domain [26]	BBa_J96000	JN204869
	Stop	Tail domain with stop codons in all three frames	BBa_J96001	JN204870
	CMV*	Cytomegalovirus immediate-early promoter	BBa_I712004	
	SV40pA*	Eukaryotic – derived from SV40 early poly A signal sequence	BBa_J52016	
	MCS1a	Multiple cloning site version 1, first frame	BBa_J96002	JN204871
	MCS1b	Multiple cloning site version 1, second frame	BBa_J96003	JN204872
	MCS1c	Multiple cloning site version 1, third frame	BBa_J96004	JN204873
	MCS2a	Multiple cloning site version 2, first frame	BBa_J96005	JN204874
	MCS2b	Multiple cloning site version 2, second frame	BBa_J96006	JN204875
	MCS2c	Multiple cloning site version 2, third frame	BBa_J96007	JN204876
	MCS3a	Multiple cloning site version 3, first frame	BBa_J96008	JN204877
	MCS3b	Multiple cloning site version 3, second frame	BBa_J96009	JN204878
	MCS3c	Multiple cloning site version 3, third frame	BBa_J96010	JN204879
Selection	FNeomycin	Resistance to G418/neomycin gene [1]	BBa_J96011	JN204880
	FPuromycin	Resistance to puromycin gene [1]	BBa_J96012	JN204881
	FHSTK	Herpes simplex thymidine kinase conferring toxicity to ganciclovir [27]	BBa_J96013	JN204882

Reporters	tdTomato	Engineered red fluorescent protein [28]	BBa_ J96029	JN204883
	EGFP	Engineered green fluorescent protein [29]	BBa_ J96031	JN204884
	Cerulean	Engineered cyan fluorescent protein [19]	BBa_ J96032	JN204885
	EBFP2	Engineered blue fluorescent protein [30]	BBa_ J96033	JN204886
	mCherry*	Engineered red fluorescent protein [28]	BBa_ J63000	
	RLuciferase	Renilla luciferase gene [31]	BBa_ J96034	JN204887
Tags	Flag	FLAG affinity tag [32]	BBa_ J96035	JN204888
	HA	HA affinity tag [33]	BBa_ J96036	JN204889
	His	His affinity tag [34]	BBa_ J96037	JN204890
	StrepII	Strep II affinity tag [35]	BBa_ J96038	JN204891
Localization	SP	Membrane or secretion: IgK leader peptide with Kozak [36]	BBa_ J96014	JN204892
	TMD	PDGF Receptor Transmembrane Domain [36]	BBa_ J96015	JN204893
	Myristoylation	Myristoylation signal sequence with Kozak [37]	BBa_ J96016	JN204894
	NLS*	Nuclear Localization Signal from SV40 [38]	BBa_ J63008	
Recombination	Loxp	Lox p sequence [39]	BBa_ J96017	JN204895
	Lox66	Directional lox sequence compatible with lox71 [40]	BBa_ J96018	JN204896
	Lox71	Directional lox sequence compatible with lox66 [40]	BBa_ J96019	JN204897
Cistron	IRES	Internal ribosomal entry site [41]	BBa_ J96040	JN204898
	P2A	Self cleaving 2A peptide [42]	BBa_ J96041	JN204899
	T2A	Self cleaving 2A peptide [42]	BBa_ J96042	JN204900

	PS3	Short DNA sequence for ribosome recruitment (mini-IRES) [43]	BBa_ J96043	JN204901
	PS4	Short DNA sequence for ribosome recruitment (mini-IRES) [43]	BBa_ J96044	JN204902
Others	Linker	24 aa flexible linker, rich in Gly and Ser.	BBa_ J96020	JN204903
	Spacer1*	Randomized DNA spacer	BBa_ J96021	
	Spacer2	Randomized DNA spacer	BBa_ J96022	JN204904
	MS2	MS2 phage coat domain binding to RNA at the MS2 binding site sequence [44]	BBa_ J96023	JN204905
	MS2BS	MS2 phage coat binding site sequence [44]	BBa_ J96024	JN204906
	LambdaN	Lambda N peptide sequence binding to RNA at the boxB binding site [45]	BBa_ J96025	JN204907
	BoxB	Lambda N peptide binding site sequence [45]	BBa_ J96026	JN204908
	TEVSite	TEV tobacco etch virus protease cleavage site [46]	BBa_ J96027	JN204909
	d1PEST	Mouse ornithine decarboxylase PEST sequence 1 hr half-life [47]	BBa_96046	JN204910
	d2PEST	Mouse ornithine decarboxylase PEST sequence 2 hr half-life [47]	BBa_96047	JN204911

To facilitate flexible construction of custom-made plasmids, all biobricks are in the same format, containing the same prefix and suffix, as described in the Methods section. Biobricks may be divided into: (i) Non-coding (not intended for containing coding sequences; lengths are often not in multiples of three, or have a stop codons in the sequence). (ii) Head domains (that contain a Kozak sequence and are used to start the translation of the protein of interest). (iii) Internal domains (the coding regions for the desired domains (e.g. reporter proteins). (iv) Tail domains (containing the stop codon used to stop the translation of the protein of interest). (v) Translation units (containing the Kozak sequence, an internal domain and a stop codon). Table 2 indicates whether a given biobrick maintains the codon frame, whether it contains a Kozak sequence or a stop codon, and shows the restriction sites present in each biobrick or plasmid backbone. In this table we explicitly present those elements that may not be used for biofusion assembly (highlighted in bold).

Table 2: Characteristics of the collection plasmids

	Name	Restriction Sites	Kozak	Stop	Backbone	Size(bp)	3n	Well
Backbones	**pSB1A3**		-	-	-	2157*	-	1A
	pSB1AK3	ClaI, HindIII, SmaI, XhoI	-	-	-	3189*	-	1B
	pSB1AC3	SacI, XhoI	-	-	-	3055*	-	1C
	pSB1AT3	BamHI, ClaI, HindIII, SalI, SphI, XhoI	-	-	-	3446*	-	1D
Construction	Kozak		YES	N	AK	12	YES	1E
	Stop		N	**YES**	AK	11	N	1F
	CMV	SacI	N	**YES**	AK	654	YES	1G
	SV40pA	SphI	N	**YES**	AK	228	YES	1H
	MCS1a	ApaI, XhoI, BamHI, HindIII	N	N	AK	32	N	2A
	MCS1b	ApaI, XhoI, BamHI, HindIII	N	N	AT	33	YES	2B
	MCS1c	ApaI, XhoI, BamHI, HindIII	N	N	AT	34	N	2C
	MCS2a	BglII, SacI, KpnI, AgeI	N	N	AT	30	YES	2D
	MCS2b	BglII, SacI, KpnI, AgeI	N	N	AT	31	N	2E
	MCS2c	BglII, SacI, KpnI, AgeI	N	N	AT	32	N	2F
	MCS3a	BsrGI, StuI, SalI, SphI, ClaI	N	N	AC	39	YES	2G
	MCS3b	BsrGI, StuI, SalI, SphI, ClaI	N	N	AC	40	N	2H
	MCS3c	BsrGI, StuI, SalI, SphI, ClaI	N	YES	AC	41	N	3A
Selection	**FNeomycin**	SphI	YES	**YES**	AK	801	YES	4A
	FPuromycin	StuI	YES	**YES**	AC	606	YES	4B
	FHSTK	ApaI, SmaI, SphI	YES	**YES**	AC	1137	YES	4C
Reporters	tdTomato		N	N	AK	1425	YES	5A

	EGFP	BsrGI	N	N	AC	714	YES	5B
	mCerulean	BsrGI, ClaI	N	N	AC	714	YES	5C
	EBFP2		N	N	AC	714	YES	5D
	mCherry		N	N	A	705	YES	5E
	RLuciferase	BsrGI, SphI	N	N	AK	930	YES	5F
Tags	Flag		N	N	AC	24	YES	6A
	HA		N	N	AC	27	YES	6B
	His		N	N	AC	18	YES	6C
	StrepII		N	N	AC	24	YES	6D
Localization	SP		YES	N	AC	69	YES	7A
	TMD		N	N	AC	147	YES	7B
	Myrist		YES	N	AC	48	YES	7C
	NLS		N	N	A	21	YES	7D
Recombina-tion	Loxp		N	N	AC	36	YES	8A
	Lox66		N	N	AC	36	YES	8B
	Lox71		N	N	AC	36	YES	8C
Cistron	**IRES**	ApaI, HindIII, KpnI	N	**YES**	AC	504	YES	9A
	P2A	SmaI	N	N	AC	42	YES	9B
	T2A		N	N	AC	54	YES	9C
	PS3	BsrGI	N	**YES**	AC	50	N	9D
	PS4		N	**YES**	AC	48	YES	9E
Others	Linker		N	N	AC	72	YES	10A
	Spacer1		N	**YES**	A	72	YES	10B
	Spacer2		N	N	AK	70	N	10C
	MS2	BglII, SacI, SalI	N	N	AK	387	YES	10D
	MS2BS	SalI	N	N	AC	39	YES	10E
	LambdaN		N	N	AC	66	YES	10F
	BoxB	ApaI	N	N	AC	21	YES	10G
	TEVSite		N	N	AC	21	YES	10H
	d1PEST		N	**YES**	AT	129	YES	11A
	d2PEST		N	**YES**	AT	129	YES	11B

Whereas the biobrick assembly workflow will work for all parts in the collection, for classical cloning it is up to the user to check which biobricks will be used to assemble a plasmid. For example, the user must check whether a given restriction enzyme site is present only once (e.g. in the MCS site region) or whether it is not unique and is also found in another biobrick or plasmid backbone.

Constructing Simple Assemblies

The biobrick assembly principle is inherently iterative, allowing longer and longer poly-biobrick constructs to be built in a stepwise fashion (Fig. 1). This is particularly useful for making related constructs with variable copies of a motif, such as a transcription factor binding sites. To illustrate this, we show how the binding sites of the bacterial MS2 phage coat protein, which can be used to repress mRNA translation in synthetic biology applications [18], can be conveniently concatenated with biobrick assembly (Fig. 1D).

One of the simplest expression plasmids that one might construct using the distribution is a plasmid such as that presented in Figure 3A; a simple EGFP expression plasmid, used to label cells, as may be observed in Figure 3B. A user will typically want to create a translation unit and insert it under a promoter with a polyadenylation sequence in the end. The translation unit will typically be composed of a head domain (in this case the Kozak sequence alone) followed by one or more internal domains (in this example the EGFP reporter coding sequence) and finished with a tail domain (here the simple stop codon biobrick).

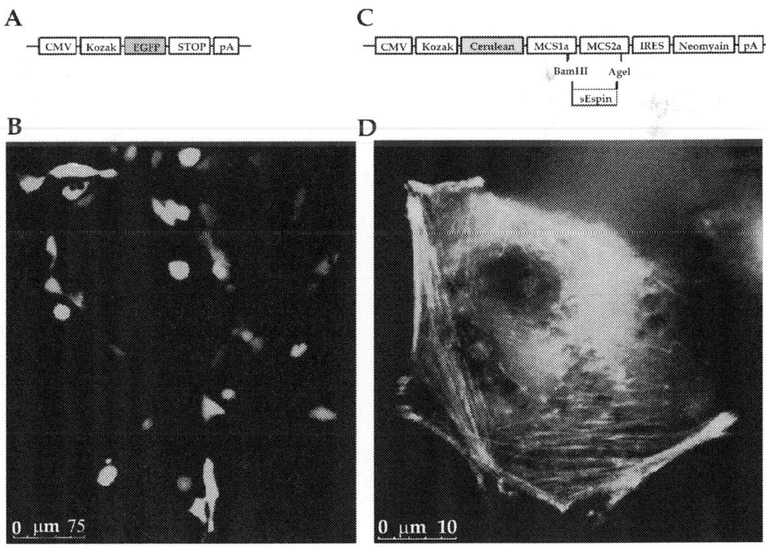

Figure 3. Examples of biobrick assemblies.

(A) Structure of a classical plasmid for EGFP expression using 5 biobricks. (B) Structure of bicistronic custom plasmid. 8 biobricks are linked together to make a construct for C-terminal fusions to the blue fluorescent protein, cerulean, using classical restriction enzyme multiple cloning sites (MCS). For illustration, the actin-bundling binding protein sEspin is cloned into the MCS, resulting in a fusion with cerulean. (C) Fluorescence microscopy image of the EGFP construct in panel A, after transient transfection into HEK293 cells. (D) A fluorescence microscopy image of the Cerulean-sEspin fusion construct in panel B, allows the visualisation of stress fiber-like structures in a HEK293T cell. Scale bars are indicated below each image.

Constructing a MCS Plasmid for Classical Cloning

Although many commercially-available plasmids contain MCSs to allow fusing proteins of interest to fluorescent proteins, many combinations of colours, N-C orientations, stable cell-line selection genes or half-life modifications are simply not available. To illustrate this we have created a more complex plasmid, containing two cistrons, that is not commercially-available. In the first cistron, we assembled the blue fluorescent protein cerulean [19] before a complex MCS (assembled using MCS1 and MCS2). We subsequently cloned an actin-binding protein, rat sEspin [17], using the BamHI and AgeI restriction sites (Figure 3C). By fluorescent microscopy, we observed Cerulean-sEspin localization in actin bundle stress fiber-like structures [17] (Fig. 3D). In the second cistron we included the neomycin resistance gene for positive selection. The example shown is just one of many customized cloning vectors that can be generated with this platform, and requires just three biobrick cloning iterations (first round: CMV-Kozak, Cerulean-MCS1a, MCS2a-IRES, Neomycin-pA; second round: CMV-Kozak-Cerulean-MCS1a, MCS2a-IRES-Neomycin-pA; third round: CMV-Kozak-Cerulean-MCS1a-MCS2a-IRES-Neomycin-pA). The use of MCSs links the biobrick format to classical cloning strategies, which should increase their appeal to researchers who simply wish to 'cut and paste' their DNA cassettes with standard restriction enzymes.

Recombinase-Mediated DNA Insertion

The collection presented here may be used to obtain stable cloning systems, for chromosomal integration in eukaryotic cells, such as Recombinase-Mediated DNA Insertion (RMDI) [20]. We implemented

RMDI by establishing a stable cell line expressing tdTomato, and containing an heteromeric lox site in its coding sequence (lox66 [21]; see scheme in Figure 4A). Through Cre-mediated recombination we inserted DNA from a donor plasmid, containing EGFP preceded by a compatible heteromeric site (lox71). Thus, red expression was converted to green expression.

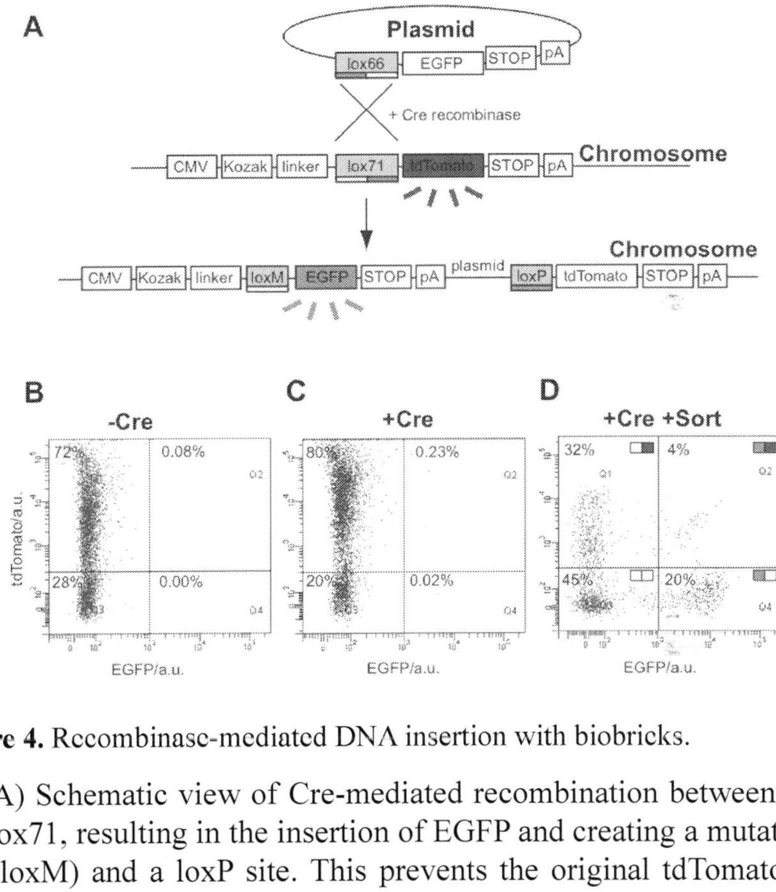

Figure 4. Recombinase-mediated DNA insertion with biobricks.

(A) Schematic view of Cre-mediated recombination between lox66 and lox71, resulting in the insertion of EGFP and creating a mutated lox site (loxM) and a loxP site. This prevents the original tdTomato from being expressed while allowing EGFP expression. (B, C and D) Flow cytometry analysis of a stable cell line expressing tdTomato, with various recombinase or cell sorting treatments. (B) The untreated cell line (−Cre) contains mainly red-positive cells (72%). (C) Upon recombinase treatment (+Cre), the amount of green cells increases in quadrants 2 and 4 (Q2, Q4). (D) By sorting the green cells (Q2 and Q4 in the previous panel), using flow cytometry (+Sort), and growing to confluency, the resulting cells are enriched for the newly-generated green cells (20%; Q4).

Since the EGFP plasmid did not contain a promoter, cells should express EGFP alone only if inserted at the correct RMDI site. Expression from random insertions next to genomic promoters (expected to be rare events) would be associated with both green and red expression. However, in the early stages after transfection, even correctly recombined cells would be expected to have some residual red expression, until the levels of tdTomato were degraded or diluted by cell division.

Indeed, after three days we observed a cell population expressing both EGFP and tdTomato by flow cytometry (0.23% for RMDI when using both the EGFP plasmid and Cre versus 0.08% for random insertion in the control with the EGFP plasmid without Cre; Fig. 4B and C). Thus, there appeared to be some Cre-induced increase in GFP fluorescence, but it was not clear whether the effects were site-specific or due to background fluorescence. Therefore, to investigate whether the tdTomato levels would subsequently fall, indicating correct RMDI, we sorted the EGFP-positive population in Figure 4C, grew them to confluency and re-analysed the final population. We thus obtained a population of cells where RMDI had indeed occurred, with high levels of EGFP and low levels of tdTomato, similar to non-fluorescent controls (20% green-only cells; Fig. 4D).

When considering only the EGFP-positive population (Q2+Q4) we note a marked shift in the Q4 population (8% of Q2+Q4 before sorting to 83% of Q2+Q4 after sorting and regrowth). This suggests that the insertion of the EGFP plasmid occurred at the intended lox site, thereby disrupting tdTomato, but that extra time was required for tdTomato levels to fall, via dilution and degradation, during regrowth of the sorted cells.

The 17% of Q2+Q4 found in Q2 is an EGFP-tdTomato double-positive population (4% of the total cell population; Fig. 4D), likely arising from cells containing multiple chromosomal copies of the initial tdTomato construct: here recombination occurred in one or more instances, but not in every copy. Random integrations of GFP near promoter regions are likely to be rare but may also contribute to this population.

We also observed tdTomato-positive EGFP-negative (32%) and tdTomato-EGFP double-negative (45%) populations. These are likely arising from cells that were unmodified and yet were carried-through the EGFP cell sorting step (false positives) or from cells that downregulated the CMV promoter [22].

DISCUSSION

In this work, we have created a distribution of DNA fragments compatible with the Standard Registry of Biological Parts [8]. This distribution is intended for the specific use of creating custom-made plasmids, with a focus on use in eukaryotic cell lines, particularly because many previous parts have focused on prokaryotic components. We have therefore constructed biobricks for elements that are routinely used in plasmids for transfection in eukaryotic cell lines. Using these elements, we provide three examples of applications of this system: the assembly of sequence motif concatemers; the assembly of vectors for transient and stable transfection; and a Cre-based Recombinase-Mediated DNA Insertion.

Although the present work makes use of existing concepts, such as the standardization of DNA parts [23] and classical cloning for expression in eukaryotic cell lines [1], the construction of the required elements into one collection and the combination of both cloning concepts makes this method a distinct new resource for preparing eukaryotic contructs. To provide this tool to users, we propose that the collection presented here be available as a stand-alone biobrick distribution exchanged in a peer-to-peer fashion, in addition to its availability through the Registry of Standard Biological Parts (http://partsregistry.org/).

A researcher may request access to the Registry to obtain a large collection of biobricks which are a valuable resource, especially to synthetic biologists. However, the size of the Registry database may be detrimental in some instances, especially for sporadic users. One example of this is the impossibility - at the moment - to search the registry database for biobricks in a specific format. In fact, special care must be taken when using biobricks from the registry since it is not always clear whether two internal protein domain biobricks may be fused in frame, because of the different biobrick formats [10], [16]. This is in part due to the definition of the registry database of biobricks as being 'compatible' with a given format when meaning they may be assembled together even if the coding frame is not maintained, whereas other users may consider 'compatibility' as an indication that the fusion conserves the codon frame.

In contrast, RFC 23 compatibility in the small collection presented here – in the sense of codon reading frame conservation – is ensured except where otherwise explicitly stated. Another advantage of having an independent small distribution for a specific function is the possibility to exchange it in 96-well plates and to be sure that one is using sequence-verified components. For example, when sequence-verifying the registry, certain discrepancies have been found [24]. The bacteria containing the plasmids can be grown and shipped in wells filled with agar medium, allowing direct sharing between peers, without need for advanced robotics or pipetting systems.

We believe that other similar biobrick libraries should be created to be self-contained. That is, all parts in such libraries should be compatible and sufficient to be used for a specific function. For instance one could envisage kinase-phosphatase or signalling cascade libraries; transcription factor libraries; metabolic enzyme libraries. Such a library-based form of distribution may be one of several solutions for the growing size of the Registry that may rapidly reach a limit in the cost-efficiency of its own distribution (the registry is >13 000 parts and growing [25]). Ultimately, we hope to stimulate debate in this growing open standard for biological engineering.

ACKNOWLEDGMENTS

The plasmid containing the rat small sEspin was a kind gift from the group of Hernán López-Schier. We would like to thank Jennifer Semple for critical reading of the manuscript.

AUTHOR CONTRIBUTIONS

Conceived and designed the experiments: MC RG MI. Performed the experiments: MC. Analyzed the data: MC RG MI. Contributed reagents/ materials/analysis tools: MC. Wrote the paper: MC RG MI.

REFERENCES

1. Sambrook J, Fritsch EF, Maniatis T (1989) Molecular Cloning: A Laboratory Manual. New York: Cold Spring Harbor Laboratory Press. J. SambrookEF FritschT. Maniatis1989Molecular Cloning: A Laboratory ManualNew YorkCold Spring Harbor Laboratory Press
2. Clontech (2005) Living Colors Fluorescent Protein Vectors. pp. 18–20.Clontech2005Living Colors Fluorescent Protein Vectors.1820Clontechniques Oct 2005. Clontechniques Oct 2005.
3. Ward TH, Lippincott-Schwartz J (2006) The uses of green fluorescent protein in mammalian cells. Methods Biochem Anal 47: 305–337. TH WardJ. Lippincott-Schwartz2006The uses of green fluorescent protein in mammalian cells.Methods Biochem Anal47305337
4. Shetty RP, Endy D, Knight TF Jr (2008) Engineering BioBrick vectors from BioBrick parts. J Biol Eng 2: 5.RP ShettyD. EndyTF Knight Jr2008Engineering BioBrick vectors from BioBrick parts.J Biol Eng25
5. Tigges M, Marquez-Lago TT, Stelling J, Fussenegger M (2009) A tunable synthetic mammalian oscillator. Nature 457: 309–312.M. TiggesTT Marquez-LagoJ. StellingM. Fussenegger2009A tunable synthetic mammalian oscillator.Nature457309312
6. Basu S, Gerchman Y, Collins CH, Arnold FH, Weiss R (2005) A synthetic multicellular system for programmed pattern formation. Nature 434: 1130–1134.S. BasuY. GerchmanCH CollinsFH ArnoldR. Weiss2005A synthetic multicellular system for programmed pattern formation.Nature43411301134
7. Guet CC, Elowitz MB, Hsing W, Leibler S (2002) Combinatorial synthesis of genetic networks. Science 296: 1466–1470.CC GuetMB ElowitzW. HsingS. Leibler2002Combinatorial synthesis of genetic networks.Science29614661470
8. Knight TF (2003) Idempotent Vector Design for Standard Assembly of BioBricks. TF Knight2003Idempotent Vector Design for Standard Assembly of BioBricksTech rep, MIT Synthetic Biology Working Group Technical Reports. Tech rep, MIT Synthetic Biology Working Group Technical Reports.
9. Goodman C (2008) Engineering ingenuity at iGEM. Nat Chem Biol 4: 13.C. Goodman2008Engineering ingenuity at iGEM.Nat Chem

Biol413

10. Phillips I, Silver PA (2006) A new biobrick assembly strategy designed for facile protein engineering. I. PhillipsPA Silver2006A new biobrick assembly strategy designed for facile protein engineering.In DSpace. : Massachussetts Institute of Technology Artificial Intelligence Laboratory, MIT Synthetic Biology Working Group. In DSpace. : Massachussetts Institute of Technology Artificial Intelligence Laboratory, MIT Synthetic Biology Working Group.
11. Canton B, Labno A, Endy D (2008) Refinement and standardization of synthetic biological parts and devices. Nat Biotechnol 26: 787–793.B. CantonA. LabnoD. Endy2008Refinement and standardization of synthetic biological parts and devices.Nat Biotechnol26787793
12. Kelly JR, Rubin AJ, Davis JH, Ajo-Franklin CM, Cumbers J, et al. (2009) Measuring the activity of BioBrick promoters using an in vivo reference standard. J Biol Eng 3: 4.JR KellyAJ RubinJH DavisCM Ajo-FranklinJ. Cumbers2009Measuring the activity of BioBrick promoters using an in vivo reference standard.J Biol Eng34
13. Rouilly V, Canton B, Nielsen P, Kitney R (2007) Registry of BioBricks models using CellML. BMC Systems Biology 1: P79.V. RouillyB. CantonP. NielsenR. Kitney2007Registry of BioBricks models using CellML.BMC Systems Biology1P79
14. Grunberg R, Ferrar TS, van der Sloot AM, Constante M, Serrano L (2010) Building blocks for protein interaction devices. Nucleic Acids Res 38: 2645–2662.R. GrunbergTS FerrarAM van der SlootM. ConstanteL. Serrano2010Building blocks for protein interaction devices.Nucleic Acids Res3826452662
15. Weeding E, Houle J, Kaznessis YN (2010) SynBioSS designer: a web-based tool for the automated generation of kinetic models for synthetic biological constructs. Brief Bioinform bbq002.E. WeedingJ. HouleYN Kaznessis2010SynBioSS designer: a web-based tool for the automated generation of kinetic models for synthetic biological constructs.Brief Bioinformbbq002
16. Anderson JC, Dueber JE, Leguia M, Wu GC, Goler JA, et al. (2010) BglBricks: A flexible standard for biological part assembly.

J Biol Eng 4: 1.JC AndersonJE DueberM. LeguiaGC WuJA Goler2010BglBricks: A flexible standard for biological part assembly.J Biol Eng41

17. Bartles JR, Zheng L, Li A, Wierda A, Chen B (1998) Small espin: a third actin-bundling protein and potential forked protein ortholog in brush border microvilli. J Cell Biol 143: 107–119.JR BartlesL. ZhengA. LiA. WierdaB. Chen1998Small espin: a third actin-bundling protein and potential forked protein ortholog in brush border microvilli.J Cell Biol143107119
18. Isaacs FJ, Dwyer DJ, Collins JJ (2006) RNA synthetic biology. Nat Biotechnol 24: 545–554.FJ IsaacsDJ DwyerJJ Collins2006RNA synthetic biology.Nat Biotechnol24545554
19. Rizzo MA, Springer GH, Granada B, Piston DW (2004) An improved cyan fluorescent protein variant useful for FRET. Nat Biotechnol 22: 445–449.MA RizzoGH SpringerB. GranadaDW Piston2004An improved cyan fluorescent protein variant useful for FRET.Nat Biotechnol22445449
20. Fukushige S, Sauer B (1992) Genomic targeting with a positive-selection lox integration vector allows highly reproducible gene expression in mammalian cells. Proc Natl Acad Sci U S A 89: 7905–7909.S. FukushigeB. Sauer1992Genomic targeting with a positive-selection lox integration vector allows highly reproducible gene expression in mammalian cells.Proc Natl Acad Sci U S A8979057909
21. Albert H, Dale EC, Lee E, Ow DW (1995) Site-specific integration of DNA into wild-type and mutant lox sites placed in the plant genome. Plant J 7: 649–659.H. AlbertEC DaleE. LeeDW Ow1995Site-specific integration of DNA into wild-type and mutant lox sites placed in the plant genome.Plant J7649659
22. Migliaccio AR, Bengra C, Ling J, Pi W, Li C, et al. (2000) Stable and unstable transgene integration sites in the human genome: extinction of the Green Fluorescent Protein transgene in K562 cells. Gene 256: 197–214.AR MigliaccioC. BengraJ. LingW. PiC. Li2000Stable and unstable transgene integration sites in the human genome: extinction of the Green Fluorescent Protein transgene in K562 cells.Gene256197214

23. Arkin A (2008) Setting the standard in synthetic biology. Nat Biotechnol 26: 771–774.A. Arkin2008Setting the standard in synthetic biology.Nat Biotechnol26771774

24. Peccoud J, Blauvelt MF, Cai Y, Cooper KL, Crasta O, et al. (2008) Targeted development of registries of biological parts. PLoS One 3: e2671.J. PeccoudMF BlauveltY. CaiKL CooperO. Crasta2008Targeted development of registries of biological parts. PLoS One3e2671

25. Galdzicki M, Rodriguez C, Chandran D, Sauro HM, Gennari JH (7005) Standard biological parts knowledgebase. PLoS One 6: e17005.M. GaldzickiC. RodriguezD. ChandranHM SauroJH Gennari7005Standard biological parts knowledgebase.PLoS One6e17005

26. Kozak M (1987) An analysis of 5′-noncoding sequences from 699 vertebrate messenger RNAs. Nucleic Acids Res 15: 8125–8148.M. Kozak1987An analysis of 5′-noncoding sequences from 699 vertebrate messenger RNAs.Nucleic Acids Res1581258148

27. Reardon JE (1989) Herpes simplex virus type 1 and human DNA polymerase interactions with 2′-deoxyguanosine 5′-triphosphate analogues. Kinetics of incorporation into DNA and induction of inhibition. J Biol Chem 264: 19039–19044.JE Reardon1989Herpes simplex virus type 1 and human DNA polymerase interactions with 2′-deoxyguanosine 5′-triphosphate analogues. Kinetics of incorporation into DNA and induction of inhibition.J Biol Chem2641903919044

28. Shaner NC, Campbell RE, Steinbach PA, Giepmans BN, Palmer AE, et al. (2004) Improved monomeric red, orange and yellow fluorescent proteins derived from Discosoma sp. red fluorescent protein. Nat Biotechnol 22: 1567–1572.NC ShanerRE CampbellPA SteinbachBN GiepmansAE Palmer2004Improved monomeric red, orange and yellow fluorescent proteins derived from Discosoma sp. red fluorescent protein.Nat Biotechnol2215671572

29. Zhang G, Gurtu V, Kain SR (1996) An enhanced green fluorescent protein allows sensitive detection of gene transfer in mammalian cells. Biochem Biophys Res Commun 227: 707–711.G. ZhangV. GurtuSR Kain1996An enhanced green fluorescent protein allows

sensitive detection of gene transfer in mammalian cells.Biochem Biophys Res Commun227707711

30. Ai HW, Shaner NC, Cheng Z, Tsien RY, Campbell RE (2007) Exploration of new chromophore structures leads to the identification of improved blue fluorescent proteins. Biochemistry 46: 5904–5910. HW AiNC ShanerZ. ChengRY TsienRE Campbell2007Exploration of new chromophore structures leads to the identification of improved blue fluorescent proteins.Biochemistry4659045910
31. Karkhanis YD, Cormier MJ (1971) Isolation and properties of Renilla reniformis luciferase, a low molecular weight energy conversion enzyme. Biochemistry 10: 317–326.YD KarkhanisMJ Cormier1971Isolation and properties of Renilla reniformis luciferase, a low molecular weight energy conversion enzyme. Biochemistry10317326
32. Einhauer A, Jungbauer A (2001) The FLAG peptide, a versatile fusion tag for the purification of recombinant proteins. J Biochem Biophys Methods 49: 455–465.A. EinhauerA. Jungbauer2001The FLAG peptide, a versatile fusion tag for the purification of recombinant proteins.J Biochem Biophys Methods49455465
33. Field J, Nikawa J, Broek D, MacDonald B, Rodgers L, et al. (1988) Purification of a RAS-responsive adenylyl cyclase complex from Saccharomyces cerevisiae by use of an epitope addition method. Mol Cell Biol 8: 2159–2165.J. FieldJ. NikawaD. BroekB. MacDonaldL. Rodgers1988Purification of a RAS-responsive adenylyl cyclase complex from Saccharomyces cerevisiae by use of an epitope addition method.Mol Cell Biol821592165
34. Hochuli E, Bannwarth W, Döbeli H, Gentz R, Stüber D (1988) Genetic Approach to Facilitate Purification of Recombinant Proteins with a Novel Metal Chelate Adsorbent. Nature Biotechnology 6: 1321–1325.E. HochuliW. BannwarthH. DöbeliR. GentzD. Stüber1988Genetic Approach to Facilitate Purification of Recombinant Proteins with a Novel Metal Chelate Adsorbent. Nature Biotechnology613211325
35. Skerra A, Schmidt TG (1999) Applications of a peptide ligand for streptavidin: the Strep-tag. Biomol Eng 16: 79–86.A. SkerraTG

Schmidt1999Applications of a peptide ligand for streptavidin: the Strep-tag.Biomol Eng167986

36. McKeown L, Robinson P, Greenwood SM, Hu W, Jones OT (2006) PIN-G–a novel reporter for imaging and defining the effects of trafficking signals in membrane proteins. BMC Biotechnol 6: 15.L. McKeownP. RobinsonSM GreenwoodW. HuOT Jones2006PIN-G–a novel reporter for imaging and defining the effects of trafficking signals in membrane proteins.BMC Biotechnol615
37. Boutin JA (1997) Myristoylation. Cell Signal 9: 15–35.JA Boutin1997Myristoylation.Cell Signal91535
38. Kalderon D, Roberts BL, Richardson WD, Smith AE (1984) A short amino acid sequence able to specify nuclear location. Cell 39: 499–509.D. KalderonBL RobertsWD RichardsonAE Smith1984A short amino acid sequence able to specify nuclear location.Cell39499509
39. Sauer B (1987) Functional expression of the cre-lox site-specific recombination system in the yeast Saccharomyces cerevisiae. Mol Cell Biol 7: 2087–2096.B. Sauer1987Functional expression of the cre-lox site-specific recombination system in the yeast Saccharomyces cerevisiae.Mol Cell Biol720872096
40. Araki K, Araki M, Yamamura K (2002) Site-directed integration of the cre gene mediated by Cre recombinase using a combination of mutant lox sites. Nucleic Acids Res 30: e103.K. ArakiM. ArakiK. Yamamura2002Site-directed integration of the cre gene mediated by Cre recombinase using a combination of mutant lox sites.Nucleic Acids Res30e103
41. Ghattas IR, Sanes JR, Majors JE (1991) The encephalomyocarditis virus internal ribosome entry site allows efficient coexpression of two genes from a recombinant provirus in cultured cells and in embryos. Mol Cell Biol 11: 5848–5859.IR GhattasJR SanesJE Majors1991The encephalomyocarditis virus internal ribosome entry site allows efficient coexpression of two genes from a recombinant provirus in cultured cells and in embryos.Mol Cell Biol1158485859
42. Szymczak AL, Workman CJ, Wang Y, Vignali KM, Dilioglou S, et al. (2004) Correction of multi-gene deficiency in vivo using a single 'self-cleaving' 2A peptide-based retroviral vector. Nat Biotechnol 22: 589–594.AL SzymczakCJ WorkmanY. WangKM VignaliS.

Dilioglou2004Correction of multi-gene deficiency in vivo using a single 'self-cleaving' 2A peptide-based retroviral vector.Nat Biotechnol22589594

43. Venkatesan A, Dasgupta A (2001) Novel fluorescence-based screen to identify small synthetic internal ribosome entry site elements. Mol Cell Biol 21: 2826–2837.A. VenkatesanA. Dasgupta2001Novel fluorescence-based screen to identify small synthetic internal ribosome entry site elements.Mol Cell Biol2128262837

44. Valegard K, Murray JB, Stockley PG, Stonehouse NJ, Liljas L (1994) Crystal structure of an RNA bacteriophage coat protein-operator complex. Nature 371: 623–626.K. ValegardJB MurrayPG StockleyNJ StonehouseL. Liljas1994Crystal structure of an RNA bacteriophage coat protein-operator complex.Nature371623626

45. Cilley CD, Williamson JR (1997) Analysis of bacteriophage N protein and peptide binding to boxB RNA using polyacrylamide gel coelectrophoresis (PACE). RNA 3: 57–67.CD CilleyJR Williamson1997Analysis of bacteriophage N protein and peptide binding to boxB RNA using polyacrylamide gel coelectrophoresis (PACE).RNA35767

46. Kapust RB, Tozser J, Copeland TD, Waugh DS (2002) The P1′ specificity of tobacco etch virus protease. Biochem Biophys Res Commun 294: 949–955.RB KapustJ. TozserTD CopelandDS Waugh2002The P1′ specificity of tobacco etch virus protease. Biochem Biophys Res Commun294949955

47. Zhao X, Duong T, Li X (2000) Comparison of Living ColorsTM EGFP and dEGFPs as Reporters. Clontechniques XV(2): 29–30.X. ZhaoT. DuongX. Li2000Comparison of Living ColorsTM EGFP and dEGFPs as Reporters.ClontechniquesXV22930

SECTION II:
PLASMID REPLICATION SYSTEMS

CHAPTER

3

SOCIOBIOLOGICAL CONTROL OF PLASMID COPY NUMBER IN BACTERIA

Mukta M. Watve, Neelesh Dahanukar, Milind G. Watve

Indian Institute of Science Education and Research, Pune, Maharashtra, India

ABSTRACT

All genes critical for plasmid replication regulation are located on the plasmid rather than on the host chromosome. It is possible therefore that there can be copy-up "cheater" mutants. In spite of this possibility, low copy number plasmids appear to exist stably in host populations. We examined this paradox using a multilevel selection model. Simulations showed that, a slightly higher copy number mutant could out-compete the wild type. Consequently, another mutant with still higher copy number could invade the first invader. However, the realized benefit of increasing intra-host fitness was saturating whereas that of inter-host fitness was exponential. As a result, above a threshold, intra-host selection was overcompensated by inter-host selection and the low copy number wild type plasmid could back invade a very high copy number plasmid. This led to a rock-paper-scissor (RPS) like situation that allowed the coexistence of plasmids with varied copy numbers. Furthermore, another type of cheater that had lost the genes required for conjugation but could hitchhike on a conjugal plasmid, could further reduce the advantage of copy-up mutants. These sociobiological interactions may compliment molecular mechanisms of replication regulation in stabilizing the copy numbers.

INTRODUCTION

Plasmids are extra chromosomal elements of circular DNA in bacteria, which replicate independent of the host genome. Plasmids exploit the machinery of host cell for their replication, but many of them carry useful or conditionally advantageous genes and therefore cannot be generalized as parasites. Many experiments have shown that the host cell has to bear a cost for carrying a plasmid [1], [2]. The cost is highly variable and ranges from undetectably small to as large as 40% [3]. The cost can be ameliorated by host parasite coevolution [4]–[6]. Although there are a number of confounding factors, the cost can be generally assumed to increase with copy number and the length of the plasmid [7], [8]. However, plasmids impart a wide range of unique features to the host by contributing to metabolic versatility and resistance to environmental factors [9], [10]. Therefore it has been argued that in the presence of positive selection for a plasmid borne gene the plasmid can be stable. However the presence of useful genes does not seem to be central to the evolution and stability of plasmids because a useful gene could ultimately be incorporated in the bacterial chromosome [11] saving the cost of carrying the plasmid. In the case of genes involved in extracellular products such as virulence factors, a 'cheater' strain that does not produce the virulence factor can invade the virulent population. Having such genes on plasmids make horizontal gene transfer possible coercing the cheaters to make the extracellular products [12]. Although such systems can be robust to bacterial cheaters, they may not be robust to plasmid cheaters with the gene for the extracellular products deleted. Such plasmids will have an equal chance of horizontal transfer but the shorter plasmids could replicate faster and replace the wild type. Therefore although many plasmids carry useful genes, that does not appear to be necessary and sufficient cause of plasmid stability.

Furthermore, it has been suggested that in the absence of selection for any plasmid borne gene a plasmid free cell should have a selective advantage and the stability of the plasmid is difficult to explain. Factors such as spread of the plasmid by conjugal transfer to compensate for the loss in host fitness, compensatory mutations, co-evolution of host and plasmid to reduce the cost [4]–[6] or plasmid addiction [13], [14] are some of the possible mechanisms responsible for maintenance of a plasmid.

Many mathematical models have focused on the problem of persistence of plasmid in the host cells [11], [15], [16]. More perplexing is the problem of stability of copy number of a plasmid in the host cell. This is one of the many paradoxes in the evolution and stability of plasmids [13]. Most of the plasmids are considerably smaller than the chromosome in terms of length of DNA. Therefore if the same machinery is being used for replication, there can be many replication cycles of the plasmid per single replication of the chromosome. This can result into rapid escalation of plasmid numbers in a cell leading to increased metabolic burden on the host cell and eventually cell death [17]. This can be prevented only by tight regulation of plasmid replication [13], [18], [19]. Ironically, all critical genes involved in the regulation of plasmid replication are on the plasmid. This raises a potential evolutionary problem. A low copy number is optimal for long term survival of the plasmid in a host cell lineage. However, any mutation that loosens the control and thereby increases the copy number has a short term advantage over the wild type although it may affect long term stability of the system. Therefore copy-up mutations can be considered as "cheaters" and such cheaters can invade the wild type and therefore low copy number is unlikely to be stable. Plasmids with copy numbers as high as 200 are observed in natural populations [20], [21] indicating that the problem is not hypothetical alone. Presence of high copy number plasmids and the stability of low copy number plasmids in spite of the potential threat of invasion is very much a real life paradox.

A number of mechanisms exist for the regulation of plasmid replication and one or more of these mechanisms can control the plasmid replication separately or in combination [9], [22]–[24]. Paulsson [16] recognized the conflict between two levels of selection in the control of plasmid copy numbers. Plasmids that systematically over-replicate relative to their cell mates have a higher chance of fixing in a cell (intra-host selection). However cells having such plasmids have to pay a greater cost and therefore are likely to be competed out by hosts with low copy numbers or ones without a plasmid (inter-host selection). Paulsson [16] suggested that intra-host selection should favor evolution of cis acting activators while inter-host selection favors evolution of retaliation by trans acting inhibitors of replication. However, such refinements in the mechanisms of replication control and reorganization of genes involved do not rule out the possibility as well as selective advantage of copy-up mutants. The demonstration of copy-up mutants in spite of multiple mechanisms of replication regulation [23] implies that the molecular mechanisms of replication regulation are not infallible and therefore unable to explain the evolutionary stability of copy numbers.

There can be multiple mechanisms by which a copy-up mutant can outcompete a low copy wild type. The effective growth rate of the mutant in terms of the number of replications per host cell cycle is greater than the wild type resulting into progressive dilution of the wild type. Since the wild type replication gets inhibited at a lower copy number and if the copy number control mechanism operates in response to the total copy number, the mutant can suppress the replication of the wild type. Also if we assume that a random copy is transferred in a conjugational event, the probability of getting transferred will be higher for a copy-up mutant. Despite a multiplicity of potential advantages of high copy numbers, plasmids with low copy numbers are surprisingly common. We therefore need to look at other mechanisms for the stability of low copy number plasmids.

It is known that host cells already bearing a plasmid are at least partially immune to conjugal intake of another plasmid [25]. This is a likely mechanism by which horizontal invasion by copy up mutants can be arrested. It has not been critically examined theoretically or experimentally whether this mechanism is necessary and sufficient to prevent high copy number cheaters. Although a host cell lineage having a low copy number plasmid may be immune to conjugational invasion by another plasmid, mutants can nevertheless arise within this population and get transmitted vertically in the cell lineage as well as horizontally to plasmid free cells. In order to test whether this mechanism is necessary to prevent invasion by high copy number mutants, the best approach would be to start with a model without any such mechanism and see whether stability of low copy numbers is possible.

We explore here the possibility that copy numbers are selected by complex sociobiological interactions and are always in a dynamic steady state emerging out of conflicting levels of selection. We use a multilevel selection approach similar to Paulsson [16] but instead of looking at the evolution of molecular mechanisms we focus on the long term population dynamics of different types of plasmid and host populations in a competitive environment.

Apart from copy numbers another type of social cheating is possible in plasmid populations. Plasmids spread by conjugal transfers and the genes for conjugal mechanisms are borne by plasmids. There is likely to be a cost to the plasmid (in terms of increased replication time) in carrying the conjugal genes and making the conjugal machinery would be a cost to the host cell. Non-conjugal plasmids can arise as cheaters in such a system and hitch-hike on conjugal plasmids during conjugal transfer. Using our model system we also examine how the two types of cheaters would interact.

METHODS

The Model

We assumed four different variants or mutants of a plasmid. The wild type or lc is a low copy number plasmid with stringent control over replication and having conjugal abilities (Table 1). Plasmid ln is a mutant with deletion of the gene cluster required for conjugation (such as the tragene complex) but retains constraint on copy number. Plasmid hc is a mutant with less constrained replication regulation resulting into higher copy number, but retaining the tra gene complex. Plasmid hn is a mutant of hc that has lost the tra gene complex. The host cells can be free of plasmid (x_0) or have one of the 4 types of plasmids (x_{lc}, x_{ln}, x_{hc}, x_{hn} respectively) or can have multiple infections (x_m, see below). The cost of carrying a plasmid which is assumed to be directly proportional to the copy number (g_{lc}, g_{ln}, g_{hc}, g_{hn}) affects the intrinsic growth rates of the five types of host populations. The cost of carrying the tra gene complex is also assumed to affect the rate of replication of the plasmids as well as the host cells carrying them. Therefore the intrinsic growth rate of a host cell type (r_{lc}, r_{ln}, r_{hc}, r_{hn}) is calculated as the baseline fitness (r_0, assumed to be 1) minus the cost of carrying a particular plasmid and the cost of conjugation if the plasmid bears the conjugal machinery. Therefore,

$$r_0 = 1$$

$$r_{lc} = 1 - \phi * g_{lc} - \lambda$$

$$r_{ln} = 1 - \phi * g_{ln}$$

$$r_{hc} = 1 - \phi * g_{hc} - \lambda$$

$$r_{hn} = 1 - \phi * g_{hn}$$

Where φ is the cost paid by the host for the intra-host fitness of plasmid, λ is the cost of conjugation for the host and g_i is the intra-host fitness of plasmid of type i (where i ∈ lc, ln, hc, hn). Fitness of the given plasmid can be given as

$$g_{lc} = 1 - \mu$$

$$g_{ln} = 1$$

$$g_{hc} = 1 + \delta - \mu$$

$$g_{hn} = 1 + \delta$$

Where μ is the cost of carrying conjugal genes and δ is the additive fitness due to extra copy number of the plasmid. We assume that the fitness δ is a linear function of plasmid copy number. The growth rates g_i reflect the competitive advantages of different types of plasmids.

In the base line model we do not assume any plasmid incompatibility and multiple infections are permitted. The four types of plasmids could result in many possible combinations of plasmids in cells with multiple infections. However, for simplicity of the model we considered only a single category x_m that represented the pooled population of all cells with multiple infections having different proportions (p_{lc}, p_{ln}, p_{hc}, p_{hn}) of the four plasmids. Although this assumption is naïve, it can be seen that it is unlikely to affect the model results qualitatively. For example if we make a limiting assumption that all types of plasmids co-occur in a single cell, the probability that a non-conjugal ln type of plasmid will get transferred by hitch-hiking on the conjugal plasmid lc can be assumed to be a function of the proportion of lc and that of ln.

Table 1: Symbols used in the model with description.

Symbol	Description
lc	Low copy number plasmid with conjugation ability
ln	Low copy number plasmid without conjugation ability
hc	High copy number plasmid with conjugation ability
hn	High copy number plasmid without conjugation ability
x_0	Frequency of plasmid free cell
x_i	Frequency of cell infected by plasmid type *i**
x_m	Frequency of cell infected by multiple types of plasmid
r_0	Intrinsic growth rate of plasmid free cell
r_i	Intrinsic growth rate of cell infected by plasmid type *i*
r_m	Intrinsic growth rate of cell infected by multiple types of plasmid
g_i	Fitness of plasmid of type *i*
p_i	Frequency of plasmid of type *i* in host cell with multiple infection
ϕ	Cost paid by the host for the fitness of plasmid
λ	Cost of conjugation for the host
μ	Cost of conjugal efficiency for the plasmid
δ	Additive fitness due to extra copy number of the plasmid
ε	Rate of plasmid curing
β_l	Conjugation frequency of the low copy number plasmid (*lc*)
β_h	Conjugation frequency of the high copy number plasmid (*hc*)

*$i \in lc, ln, hc, hn$.

doi:10.1371/journal.pone.0009328.t001

Thus the transfer rates can be given as $\beta_l\ p_{lc}\ p_{ln}$. If on the other hand we assume that plasmid types co-occur only in pairs then a non-conjugal plasmid can get hitch hiked only when it pairs with a conjugal plasmid. The probability of the two co-occurring in a host cell will be a function of the proportion of conjugal plasmid and that of non-conjugal plasmid which can be written as $p_{ln}\ p_{lc}$ and the rate of plasmid transfer as $\beta_l\ p_{lc}\ p_{ln}$. Thus both the limiting assumptions give rise to very similar mathematical forms. Therefore we believe that treating multiple infections as a single category is unlikely to bias the results. Treating every possible combination of plasmids in mixed infection will add 11 more population types making the model highly complex and therefore to avoid the complexity we merged them into a single category. The fitness of multiple infected cell (r_m) was calculated by taking the weighted average cost of all plasmids. Plasmid curing was assumed to take place spontaneously with a constant rate ε. The assumption that plasmid curing occurred at a constant rate was not completely arbitrary. For plasmids with higher copy number the probability of curing due to stochastic unequal segregation is low by chance alone while for plasmids with low copy number selection favors evolution of efficient partition system reducing the chances of spontaneous curing [16]. As a result, the plasmid curing rate need not scale with copy number.

Conjugation and curing is assumed to result in the transformation of one type of cell to another as given in Figure 1. We assume that only one type of plasmid is transferred or cured at a time so that a cell does not go from an uninfected state to a multiple infection state in a single step. Also, a multiple infected cell does not become completely cured of plasmids in single time step.

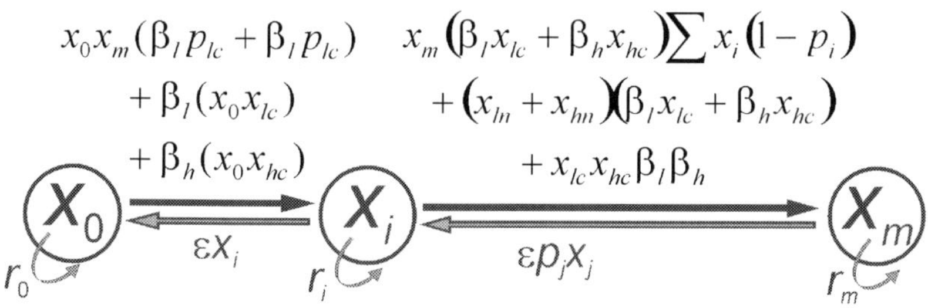

Figure 1: Transformations of cell types due to conjugation and segregation of plasmids.

Only one plasmid is assumed to get transferred or get cured at a time. Therefore x_0 cannot be directly transformed into x_m and vice versa. Black arrows indicate conjugation and grey straight arrow indicates curing of plasmid. Curved grey arrows are intrinsic growth rates of respective cell types. Symbols used for the model are explained in Table 1.

Let β_l and β_h be the conjugation frequency of the low copy number and high copy number plasmids (plasmid lc and hc respectively). The dynamics of transfer of one host to another host can be realized from Table 2. Based on Table 2 we can write the differential equations which govern the change in the frequency of different types of cells by adding all the terms that contribute to the change in the cell type. For instance, the differential equation for a plasmid free cell can be given as,

$$\dot{x}_0 = x_0 r_0 + \varepsilon(x_{lc} + x_{\ln} + x_{hc} + x_{hn}) - (x_0 x_m)(\beta_l p_{lc} + \beta_h p_{hc})$$
$$- \beta_l(x_0 x_{lc}) - \beta_h(x_0 x_{hc}).$$

Similarly, the dynamics of change in the proportions of plasmids (p_{lc}, p_{ln}, p_{hc}, p_{hn}) in the case of multiple infections can be written based on Table 3.

Table 2: Dynamics of change in the population of cell types.

Cell type	Cell growth	Addition due to curing of plasmids of other hosts	Decrease in the population due to curing of plasmid	Transfer to other cell type due to receiving plasmid from conjugation	Transferred from other cell type to current cell type due to receiving plasmid from conjugation
x_0	$x_0 r_0$	$\varepsilon (x_{lc}+x_{ln}+x_{hc}+x_{hn})$	NA	$-x_0 x_m(\beta_l p_{lc}+\beta_h p_{hc})$ $-\beta_l x_0 x_{lc}-\beta_h x_0 x_{hc}$	NA
x_{lc}	$x_{lc} r_{lc}$	$\varepsilon (p_{lc} x_m)$	$-\varepsilon x_{lc}$	$-x_{lc} x_m(1-p_{lc})(\beta_l p_{lc}+\beta_h p_{hc})$ $-\beta_h x_{lc} x_{hc}$	$x_0 x_m p_{lc} (\beta_l p_{lc}+\beta_h p_{hc})$ $+\beta_l x_0 x_{lc}$
x_{ln}	$x_{ln} r_{ln}$	$\varepsilon (p_{ln} x_m)$	$-\varepsilon x_{ln}$	$-x_{ln} x_m (1-p_{ln})(\beta_l p_{lc}+\beta_h p_{hc})$ $-x_{ln} (\beta_l x_{lc}+\beta_h x_{hc})$	$x_0 x_m p_{ln} (\beta_l p_{lc}+\beta_h p_{hc})$
x_{hc}	$x_{hc} r_{hc}$	$\varepsilon (p_{hc} x_m)$	$-\varepsilon x_{hc}$	$-x_{hc} x_m(1-p_{hc})(\beta_l p_{lc}+\beta_h p_{hc})$ $-\beta_l x_{lc} x_{hc}$	$x_0 x_m p_{hc} (\beta_l p_{lc}+\beta_h p_{hc})$ $+\beta_h x_0 x_{hc}$
x_{hn}	$x_{hn} r_{hn}$	$\varepsilon (p_{hn} x_m)$	$-\varepsilon x_{hn}$	$-x_{hn} x_m(1-p_{hn})(\beta_l p_{lc}+\beta_h p_{hc})$ $-x_{hn} (\beta_l x_{lc}+\beta_h x_{hc})$	$x_0 x_m p_{hn} (\beta_l p_{lc}+\beta_h p_{hc})$
x_m	$x_m(p_{lc} r_{lc}+p_{ln} r_{ln}$ $+p_{hc} r_{hc}+p_{hn} r_{hn})$	NA	$-\varepsilon x_m$	NA	$x_m (\beta_l x_{lc}+\beta_h x_{hc}) \{x_{lc} (1-p_{lc})$ $+x_{ln} (1-p_{ln})+x_{hc} (1-p_{hc})$ $+x_{hn} (1-p_{hn})\}+x_{lc} x_{hc} \beta_l \beta_h$ $+(x_{ln}+x_{hn})(\beta_l x_{lc}+\beta_h x_{hc})$

doi:10.1371/journal.pone.0009328.t002

Table 3: Dynamics of change in the proportion of different plasmids in multi-plasmid infected cell.

Plasmid type	Cell growth	Replication of plasmid within host	Addition due to conjugal transfer of the plasmid	Plasmid lost by curing
p_{lc}	$p_{lc} x_m$	$p_{lc} x_m g_{lc}$	$p_{lc} x_m (x_{ln} + x_{hc} + x_{hn})(\beta_l p_{lc} + \beta_h p_{hc}) + \beta_l p_{lc} (x_{ln} + x_{hc} + x_{hn}) + \beta_h x_{lc} x_{hc}$	$-\varepsilon x_m p_{lc}$
p_{ln}	$p_{ln} x_m$	$p_{ln} x_m g_{ln}$	$p_{ln} x_m (x_{lc} + x_{hc} + x_{hn})(\beta_l p_{lc} + \beta_h p_{hc}) + x_{ln} (x_{lc} + x_{hc} + x_m (1-p_{ln}))(\beta_l p_{lc} + \beta_h p_{hc})$	$-\varepsilon x_m p_{ln}$
p_{hc}	$p_{hc} x_m$	$p_{hc} x_m g_{hc}$	$p_{hc} x_m (x_{lc} + x_{ln} + x_{hn})(\beta_l p_{lc} + \beta_h p_{hc}) + \beta_h p_{hc} (x_{lc} + x_{ln} + x_{hn}) + \beta_l x_{lc} x_{hc}$	$-\varepsilon x_m p_{hc}$
p_{hn}	$p_{hn} x_m$	$p_{hn} x_m g_{hn}$	$p_{hn} x_m (x_{lc} + x_{ln} + x_{hc})(\beta_l p_{lc} + \beta_h p_{hc}) + x_{hn} (x_{lc} + x_{hc} + x_m (1-p_{hn}))(\beta_l p_{lc} + \beta_h p_{hc})$	$-\varepsilon x_m p_{hn}$

doi:10.1371/journal.pone.0009328.t003

Simulations

Analytical solutions were possible when only one type of plasmid or two types of host cells were present. Considering all plasmid types and cell types was difficult to study analytically owing to the inherent complexity of the model. Therefore we analyzed the model numerically. All simulations began with a plasmid free cell population to which all plasmid types were invaders. We assumed that any of the plasmid types can arise by mutation and a low frequency of mutation selection balance was maintained for all the types so that they had a fair chance of back invasion at any point in time. The total population of host cells as well as that of plasmids in x_m was normalized to unity at the end of each time unit.

To study the effect of copy number increase with small increments, we adopted the following approach. If an lc plasmid was successfully invaded by an hc mutant, the parameters of the hcinvader were made the parameters of lc in the following round of simulations and a new hc with slightly higher copy number was introduced. At each turn of the simulation cycle back invasion by the initial low copy number wild type was also attempted. A similar approach was adopted to study the effects of other parameters such as conjugation efficiency.

Results

When only the wild type (lc) plasmid was present, equilibrium was established between the plasmid free and plasmid bearing cells owing to a balance between the greater inter-host fitness of plasmid free population, plasmid transfer by conjugation and spontaneous plasmid curing. The differential equations governing the change in population of plasmid free cells and lc plasmid containing cells can be given as

$\dot{x}_0 = x_0 r_0 + \varepsilon\, x_{lc} - \beta_l (x_0 x_{lc})$ and $\dot{x}_{lc} = x_{lc} r_{lc} - \varepsilon\, x_{lc} + \beta_l (x_0 x_{lc})$

At equilibrium,

$$\frac{\dot{x}_0}{x_0} = \frac{\dot{x}_{lc}}{x_{lc}}, \text{ implies}$$

$$\frac{x_0 r_0 + \varepsilon\, x_{lc} - \beta_l (x_0 x_{lc})}{x_0} = \frac{x_{lc} r_{lc} - \varepsilon\, x_{lc} + \beta_l (x_0 x_{lc})}{x_{lc}}$$

It can be seen that as $x_{lc}+x_0=1$, the stable ratio at equilibrium would be

$$\frac{x_{lc}}{x_0} = \frac{\beta_l + r_{lc} - r_0 - \varepsilon}{\varepsilon}$$

Numerical simulations matched the analytical solution.

When a mutant with a higher copy number (hc plasmid) was introduced in the presence of wild type (lc) plasmid, provided the difference in copy number was not too large, the wild type was driven to extinction or very low frequencies by the invader (Figures 2A and 2B). This was obvious as the mutant had more intra-host fitness than the wild type plasmid. The mutant with a higher copy number was vulnerable to invasion by another mutant that had higher copy number, which in turn was vulnerable to invasion by another mutant with still higher copy number and so on. Thus, mutants with slightly higher fitness invaded the ones with lower fitness. This process could increase the copy number to a very large value especially when the host cost of plasmid fitness (i.e. cost of carrying additional plasmid copies) was small. However, at some stage the wild type plasmid with low copy number could back invade the high copy number population (Figures 3A and 3B). This was owing to the large metabolic burden on the host cell carrying the high copy number plasmid. The back invasion was possible only when the copy number of the invading variant was substantially smaller. The back invading low copy plasmid was eventually susceptible to invasion by slightly high copy number variant. Therefore the system did not have a unique stable equilibrium.

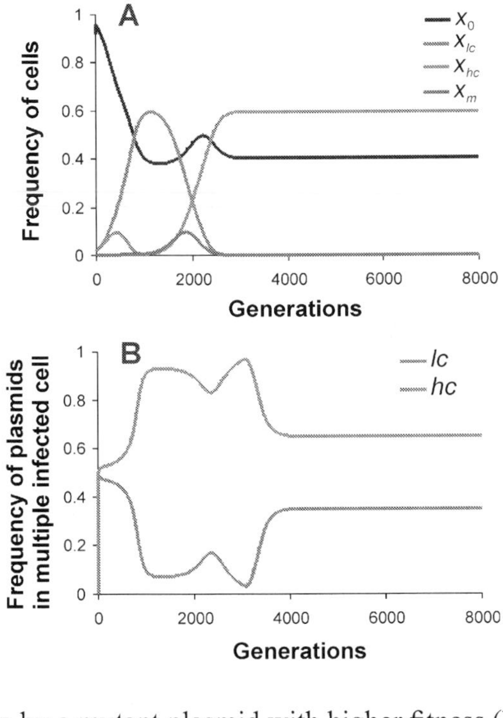

Figure 2: Invasion by a mutant plasmid with higher fitness (hc plasmid).

(A) Populations of host cells and (B) proportions of plasmids in host cells with multiple infection. Parameter values for the figure are $\mu=0.2$, $\lambda=0.02$, $\varphi=0.004$, $\varepsilon=0.01$, $\beta_l=0.05$, $\beta_h=0.05$, $\delta=0.5$.

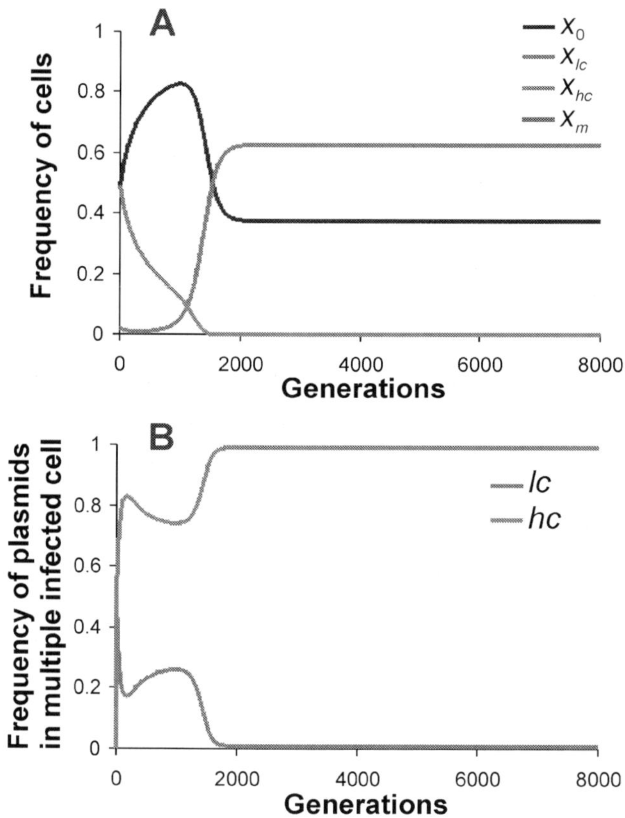

Figure 3: Mutant plasmid with exorbitantly high copy number is invaded back by wild type plasmid.

(A) Populations of host cells and (B) proportions of plasmids in host cells with multiple infection. $\delta=5$ all other parameters as in Fig 2.

The major reason for this dynamics was the difference in the effects of inter and intra-host fitness. In order to differentially study and compare the effects of the two we defined realized fitness as the net increment in the total (including all host cells) plasmid proportion over time t (t=100 in the simulations). A common problem with multi-level selection is that the different levels of selection work with different net effects on the realized fitness. We took the approach of studying the nature of the relationship between fitness levels with realized fitness taking one level at a time. To analyze the effects of intra-host fitness on realized fitness, inter-host fitness difference was set to zero by making $\varphi=0$ and the increment in the net population of plasmid over time t was monitored at different values of δ. Inversely to study the realized inter-host fitness in isolation, δ was set to zero and the increment in the net population of the plasmid over time t was monitored at increasing inter-host fitness. It was observed that with increasing δ the realized fitness increased in a saturation curve. On the other hand, the realized fitness increased exponentially with increasing inter-host fitness difference as shown schematically in Figure 4A and B. The saturating nature of the realized fitness curve with increasing intra-host fitness was owing to the inherent limitation on growth within a host cell and a constant and limiting rate of transfer across cells. Although a plasmid with a very high copy number could replace the wild type very rapidly within a host cell clone, its spread to other cells was not equally rapid being limited by the rate of conjugal transfer which was assumed to be constant. As a result of the two different natures of the curves, with a small increment in copy number the intra-host fitness advantage dominated over the inter-host fitness but with a larger difference, the inter-host fitness increased more sharply to overcompensate for the saturating effects of intra-host fitness and resulted in a net loss to the high copy number plasmid.

A

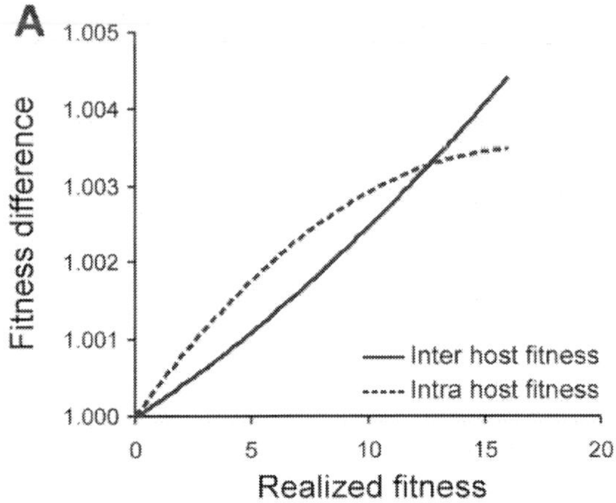

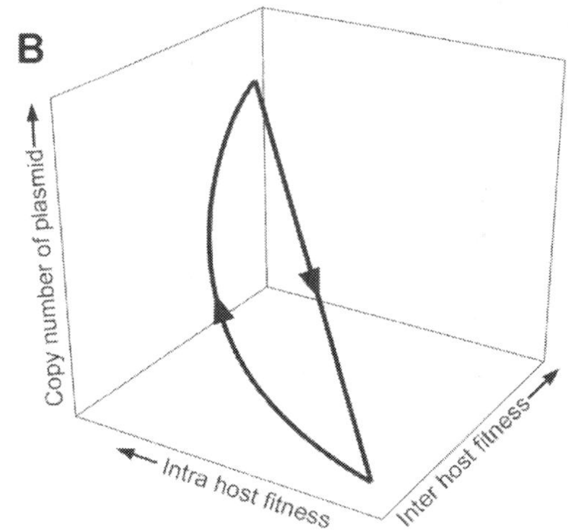

Figure 4: A conceptual diagram based on generalization from many numerical results, depicting the effect of intra and inter-host fitness on realized fitness.

(A) The realized fitness is a saturating function of intra-host fitness and an exponential function of inter-host fitness. As a result of the different nature of the two curves, RPS like dynamics results which is depicted schematically in (B). Arrows indicate the direction of evolutionary change. Mutants with slightly higher copy numbers can invade lower copy number plasmids owing to higher intra-host fitness. As a result the mean copy number increases, however the effect of intra-host fitness is saturating and therefore at a certain stage the reducing positive increment in realized fitness is over-compensated by increasing negative contribution of inter-host fitness difference. At this stage a low copy number plasmid can back invade resulting into a cyclic evolutionary dynamics.

The resulting dynamics were similar to Rock-Paper-Scissors (RPS) game [26]–[28] where plasmids with low copy number were successively invaded by plasmids with higher copy number due to the increase in the intra-host fitness till a point where the intrinsic growth rate of the host was greatly reduced, due to increased metabolic burden, and owing to the inter-host fitness difference the plasmids with lower copy number could back invade the population and the cycle started again (Figure 4B). It is well known that RPS game leads to oscillating or stable coexistence of all the three strategies [26]–[28]. Therefore we expect that low intermediate and high copy number plasmids could coexist in a host species in a dynamic equilibrium. The only difference between the classical RPS game and the plasmid dynamics is that the mean copy numbers of plasmids can vary on a single continuous scale and therefore have a number of intermediates while classically rock-paper-scissor are three qualitatively distinct and discrete strategies.

Introduction of the second type of cheating in the form of ln and hn plasmids changed the picture considerably. Plasmids ln and hn were assumed not to carry the tra gene complex required for conjugation but in the presence of lc or hc plasmids they could get transferred to other hosts through conjugal hitchhiking. The absence of tra gene complex was assumed to reduce cost and thereby increase the fitness of the plasmid as well as the host carrying it. This led to an increase in efficient vertical transfer of the plasmid but the horizontal transfer was restricted from host cells co-infected with lc and/or hc and ln and/or hn. In pair wise interactions ln plasmid could invade (but not replace) lc plasmid but not the hc plasmid. The invasion was restricted and led to a stable coexistence of the wild type and cheater. Since the cheater depended upon the wild type for conjugal transfer, it could not drive the wild type to extinction. Plasmid hn could invade (but not replace) lc as well as hc plasmids and also outcompeted lninvariably.

When all the four types were simultaneously included in the simulations there were interesting consequences of the interaction between the two types of cheaters. Whenever the hn plasmid successfully established in the population it reduced the fitness advantage of hc. Since hnshared the benefit of higher copy number with hc and also had a lower cost, it had a higher intra-host fitness then hc. In a typical result (Figure 5A and 5B) hc rapidly increased in frequency but was soon followed by hn which largely replaced hc in co-infected host. The reduction in the frequency of hc was followed by back invasion by lc. Since hn had a limited capacity to invade lc, both lc and hn coexisted in a stable steady state where the pure plasmid host population was dominated by lc and co-infected host population was dominated by hn. Keeping other parameters the same omission of hn resulted in invasion by hc and near extinction of lc (Figures 5C and 5D).

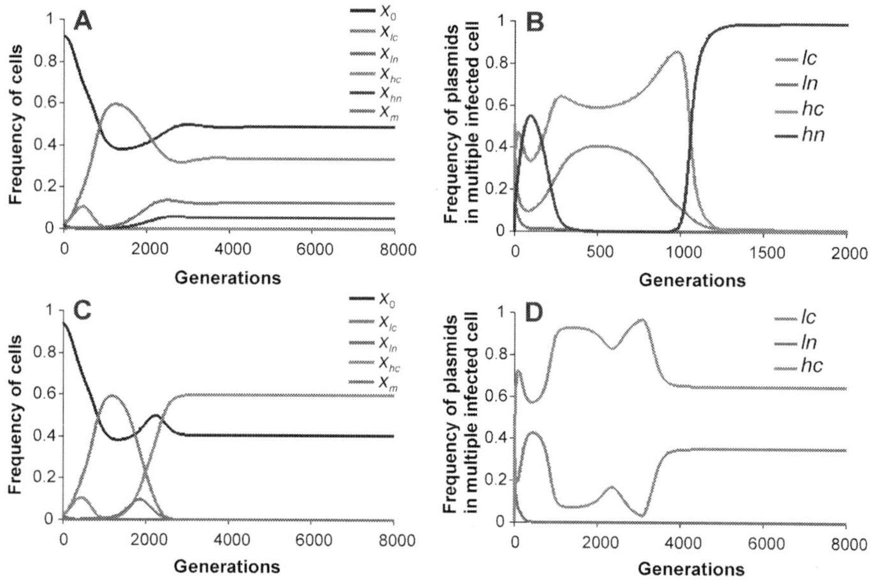

Figure 5. The interaction of the two types of cheaters.

Successful invasion of hc plasmid by hn plasmid in cells with multiple infection brings the wild type back (A, B). Parameter values for the figure are as per Fig. 2. In the absence of hn plasmid, for the same parameters, hc plasmid dominates the population (C and D).

Across a wide range of costs of the plasmid (Figure 6) the non-conjugal plasmids were observed to restrict the width of copy numbers over which the RPS situation worked. The low copy number wild type was able to back invade the higher copy mutants at a much smaller copy number when non-conjugal plasmids were present. The two modes of cheating thus had antagonistic effects resulting into effective arrest of escalation. The RPS dynamics itself was robust to changes in other parameters. Throughout the range of parameters that did not result into extinction of all types of plasmids, low copy number plasmids could exist either stably or in RPS dynamics. The RPS limit lines (Figure 6) were unaffected by changes in ε as long as it was independent of the copy number. Both μ and λ pushed down the limit line marginally in the absence of cheaters. However, in presence of non-conjugal cheaters μ exerted a strong effect in lowering the RPS limit whereas λ had a marginal effect. In other words increase in μ helped the stability of low copy number plasmids in presence of non-conjugal cheaters.

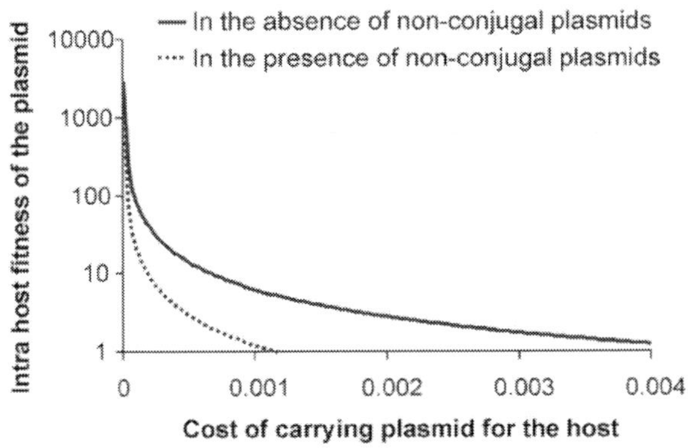

Figure 6. The maximum intra-host plasmid fitness as a function of plasmid cost.

The lines show the limit of intra-host fitness above which the wild type can back invade. In presence of the non-conjugal cheaters the limit is substantially reduced. All other parameters as in Fig 2.

Unlike the fitness effect of copy number, the conjugation efficiencies β_l and β_h had a linear effect on fitness. Plasmid with higher conjugation efficiency was selected over one with lower efficiency when other parameters were equal. Therefore although conjugal and non-conjugal plasmids coexisted, coexistence of different conjugation efficiencies was not possible. Since selection for conjugation efficiencies was linear, there was a unidirectional selection for highest conjugation efficiency. We must assume then that the rate of conjugation is limited by physical factors and availability of recipient cells rather than the efficiency of conjugation machinery.

Reducing the probability of conjugal transfer to any cell with pre-existing plasmid reduced the frequency of co-infection. This could effectively make intra-host selection weaker and as a result, low copy number plasmids increased substantially in frequency. Prevention of co-infection can therefore enhance the stability of low copy numbers. However, the model showed that preventing co-infection was not necessary for the stability of low copy numbers.

DISCUSSION

The model shows that low copy number plasmids could be maintained in spite of a threat by copy-up mutants owing to a continuous RPS like relationship in mutants with low, intermediate and high copy numbers. Although multi-level selection in plasmids [16] and other host parasite interactions [29], [30] has been examined before, co-existence mediated by an RPS like situation was not predicted so far. Paulsson [16] realized that a selfish plasmid was at a net advantage "...as long as it is not too selfish" but failed to extend this logic to visualize the RPS like consequence of it. The RPS dynamics results from the two types of non-linear effects of intra and inter-host fitness. Inter-host fitness working in an exponentially growing host population has an exponential effect on the realized fitness. On the other hand, intra-host fitness has an inherent upper limit since they apply only to the currently infected host. The saturation is also owing to the rate limiting nature of conjugation. Interaction of these two types of curves depicted in Figure 4 is fundamental to the central result of the model.

The assumption of rate limiting nature of conjugation is not arbitrary. The efficiency of conjugation does not bear a saturating relationship with realized fitness and therefore both low and high copy number plasmids should evolve conjugation machinery with high efficiency. The conjugation rate then should be limited by physical factors and the probability of finding recipient cells. Therefore the conjugation efficiency can be assumed to be constant independent of the intra-host fitness.

The results of this model may be generalized to any situation of conflict between intra-group and inter-group selection. It is likely to work similarly and give rise to RPS mediated coexistence if only one basic requirement is met with, that is the intra-group fitness advantage needs to be a saturating function. It is natural for the inter-group fitness advantage to be exponential since most biological populations are assumed to replicate giving rise to exponential kinetics. Generalizing the model is likely to give us more insights into the dynamics of group selection and conflicts between levels of selection.

If cells with preexisting plasmids refuse to accept a conjugal transfer the frequency of co-infection will be reduced making inter-host fitness stronger. This strategy however is not fool proof since co-infections can also arise by mutations within cells. Secondly the non-conjugal plasmids should evolve to welcome co-infection by accepting conjugal transfer since they can increase their fitness in co-infected cells. Therefore prevention of co-infection by incompatibility or rendering conjugation immunity to the host cell is unlikely to be a stable and effective strategy by itself. Our model shows, on the other hand that low copy numbers can be effectively maintained even without prevention of co-infection. Therefore this mechanism appears neither necessary nor sufficient but nevertheless helpful for the stability of low copy numbers. An important testable prediction of the model is that low copy number conjugal plasmids should confer immunity to the host against conjugal invasion whereas high copy number non-conjugal plasmids should welcome conjugal invasion by other plasmids. Another testable prediction of the model is that plasmids with different copy numbers would be present in any given host species within a single population or metapopulations. Metapopulation dynamics is likely to affect coexistence further which has not been examined as yet but which is beyond the scope of this paper. Nevertheless the model predicts variability in plasmids in a host population and variations in copy number and associated cost have indeed been shown by plasmids over different host strains [3] and over time [4]–[6]. It would also be possible to study the long term dynamics of plasmid varieties in a sufficiently large host population using a continuous culture system and using carefully selected plasmid types in pairs and in complex combinations.

The further twist to the story added by the second type of cheating is not hypothetical alone. It is well known that non-conjugal plasmids are common and they can undergo conjugal transfer in presence of conjugal plasmids [9], [31]–[33]. Two mechanisms can be employed by the non-conjugal plasmids for hitch-hacking. First method is called mobilization, a process by which plasmids achieve transfer by 'borrowing' the gene products of a conjugal plasmid, as in the case of ColE1 and RSF1010 plasmids of Escherichia coli. The second method is called conduction, which involves a physical association with conjugal plasmid forming co-integrate that may resolve into two plasmids in the recipient. The model predicts co-existence of conjugal and non-conjugal plasmids along with low and high copy number plasmids and further that non-conjugal plasmids with very low copy numbers are unlikely to be stable since ln plasmids were always invaded by hn. Therefore all non-conjugal plasmids should have high copy numbers. This can be taken as another testable prediction of the model. The interaction between the two types of cheating is important for the maintenance of plasmid diversity along both the axes.

While evolution would have shaped both molecular and sociobiological processes, much research has focused on the molecular mechanisms of plasmid replication dynamics and the sociobiological interactions have largely been ignored. We tried to develop some insights in the possible sociobiological factors in plasmid dynamics. It would be interesting to further elucidate the interplay between molecular and sociobiological mechanisms in plasmid dynamics.

AUTHOR CONTRIBUTIONS

Conceived and designed the experiments: MMW MGW. Performed the experiments: MMW MGW. Analyzed the data: MMW ND MGW. Wrote the paper: ND MGW.

REFERENCES

1. Freter R, Freter RR, Brickner H (1983) Experimental and mathematical models of Escherichia coli plasmid transfer in vitro and in vivo. Infection Immunity 39: 60–84.R. FreterRR FreterH. Brickner1983Experimental and mathematical models of Escherichia coli plasmid transfer in vitro and in vivo.Infection Immunity396084
2. Gelder LD, Ponciano JM, Abdo Z, Joyce P, Forney LJ, Top EM (2004) Combining mathematical models and statistical methods to understand and predict dynamics of antibiotic-sensitive mutants in a population of resistant bacteria during experimental evolution. Genetics 168: 1131–1144.LD GelderJM PoncianoZ. AbdoP. JoyceLJ ForneyEM Top2004Combining mathematical models and statistical methods to understand and predict dynamics of antibiotic-sensitive mutants in a population of resistant bacteria during experimental evolution.Genetics16811311144
3. Gelder LD, Ponciano JM, Joyce P, Top EM (2007) Stability of a promiscuous plasmid in different hosts: no guarantee for a

long-term relationship. Microbiol 153: 452–463.LD GelderJM PoncianoP. JoyceEM Top2007Stability of a promiscuous plasmid in different hosts: no guarantee for a long-term relationship. Microbiol153452463
4. Lenski RE, Simpson SC, Nguyen TT (1994) Genetic analysis of a plasmid-encoded, host genotype-specific enhancement of bacterial fitness. J Bacteriol 176: 3140–3147.RE LenskiSC SimpsonTT Nguyen1994Genetic analysis of a plasmid-encoded, host genotype-specific enhancement of bacterial fitness.J Bacteriol17631403147
5. Dahlberg C, Chao L (2003) Amelioration of the cost of conjugative plasmid carriage in Eschericha coli K12. Genetics 65: 1641–1649.C. DahlbergL. Chao2003Amelioration of the cost of conjugative plasmid carriage in Eschericha coli K12.Genetics6516411649
6. Dionisio F, Conceicao IC, Marques ACR, Fernandes L, Gordo I (2005) The evolution of a conjugative plasmid and its ability to increase bacterial fitness. Biol Lett 1: 250–252.F. DionisioIC ConceicaoACR MarquesL. FernandesI. Gordo2005The evolution of a conjugative plasmid and its ability to increase bacterial fitness. Biol Lett1250252
7. Smith MA, Bidochka MJ (1998) Bacterial fitness and plasmid loss: importance of culture conditions and plasmid size. Canadian J Microbiol 44: 351–355.MA SmithMJ Bidochka1998Bacterial fitness and plasmid loss: importance of culture conditions and plasmid size.Canadian J Microbiol44351355
8. Corchero JL, Villaverde A (1998) Plasmid maintenance in Escherichia coli recombinant cultures is dramatically, steadily and specifically influenced by features of the encoded proteins. Biotechnol Bioengineering 58: 625–632.JL CorcheroA. Villaverde1998Plasmid maintenance in Escherichia coli recombinant cultures is dramatically, steadily and specifically influenced by features of the encoded proteins.Biotechnol Bioengineering58625632
9. Summers DK (1996) The biology of plasmids. London: Blackwell Sciences. DK Summers1996The biology of plasmidsLondonBlackwell Sciences
10. Williams PA (2004) Catabolic plasmids: fast-track bacterial evolution to combat pollution. Microbiol Today 31: 168–170.PA

Williams2004Catabolic plasmids: fast-track bacterial evolution to combat pollution.Microbiol Today31168170

11. Bergstrom CT, Lipsitch M, Levin BR (2000) Natural selection, infectious transfer and the existence conditions for bacterial plasmids. Genetics 155: 1505–1519.CT BergstromM. LipsitchBR Levin2000Natural selection, infectious transfer and the existence conditions for bacterial plasmids.Genetics15515051519
12. Smith J (2001) The social evolution of bacterial pathogenesis. Proc Royal Soc B 268: 61–69.J. Smith2001The social evolution of bacterial pathogenesis.Proc Royal Soc B2686169
13. Yarmolinsky MB (2000) A pot-pourri of plasmid paradoxes: effects of a second copy. Mol Microbiol 38: 1–7.MB Yarmolinsky2000A pot-pourri of plasmid paradoxes: effects of a second copy.Mol Microbiol3817
14. Zielenkiewicz U, Cegłowski P (2001) Mechanisms of plasmid stable maintenance with special focus on plasmid addiction systems. Acta Biochimica Polonica 48: 1003–1023.U. ZielenkiewiczP. Cegłowski2001Mechanisms of plasmid stable maintenance with special focus on plasmid addiction systems.Acta Biochimica Polonica4810031023
15. Lili LN, Britton NF, Feil EJ (2007) The persistence of parasitic plasmids. Genetics 177: 399–405.LN LiliNF BrittonEJ Feil2007The persistence of parasitic plasmids.Genetics177399405
16. Paulsson J (2002) Multilevel selection on plasmid replication. Genetics 161: 1373–1384.J. Paulsson2002Multilevel selection on plasmid replication.Genetics16113731384
17. Velmurugan S, Mehta S, Uzri D, Jayaram M (2003) Stable propagation of 'selfish' genetic elements. J Biosci 28: 623–636.S. VelmuruganS. MehtaD. UzriM. Jayaram2003Stable propagation of 'selfish' genetic elements.J Biosci28623636
18. Abhyankar MM, Reddy JM, Sharma R, Bullesbach E, Bastia D (2004) Biochemical investigation of control of replication initiation of plasmid R6K. J Biol Chem 279: 6711–6719.MM AbhyankarJM ReddyR. SharmaE. BullesbachD. Bastia2004Biochemical investigation of control of replication initiation of plasmid R6K.J Biol Chem27967116719

19. del Solar G, Giraldo R, Ruiz-Echevarría MJ, Espinosa M, Díaz-Orejas R (1998) Replication and control of circular bacterial plasmids. Microbiol Mol Biol Rev 62: 434–464.G. del SolarR. GiraldoMJ Ruiz-EchevarríaM. EspinosaR. Díaz-Orejas1998Replication and control of circular bacterial plasmids.Microbiol Mol Biol Rev62434464

20. Projan SJ, Monod M, Narayanan CS, Dubnau D (1987) Replication properties of pIM13, a naturally occurring plasmid found in Bacillus subtilis, and of its close relative pE5, a plasmid native to Staphylococcus aureus. J Bacteriol 169: 5131–5139.SJ ProjanM. MonodCS NarayananD. Dubnau1987Replication properties of pIM13, a naturally occurring plasmid found in Bacillus subtilis, and of its close relative pE5, a plasmid native to Staphylococcus aureus.J Bacteriol16951315139

21. Acebo P, Alda MT, Espinosa M, del Solar G (1996) Isolation and characterization of pLS 1 plasmid mutants with increased copy numbers. FEMS Microbiol Lett 140: 85–91.P. AceboMT AldaM. EspinosaG. del Solar1996Isolation and characterization of pLS 1 plasmid mutants with increased copy numbers.FEMS Microbiol Lett1408591

22. Chattoraj D, Cordes K, Abeles A (1984) Plasmid P1 replication: Negative control by repeated DNA sequences. Proc Natl Acad Sci USA 81: 6456–6460.D. ChattorajK. CordesA. Abeles1984Plasmid P1 replication: Negative control by repeated DNA sequences.Proc Natl Acad Sci USA8164566460

23. Das N, Valjavec-Gratian M, Basuray AN, Fekete RA, Papp PP, et al. (2005) Multiple homeostatic mechanisms in the control of P1 plasmid replication. Proc Natl Acad Sci USA 102: 2856–2861.N. DasM. Valjavec-GratianAN BasurayRA FeketePP Papp2005Multiple homeostatic mechanisms in the control of P1 plasmid replication. Proc Natl Acad Sci USA10228562861

24. Paulsson J, Ehrenberg M (2001) Noise in a minimal regulatory network: plasmid copy number control. Quart Rev Biophysics 34: 1–59.J. PaulssonM. Ehrenberg2001Noise in a minimal regulatory network: plasmid copy number control.Quart Rev Biophysics34159

25. Novick RP (1987) Plasmid incompatibility. Microbiol Rev 51:

381–395.RP Novick1987Plasmid incompatibility.Microbiol Rev51381395

26. Sinervo B, Lively CM (1996) The rock-paper-scissors game and the evolution of alternative male strategies. Nature 308: 240–243.B. SinervoCM Lively1996The rock-paper-scissors game and the evolution of alternative male strategies.Nature308240243
27. Hauert C, De Monte D, Hofbauer J, Sigmund K (2002) Volunteering as red queen mechanism for cooperation in public goods game. Science 296: 1129–1132.C. HauertD. De MonteJ. HofbauerK. Sigmund2002Volunteering as red queen mechanism for cooperation in public goods game.Science29611291132
28. Kerr B, Riley MA, Feldman MW, Bohannan BJM (2002) Local dispersal promotes biodiversity in a real-life game of rock–paper–scissors. Nature 418: 171–174.B. KerrMA RileyMW FeldmanBJM Bohannan2002Local dispersal promotes biodiversity in a real-life game of rock–paper–scissors.Nature418171174
29. Nowak MA, May RM (1994) Superinfection and the evolution of parasite virulence. Proc Royal Soc B 255: 81–89.MA NowakRM May1994Superinfection and the evolution of parasite virulence. Proc Royal Soc B2558189
30. Mideo N, Alizon S Day T (2008) Linking within- and between-host dynamics in the evolutionary epidemiology of infectious diseases. Trends Ecol Evol 23: 511–517.N. MideoT. Alizon S Day2008Linking within- and between-host dynamics in the evolutionary epidemiology of infectious diseases.Trends Ecol Evol23511517
31. Andrup L, Smidt L, Andersen K, Boe L (1998) Kinetics of conjugative transfer: a study of the plasmid pxo16 from Bacillus thuringiensis subsp israelensis. Plasmid 40: 30–43.L. AndrupL. SmidtK. AndersenL. Boe1998Kinetics of conjugative transfer: a study of the plasmid pxo16 from Bacillus thuringiensis subsp israelensis.Plasmid403043
32. Flett F, Humphrey GO, Saunders JR (1981) Intraspecific and intergeneric mobilization of non-conjugative resistance plasmids by a 24.5 megadalton conjugative plasmid of Neisseria gonorrhoeae. J Gen Microbiol 125: 123–129.F. FlettGO HumphreyJR Saunders1981Intraspecific and intergeneric mobilization of non-

conjugative resistance plasmids by a 24.5 megadalton conjugative plasmid of Neisseria gonorrhoeae.J Gen Microbiol125123129

33. Udo EE, Jacob LE, Mokadas EM (1997) Conjugative transfer of high-level mupirocin resistance from Staphylococcus haemolyticus to other staphylococci. Antimicrobial Agents Chemotherapy 41: 693–695.EE UdoLE JacobEM Mokadas1997Conjugative transfer of high-level mupirocin resistance from Staphylococcus haemolyticus to other staphylococci.Antimicrobial Agents Chemotherapy41693695

CHAPTER

4

MODELING SRNA-REGULATED PLASMID MAINTENANCE

Chen Chris Gong[1,2], Stefan Klumpp[1,3]

[1]Department of Theory and Bio-Systems, Max Planck Institute of Colloids and Interfaces, Potsdam, Germany,

[2]Institute of Physics and Astronomy, University of Potsdam, Potsdam, Germany,

[3]Institute for Nonlinear Dynamics, Georg August University of Go¨ttingen, Go¨ttingen, Germany

ABSTRACT

We study a theoretical model for the toxin-antitoxin (hok/sok) mechanism for plasmid maintenance in bacteria. Toxin-antitoxin systems enforce the maintenance of a plasmid through post-segregational killing of cells that have lost the plasmid. Key to their function is the tight regulation of expression of a protein toxin by an sRNA antitoxin. Here, we focus on the nonlinear nature of the regulatory circuit dynamics of the toxin-antitoxin mechanism. The mechanism relies on a transient increase in protein concentration rather than on the steady state of the genetic circuit. Through a systematic analysis of the parameter dependence of this transient increase, we confirm some known design features of this system and identify new ones: for an efficient toxin-antitoxin mechanism, the synthesis rate of the toxin's mRNA template should be lower that of the sRNA antitoxin, the mRNA template should be more stable than the sRNA antitoxin, and the mRNA-sRNA complex should be more stable than the sRNA antitoxin. Moreover, a short half-life of the protein toxin is also beneficial to the function of the toxin-antitoxin system. In addition, we study a therapeutic scenario in which a competitor mRNA is introduced to sequester the sRNA antitoxin, causing the toxic protein to be expressed.

INTRODUCTION

Small regulatory RNA (sRNA) plays an important role in gene regulation in organisms from bacteria to mammals by controlling for example translation and/or mRNA stability and the list of known RNA regulation systems keeps increasing at a rapid pace [1–3]. sRNA regulation possesses characteristics that are distinct from protein regulation, in particular a threshold-linear response, which provides an ultrasensitive mechanism for regulatory switching, and the possibility of hierarchical crosstalk, which allows prioritizing of expression [4].

Some of the best known sRNA regulation systems are related to plasmid replication and plasmid maintenance in bacteria. In the first case, an sRNA controls whether synthesis of a replication primer proceeds to plasmid replication [5–7]. In the second, the plasmid encodes a (type I) toxin-antitoxin system such as the hok/sok system ("host-killing/suppression-of-killing"), encoding a protein toxin and an antisense RNA which acts as an antitoxin by binding to the toxin mRNA and blocking ribosome access, thus preventing toxin synthesis [8]. (Other types of toxin-antitoxin systems have different functions [9], in particular related to the formation of persister cells [10].)

The toxin-antitoxin system, which is also known as an "addiction module", maintains the plasmid number through post-segregational killing of plasmid-free progeny due to differential stability of the toxin and antitoxin RNAs. The killing is done by a potent protein toxin that irreversibly damages the cell membrane [11]. In a steady state with a stable plasmid concentration, sRNA antitoxin exists in excessive molar amount compared to target mRNA, such that the latter is entirely sequestered in translationally inactive sRNA-mRNA complexes [12]. However, when a progeny cell becomes plasmid-free through cell division, synthesis of both toxin mRNA and antitoxin sRNA are stopped. The sRNA, which has a very short half-life, is rapidly depleted and the more stable target mRNA can be translated into toxic protein, killing the cell (Fig 1).

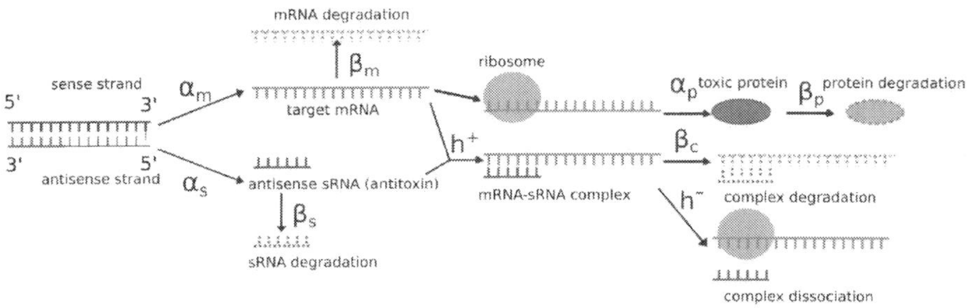

Figure 1: Simplified drawing of the sRNA regulated toxin-antitoxin mechanism for plasmid maintenance.

Structural transitions and processing of the mRNA have been omitted [8]. Parameters of the system are labeled in blue: α_m, α_s, α_p are the synthesis rates of mRNA, sRNA and protein (transcription and translation rates, respectively); β_m, β_s, β_c, β_p are the degradation rates of mRNA, sRNA, the mRNA-sRNA complex and protein, respectively; h^+ and h^- are the binding and unbinding rate of the complex.

To understand the design of the gene circuit encoding the toxin-antitoxin mechanism of post-segregational killing, here we analyze a theoretical model for the dynamics of such a circuit. The model is related to and extends previous models for sRNA-based post-transcriptional regulation [4, 13, 14]. It allows us to address the parameter dependence of the genetic circuit to identify essential features and criteria for its efficient function, for example, whether there are any criteria beyond the difference between toxin and antitoxin RNA lifetime. It also allows us to test whether the system can be designed in such a way that only complete loss of the plasmids triggers killing and not a low but non-zero plasmid copy number. We also use the model to study a proposed antibacterial strategy [15, 16] making use of regulatory crosstalk between RNAs. To answer these questions we performed extensive parameter sweeps, scanning all parameters of the model individually as well as extensively testing randomly chosen parameter sets, thus varying all parameters simultaneously. Overall, the model confirms the known principles for the function of the toxin-antitoxin mechanism as the dominant ones, in particular, the difference in RNA lifetime and the threshold condition on the synthesis rates. In addition, it also indicates a few new design features. Specifically, our analysis of the parameter dependence shows that the stability of the mRNA-sRNA complex plays an important role as well and should exceed the stability of the free antitoxin. Moreover, an unstable toxic protein is superior to a stable one for induction of killing upon plasmid loss.

MODEL AND METHODS

Dynamic Equations and Analytical Solution of Steady State Concentrations

The dynamics of a toxin-antitoxin system as described in Fig 1, is modeled with four coupled ordinary differential equations for four dynamical variables, the concentrations of the toxin mRNA (m), the antitoxin sRNA (s), the toxic protein (p) and of the mRNA-sRNA complex (c), in which the mRNA is silenced. All concentrations are expressed in units of number of molecules per volume of a cell. The four equations describe the synthesis and degradation of the mRNA, the sRNA, and the protein, as well as the formation and dissociation of the mRNA-sRNA complex:

$$\dot{m} = \alpha_m \cdot g - \beta_m \cdot m - h^+ \cdot m \cdot s + h^- \cdot c \tag{1a}$$

$$\dot{s} = \alpha_s \cdot g - \beta_s \cdot s - h^+ \cdot m \cdot s + h^- \cdot c \tag{1b}$$

$$\dot{c} = h^+ \cdot m \cdot s - h^- \cdot c - \beta_c \cdot c \tag{1c}$$

$$\dot{p} = \alpha_p \cdot m - \beta_p \cdot p \tag{1d}$$

The 10 parameters of these equations are as follows: α_m, α_s, α_p are the synthesis rates of mRNA, sRNA and protein (transcription and translation rates, respectively); β_m, β_s, β_c, β_p are the degradation rates of mRNA, sRNA, the mRNA-sRNA complex and protein, respectively; h^+ and h^- are the binding and unbinding rate of the complex, and g is the plasmid copy number per cell volume.

A few remarks are in order with respect to the model's generality. First, for many toxin-antitoxin systems, including hok/sok and the pAD1 par system, there exist intramolecular structures of the toxin mRNA that prohibit the translation of protein by partially restricting ribosome access. Such structural states of the mRNA are not included in our model. However, averaging over the active and inactivated configurations of the mRNA, we obtain an effective synthesis rate of the protein and an effective degradation rate of the mRNA, which do not qualitatively alter the model. Averaging requires however that the mRNA switches between those configurations sufficiently rapidly. Second, saturation of the toxin-antitoxin binding naturally occurs within this model (provided that binding is sufficiently strong) and is accounted for in the nonlinear binding term $h^+ \cdot m \cdot s$. As a consequence an excess of antitoxin sRNA over toxin mRNA results in all toxin mRNA being sequestered in complexes, as discussed in more detail below. Third, the effect of dilution by cell growth can be seen as included in the degradation rates. Dilution by cell growth sets a lower limit on all degradation rates, so proteins or RNA effectively cannot have half-lives exceeding the doubling time of the cell (even though their true degradation rate, due to proteolysis and RNA degradation could be lower).

The steady state solution is obtained by setting the time derivatives on the left hand side of the equations to zero. The steady state for this system can be explicitly given, because the nonlinear terms cancel each other in Eqs (1a) and (1b), leaving a quadratic equation which can be solved analytically. This leads to the steady state sRNA concentration:

$$s^* = \frac{-A \pm \sqrt{A^2 + 4\beta_s \beta_m \alpha_s g \frac{h^+ \beta_c}{h^- + \beta_c}}}{2\beta_s \frac{h^+ \beta_c}{h^- + \beta_c}},$$

where

$$A = (\alpha_m g - \alpha_s g)\frac{h^+ \beta_c}{h^- + \beta_c} + \beta_s \beta_m.$$

The steady state concentrations of the other three components can be in turn given as:

$$m^* = \frac{\alpha_m g - \alpha_s g + \beta_s s^*}{\beta_m}$$

$$c^* = \frac{m^* s^* h^+}{h^- + \beta_c}$$

$$p^* = \frac{\alpha_p m^*}{\beta_p}$$

(2)

This result includes some limiting cases that have been studied in earlier work on sRNA-dependent gene regulation. In the limit of large binding and unbinding rates h^+ and h^- (i.e. the limit of "rapid equilibrium"), the steady state concentration of the RNA complex concentration becomes $c^* = (m^* s^* h^+)/h^-$, as previously obtained by Legewie et al. [13]. The limit of irreversible binding, $h^- = 0$, in which only the equations for sRNA and mRNA concentrations need to be considered, was previously studied by Levine et al. [4] and Mitarai et al. [14].

Numerical Methods

To study the dynamic behavior, Eqs (1a–1d) are numerically integrated using the 4th order Runge-Kutta method with an integration time step on the order of 1×10^{-4} min, for a time span of 300 min. We note that the time unit of the dynamics could be made dimensionless by rescaling all rates relative to one rate that determines the time unit. Correspondingly, the results presented below will typically depend on ratios of time scales or rates. To achieve both stability and speed, an adaptive time step for the numerical integration is implemented, such that when the integration becomes numerically unstable, a smaller time step is used. A tell-tale sign for numerical instability is the appearance of negative concentration values of one or more components. The analytical results for the steady state concentrations are in good agreement (up to floating point error) with the results obtained from the numerical integration.

RESULTS AND DISCUSSION

Qualitative Description of the Dynamics of the Genetic Circuit

For the model described above, the following scenario is considered: the dynamics of the toxin-antitoxin system begin with a cell which has just acquired one or more copies of the plasmid, but has not yet synthesized any of its products, i.e. we start with m = s = c = p = 0. The dynamics are numerically integrated for a sufficiently long time, so that a steady state of the system is reached. At a certain time point (here at t = 150 min of a total t = 300 min), the plasmid copy number is set to 0 to mimic plasmid loss, e.g. due to insufficient replication or unequal partitioning of the plasmid during cell division. To illustrate the dynamics, realistic values of the parameters are used, which are estimated from the literature (Table 1). Eqs (1a–1d) are numerically integrated to produce Fig 2.

Table 1: Table of parameter values based on the literature for toxin-antitoxin systems and related sRNA regulation systems

	α_m[gene copy^{-1}min^{-1}]	β_m[min^{-1}]	α_s[gene copy^{-1}min^{-1}]	β_s[min^{-1}]	α_p[min^{-1}]	β_p[min^{-1}]
Range of values	0.1–15.8	0.034–0.693	0.1–15.8	0.35–1.4	1–7	0.0346
References	[4, 17, 18]	[19, 20]	[4, 17, 18, 21]	[8, 19, 22, 23]		[8]
Default parameters used here	1.0	0.2	6.0	1.0	5.0	0.035

doi:10.1371/journal.pone.0169703.t001

For some parameters, the values are known to depend on growth conditions, which is not explicitly incorporated into our model. For components with long half-life, the degradation rate is dominated by dilution through cell growth and division, which provides a lower limit on the possible (effective) degradation rates.

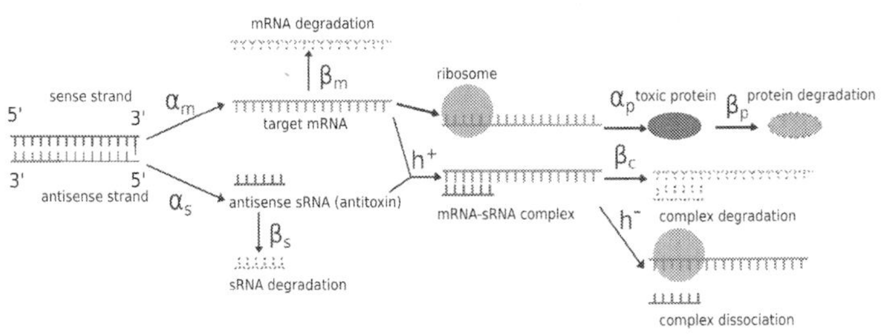

Figure 2: Simulated dynamics of the toxin-antitoxin gene circuit.

The different colors show the concentrations (in molecules per cell volume) of mRNA (orange), sRNA (blue), mRNA-sRNA complex (red) and protein (green). At t = 150 min, the plasmid is lost, described by resetting the copy number g to zero, inducing a transient peak in the protein concentration. The protein concentration peak fold-increase R is 8.6 and the peak width T_p equals 48 min. The parameters are as listed in the last row of Table 1 and g = 6.0, $h^+ = 20.0$, $h^- = 1.0$, $\beta_c = 0.1$.

In Fig 2, all four concentrations increase initially until the first steady state is reached. At t = 150 min, due to the loss of all plasmid copies, the synthesis of new RNA molecules stops. Nevertheless, a transient increase in the free mRNA concentration is observed, which results in a transient increase of the protein concentration. The transient increase in mRNA is a result of sRNA degrading much faster than mRNA, so when a mRNA-sRNA complex dissociates, there is a surplus amount of free mRNA molecules released, which in turn allows the synthesis of toxic protein. Thus, the main function of the toxin-antitoxin system, plasmid maintenance via the synthesis of a toxin upon plasmid loss, which results in the removal of plasmid-free cells from the population, is dependent on a transient dynamics rather than on a steady state. This behavior is in contrast to other sRNA-based regulation systems, which are based on similar mechanisms, but control the steady state concentration of mRNA [4].

For a quantitative characterization of the dynamics, specifically that of the protein concentration change after the loss of plasmids, two quantities are measured for each simulated scenario: (i) the peak fold-increase of the protein concentration, R, defined as the ratio of the maximum of the protein concentration after plasmid loss and the steady state protein concentration before the plasmid loss, $R = p_{peak}/p^*$; and (ii) the transient peak width T_p, defined as the width of the peak at half maximum.

The effectiveness of the toxin-antitoxin mechanism is partially determined by the protein concentration peak fold-increase R, as opposed to by the absolute concentration of the toxin. A high peak fold-increase allows a toxicity threshold to be set, such that the fluctuations of the steady state protein concentration are very unlikely to cross such a threshold and thereby “accidentally” lead to cell death, while simultaneously permitting the transient increase of protein concentration when the plasmid number is not being properly maintained to easily overcome such a threshold.

For example, if the cell will be killed by a 10-fold-increase in toxin concentration (10 being the killing threshold), and if there is little chance of an accidental 10-fold protein concentration increase due to other factors and noise, the dynamics which can produce $R > 10$ are considered effective in exerting a robust control on the host cell.

Dependence of the Protein Dynamics on Individual Model Parameters

We first investigate the parameter dependence of the model by varying each parameter in isolation, holding the other 9 fixed at constant values. In all cases, the peak fold-increase R and the peak width T_p are determined as functions of the modulated parameters.

First, the dependence of R and T_p on the plasmid copy number g (Fig 3) is investigated. An increase of g is equivalent to increasing the synthesis rates of the sRNA and mRNA molecules by the same ratio. R and T_p are both positively affected when g is increased. For the default parameters (Fig 3(a)) as well as for the rapid equilibrium case (Fig 3(b)), the rates of change of both R and T_p (i.e. the slope of the curves) are higher when g is low, and lower when g is high. In the case of rapid equilibrium, i.e. when both h^+ and h^- are high, the changes in R and T_p at the loss of one or two plasmid copies are more substantial than in the cases of lower binding and unbinding rates. This is to say, in the case of rapid equilibrium, the model could be effective in triggering killing even when the plasmids are not completely lost.

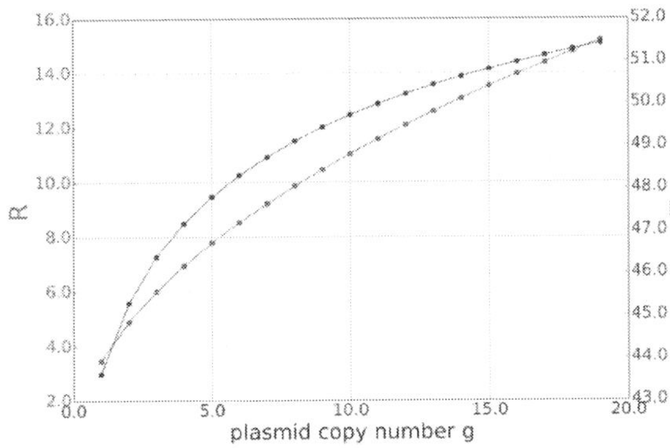

(a)

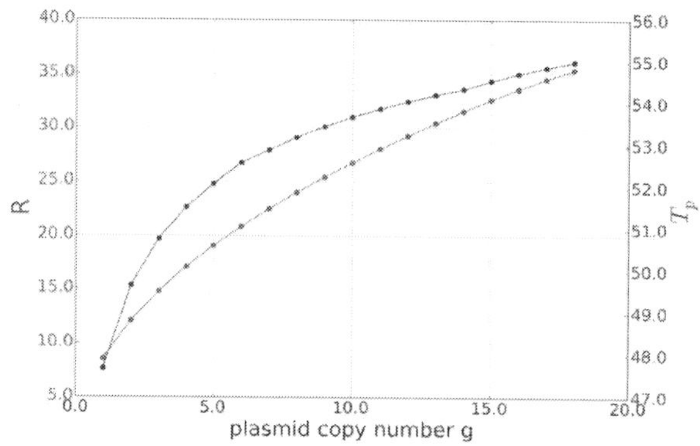

(b)

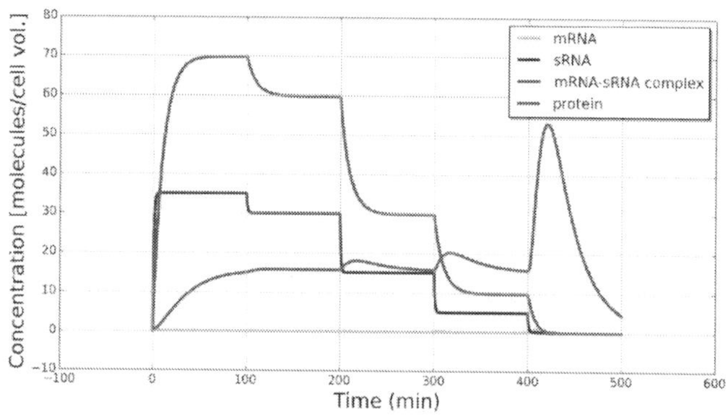

(c)

Figure 3: Dependence on plasmid copy number g.

(a,b) protein concentration peak fold-increase R and width T_p as a function of the plasmid copy number g for our default binding and unbinding rates $h^+ = 20$, $h^- = 1$ (a), and for rapid equilibrium with $h^+ = 1000$, $h^- = 10$ (b). (c) the simulation of sequential plasmid loss under default parameters. The plasmid copy number is reduced from $7 \rightarrow 6 \rightarrow 3 \rightarrow 1 \rightarrow 0$. A substantial increase in toxin concentration only occurs when the last plasmid is lost.

Another way to think about plasmid maintenance is shown in the simulation of sequential loss of plasmid copies under default parameters (Fig 3(c)). The change in plasmid number does not result in the sudden release of large amount of toxin until the last plasmid is lost. Comparing to when all plasmids are lost at once (Fig 2), sequential plasmid loss will result in a less prominent transient peak when the last plasmid is lost. This observation is consistent with the function of the toxin-antitoxin system to enforce plasmid maintenance, i.e. the host cells are forced to retain at least one copy of the plasmid, but there is little dependence on whether there are more or fewer copies.

The dependence of R and T_p on the other 9 parameters are shown in Fig 4. In general, two conditions for the existence of a transient peak of the toxic protein concentration can be extracted from these parameter dependence:

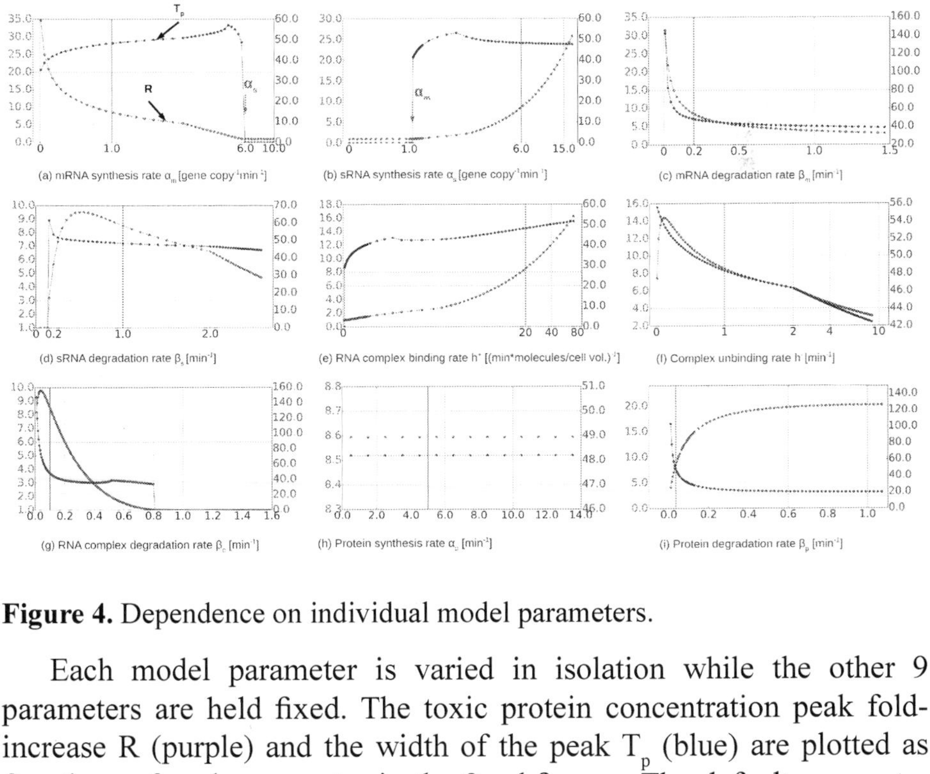

Figure 4. Dependence on individual model parameters.

Each model parameter is varied in isolation while the other 9 parameters are held fixed. The toxic protein concentration peak fold-increase R (purple) and the width of the peak T_p (blue) are plotted as functions of each parameter in the 9 subfigures. The default parameter values (from Table 1 and as in Fig 2) are marked by red lines in the plots.

First, the synthesis rate of sRNA (α_s) must exceed the synthesis rate of mRNA (α_m). This threshold condition can be observed by comparing the varying values to the default values for mRNA and sRNA synthesis rates. In Fig 4(a), when the synthesis rate of mRNA α_m exceeds the default synthesis rate of sRNA α_s at 6/min/gene copy, the peak disappears.

Conversely in Fig 4(b), when α_s becomes larger than α_m at a default value of 1/min/gene copy, the protein concentration peak starts to appear (R > 1).

This threshold condition is well-known for sRNA regulation [4]. It can be explained as follows. Increasing the number of mRNA molecules while keeping the number of sRNA molecules constant increases both the protein concentration peak p_{peak} as well as the first steady state protein concentration p*. However, p* increases faster than p_{peak} as α_m increases. Therefore the relative increase of the protein concentration R diminishes as α_m increases. A high level of free mRNA before plasmid loss will also mean that plasmid-containing cell might "poison itself" without any loss of plasmids. When $\alpha_m > \alpha_s$, there will be more free mRNA than what can be "neutralized" by complex formation, and the toxic protein will be expressed at a high level before the loss of plasmids. The toxin-antitoxin mechanism relies on the low amount of surplus of free mRNA after the loss of plasmids, so that when α_m is higher than α_s, this effect is lost.

Second, the degradation rate of sRNA must exceed the degradation rate of mRNA: $\beta_s > \beta_m$. This condition ensures that after the loss of plasmids, a pool of free mRNA builds up, since the sRNA is degraded more rapidly. When sRNA is more long-lived than mRNA, that is, when β_m exceeds β_s at 1/min, or when β_s is lower than β_m at 0.2/min (Fig 4c and 4d), a pronounced protein concentration peak is not observed. In the following section, when we discuss the results of random sampling of the parameter space, we will show that this condition on the degradation rates of the RNAs is not as strict as the threshold condition on the synthesis rates. This can be seen in Fig 5(a), where some parameter combination exhibit a protein peak despite a ratio $\beta_m/\beta_s > 1$, while there never is a peak for $\alpha_m/\alpha_s > 1$.

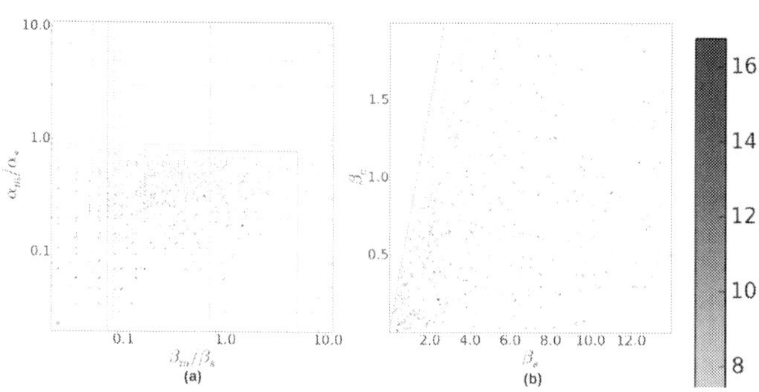

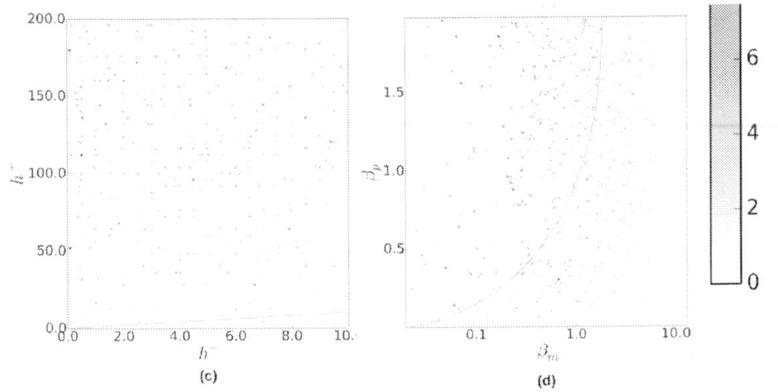

Figure 5: Parameter dependence found by random parameter sampling.

The color gradient shows the magnitude of the protein peak fold-increase R. In Fig (a), the results are projected onto α_m/α_s vs. β_m/β_s on a log-log scale. In (b) the simulation results that fall within the subregion $\alpha_m/\alpha_s < 0.8$ and $\beta_m/\beta_s < 4$ (marked by lines in Fig (a)) are projected onto β_c vs. β_s. In (c,d) the parameter sets are further restricted to the subregion $\alpha_m/\alpha_s > 0.8$, $\beta_m/\beta_s > 4$ and $\beta_c/\beta_s < 2/3$ (marked by a dark line in Fig (b)) and the results are projected onto h^+ vs. h^- in Fig (c), and β_p vs. β_m in Fig (d), respectively. The color scale values are taken to be $\log_{1.4}R$ for ease of viewing. Dark lines mark the rough division between the region with more high peaks and the region with fewer. In (c) and (d) the dark lines mark $h^+ = h^-$ and $\beta_p = \beta_m$ respectively.

Besides these main two conditions, high binding rate, low unbinding rate and sufficient stability of the complex are also needed as shown in Fig 4e–4g. This is consistent with the previous knowledge that the binding rate strongly influences the effectiveness of the toxin-antitoxin mechanism [24, 25]. Finally in Fig 4(i), when the protein degradation rate β_p is very low (equivalent to having a half life of longer than 5 min), R decreases drastically and eventually drops to a constant around 4, where no protein is being degraded. Thus, the proteolysis of the toxic protein contributes to the efficiency of the toxin-antitoxin mechanism for plasmid maintenance. The translation (protein synthesis) rate, on the other hand, has no effect on the ratio R as expected (Fig 4(h)).

The parameter dependence study shows two features of the toxin-antitoxin system that were not known or have not been demonstrated via theoretical studies. Firstly, it is known that the antitoxin sRNA in plasmid number maintenance is very short-lived [26] (which is indeed true for many types of sRNA [27, 28]), while the toxin mRNA has an unusually long half-life [26]. However, in the numerical study above we find when the degradation rate of the sRNA exceeds that of the template mRNA, in some range, the mechanism becomes actually less effective: Rdecreases in Fig 4(d) after β_s becomes larger than around 0.5. However, we did not see this effect clearly in the random sampling of the parameter space in the following section. It is therefore possible this effect only applies to situations where some other parameters, for example the synthesis rates of the RNA molecules, take on certain values.

Secondly, in some well-studied cases, the binding of mRNA with sRNA leads to the rapid degradation of the complex [29]. However, this cannot be the case here. As shown in Fig 4(g), when the RNA complex is degraded at a rate higher than 0.8 there is no transient toxin peak at all (R = 0). This can be explained by the fact that mRNA must be released from the complex upon plasmid loss, which requires a sufficiently high concentration of the complex. If the complex degradation rate β_c is too high, too few mRNA templates will be left for the translation of the toxin. Nevertheless, complex stability is an important parameter beyond this obvious feature, as will be shown by the second parameter dependence study conducted in the following section "Random Sampling of the Parameter Space" Fig 5(b). Indeed, a stabilization of RNAs in the complex has been demonstrated in at least one toxin-antitoxin system, the par system in the E. faecalis [30].

Random Sampling of the Parameter Space

We have seen that the dependence of the toxin-antitoxin circuit on the individual parameters exhibit expected but also surprising behaviors. However, due to the fact that in our single parameter scans only one parameter was varied at a time, while the other 9 are fixed, the generality of our conclusions is up for debate. To find the region of the 10-dimensional parameter space where the toxin-antitoxin mechanism works effectively, and to explore the joint conditions on the parameters, we randomly sample the parameter space. Random scans of multidimensional parameter spaces have previously been used in various contexts including cytoskeletal dynamics [31] and signaling cascades [32].

We run a total of 4025 simulations of the toxin-antitoxin dynamics with randomly generated parameters. The only restrictions imposed on the random parameter sampling are $\alpha_m > \beta_m$ and $\alpha_s > \beta_s$, to make sure that average molecule numbers exceed 1 and the differential equations approach is appropriate. The algorithm generates random values for each of the 9 parameters (g = 6 remains fixed) within the range given in Table 2 and with the aforementioned restriction, at an appropriate sampling resolution. The resolution for α_m, β_m, α_s, β_s are chosen such that the log of the ratios α_m/α_s and β_m/β_s are uniformly distributed. The sampling is therefore random and to a large degree uniform, or log-uniform.

Table 2: Parameter space being sampled and the rates of sampling

Parameter name	α_m	β_m	α_s	β_s	h^+	h^-	β_c	α_p	β_p
Maximum Value	20	14	20	14	200	10	2	30	2
Minimum Value	0.001	0.001	0.001	0.001	0.1	0.001	0.001	0.01	0.001
Sampling Resolution	0.0005	0.0005	0.0005	0.0005	4	0.005	0.001	0.9	0.02

doi:10.1371/journal.pone.0169703.t002

The value of the protein concentration peak fold-increase R for each simulation is recorded and projected onto the space spanned by α_m/α_s and β_m/β_s (Fig 5(a)). The observations obtained by random sampling are consistent with the results of the single parameter dependence study. The results confirm that $\alpha_m < \alpha_s$ is indeed a strong criterion for the existence of a high protein concentration peak. Fig 5(a) shows when $\alpha_m/\alpha_s > 0.8$ there is almost no protein concentration peak after the loss of plasmids. $\beta_m < \beta_s$ is also a very important criteria, but exceptions are allowed: The protein level decreases on average when $\beta_m > \beta_s$, nevertheless, there are distinct protein concentration peaks beyond $\beta_m/\beta_s > 1$. However, the majority of the simulations with $\beta_m/\beta_s > 1$ exhibit no peaks. These exceptions could be due to some other competing effects from the dynamics of the mRNA-sRNA complex or from the dynamics of the RNA molecules.

Next, we focus on the cases that satisfy $\alpha_m/\alpha_s < 0.8$ and $\beta_m/\beta_s < 4$, and plot those as a function of the degradation rate of the complex β_c and that of the sRNA β_s. This will eliminate toxin-antitoxin dynamics with parameter combinations that are inefficient in producing a transient peak due to the criteria discussed above, allowing new effects to be discovered more easily. Fig 5(b) shows that no peaks occur for $\beta_c \gtrsim \beta_s$, in contrast to the rest of the region. In fact, the peaks are visually prominent only when $1.5 * \beta_c \lesssim \beta_s$. A likely explanation is as follows: when β_c is low, there is a higher level of total mRNA molecules in bound form both before and after the plasmid loss. Because the DNA gene copies are removed when the plasmids are lost, the transient increase of protein synthesis is almost entirely due to the mRNA released from the complex. Therefore, the complex degradation (as opposed to mRNA degradation) is the dominant reason for total mRNA loss. At the same time, lower β_c does not increase mRNA expression before plasmid loss, but merely increases the amount of bound mRNA molecules, and will therefore not lead the plasmid-containing cell to "poison itself". If this loss of mRNA due to complex degradation is slower than the loss of sRNA (the loss of sRNA is equal to the gain of the free mRNA when the plasmids are lost), the toxic protein can be synthesized from the surplus free mRNA. Therefore, if antitoxin sRNA is less stable than the total mRNA, there will be a transient peak of mRNA, and correspondingly one for the toxic protein, after plasmid loss. However, when a higher level of free mRNA is present before plasmid loss, due to faster degradation of sRNA, i.e. when β_s is large, R is in turn negatively affected. Considering this negative impact, $\beta_c \lesssim \beta_s$ therefore does not always guarantee high peaks, which is shown from a wide range of variations in R within this region, and there is no increase in R when β_c is increasingly small compared to β_s.

It is surprising that the stability of the complex plays such an important role in the toxin-antitoxin mechanism. It is secondary only to the threshold condition on the synthesis rates and the differential stability of the RNA molecules. This highlights the crucial role played by the RNA complex in the dynamics. In connection to the single parameter study conducted before, this demonstrates that the two known effects of unstable sRNA [26] and unstable RNA complexes [29] need to be understood in relation to each other and not in isolation. Unstable RNA complexes are only useful for the functioning of the toxin-antitoxin system if the sRNA is more unstable than the RNA complex.

Next, we restrict our set of parameter combinations further, by imposing $\beta_c/\beta_s \lesssim 2/3$, $\alpha_m/\alpha_s < 0.8$ and $\beta_m/\beta_s < 4$. Fig 5(c) shows that for a prominent protein concentration peak to occur, h^-/h^+ should not be larger than one. Projections onto β_p and β_m show that when β_p is close to 0 there is no major peak (Fig 5(d)). This means that by making extremely stable toxic proteins we cannot increase the relative increase in protein concentration. It also shows generally $\beta_m < \beta_p$ result in less pronounced peaks than $\beta_m > \beta_p$. This is usually satisfied in real situations, since mRNA molecules are typically less stable than proteins.

The fact that high peaks do not occur in the subregion with $h^-/h^+ > 1$ tells us that the binding rate needs to be higher than the unbinding rate, i.e. the binding needs to be strong. However, it also cannot be infinitely strong, i.e. irreversible, because in that case there will be no free mRNA available for protein synthesis after plasmid loss.

Consistent with the results from the single parameter dependence study, the synthesis rate α_p does not have an obvious effect on the protein concentration peak. This shows that the protein concentration peak is mostly a result of the dynamics between the mRNA, sRNA and the RNA complex, characterized for example by relations between β_m and β_s, α_m and α_s, β_c and β_s, and not of a high synthesis rate of the protein. This also emphasizes the importance of understanding sRNA regulation purely from the point of view of RNA dynamics instead of protein dynamics, which is traditionally viewed as playing the dominant role in cellular regulation.

To summarize, by randomly sampling the parameter space as well as performing a single parameter dependence study, we determine that the conditions for an efficient toxin-antitoxin mechanism are as follows:

- The synthesis rate of mRNA should be lower than the synthesis rate of sRNA;
- The degradation rate of mRNA should be relatively low, compared to the degradation rate of sRNA;
- The stability of the complex should be higher than the stability of the sRNA antitoxin;
- A high affinity and irreversibility (but not complete irreversibility) in RNA complex formation and some degrees of protein instability are also helping factors, but are of less significance.
- Protein synthesis rate does not contribute to the forming of a transient protein concentration peak.

Despite the nonlinear nature of our system, by individually controlling the available parameters, a genetic circuit could be engineered to produce specific effects, such as a higher increase in toxin concentration after the loss of plasmids for an effective duration (from a few minutes up to an hour).

Analytical approximation to the transient protein concentration peak

Using a simplified version of the differential equations after plasmid loss, an analytical expression for the transient peak can be obtained, which qualitatively describes the dynamics. For this approximation, we assume that after the plasmid copies are lost, all mRNA molecules are immediately available in free form. That is, the number of free mRNA molecules after plasmid loss equals the sum of the mRNA steady state concentration m* and the complex steady state concentration c*. The differential equations which describe the dynamics are as follows:

$$\begin{aligned} \dot{m} &= \beta_m \cdot m \\ \dot{p} &= \alpha_p \cdot m - \beta_p \cdot p. \end{aligned} \tag{3}$$

The initial conditions and integration constants required to solve the differential Eq (3) are provided by the steady state concentrations before and after the loss of plasmids, respectively. Explicit expressions for mRNA and protein concentration as a function of time can then be obtained:

$$m = m_2^* + (m^* + c^* - m_2^*)e^{-\beta_m t} \tag{4a}$$

$$p = p_2^* - \frac{(m^* + c^* - m_2^*)\alpha_p}{\beta_m - \beta_p} e^{-\beta_m t} + \left(p^* - p_2^* + \frac{(m^* + c^* - m_2^*)\alpha_p}{\beta_m - \beta_p} \right) e^{-\beta_p t}, \tag{4b}$$

where m*, c* and p* are the steady state concentrations for mRNA, complex and protein before plasmid loss, and m_2^* and p_2^* are the steady state concentrations after plasmid loss (see Section "Dynamic Equations and Analytical Solution of Steady State Concentrations"). In the case that all plasmid copies are lost, the steady state concentrations m_2^* and p_2^* are both zero and the above expressions can be simplified to:

$$m = m^* e^{-\beta_m t} \tag{5a}$$

$$p = -\frac{(m^* + c^*)\alpha_p}{\beta_m - \beta_p} e^{-\beta_m t} + \left(p^* + \frac{(m^* + c^*)\alpha_p}{\beta_m - \beta_p}\right) e^{-\beta_p t}. \quad \textbf{(5b)}$$

These expressions qualitatively describe the transient peak of the protein concentration after plasmid loss, however the height of the peak is strongly overestimated (Fig 6). In the full model, mRNA is released slowly from the complex, which reduces the peak height. This behavior can be mimicked within the approximation by the sudden release of only a fraction of the mRNAs (replacing m* by an effective mRNA concentration somewhere between m* and c*).

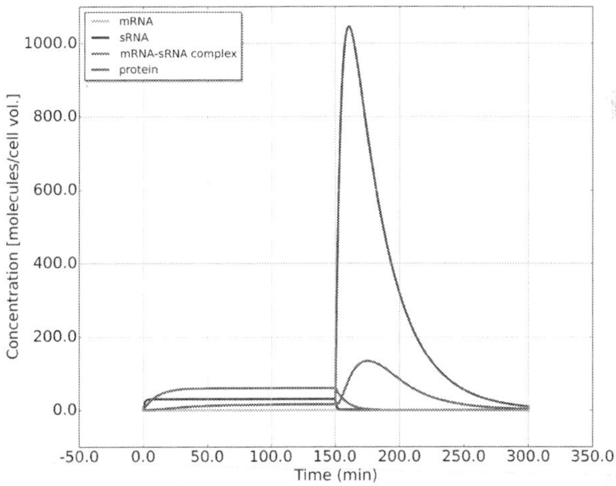

Figure 6: Analytical approximation of the dynamics after plasmid loss.

With the same parameter values as in Fig 2, solution Eq (5b) gives a very large peak in approximated protein concentration (purple) compared to the actual simulated protein concentration change (green). It should be noted that this approximation considers no reaction time delay after plasmid loss.

Despite these quantitative shortcomings of the approximation, it is useful to understand some properties of the dynamics. For example, one can easily see from the solution (Eq (5b)), that the translation rate does not affect the peak fold-increase R, because both terms are proportional to α_p, while the steady state concentration before plasmid loss, p*, is also proportional to α_p. Thus, R being the ratio of the two is independent of α_p. This result should also be true for the full dynamics since the analytical approximation used here is only considering the RNA dynamics and assumptions about the protein dynamics have not been made. The dependence on $\beta_m - \beta_p$ is also important: An increase of this quantity is equivalent to an increase of p*, consistent with the tendency observed in the random parameter sampling (Fig 5(d)).

Introduction of a Competitor mRNA to Increase Toxin Levels

Regulation by small RNAs typically displays crosstalk with multiple mRNAs under the control of the same sRNA [33]. In the case of toxin-antitoxin systems, this can be exploited as an antibacterial strategy [15, 16]. By inducing a gene encoding a competitor mRNA (on the same or another plasmid or on the chromosome) that can bind to the antitoxin sRNA and hence allows toxin mRNA to exist in free form, a similar increase in toxin concentration can be induced as with plasmid loss, which may also lead to the killing of the cell. To describe the dynamics of this scenario, the equations from above are extended to include a second type of mRNA, the competitor, and the corresponding mRNA-sRNA complex. The full dynamics is then described by the following equations:

$$\begin{aligned}
\dot{m} &= \alpha_m \cdot g - \beta_m \cdot m - h^+ \cdot m \cdot s + h^- \cdot c \\
\dot{s} &= \alpha_s \cdot g - \beta_s \cdot s - h^+ \cdot m \cdot s + h^- \cdot c - k^+ \cdot m_2 \cdot s + k^- \cdot c_2 \\
\dot{c} &= h^+ \cdot m \cdot s - h^- \cdot c - \beta_c \cdot c \\
\dot{p} &= \alpha_p \cdot m - \beta_p \cdot p \\
\dot{m}_2 &= \alpha_2 \cdot g - \beta_2 \cdot m_2 - k^+ \cdot m_2 \cdot s + k^- \cdot c_2 \\
\dot{c}_2 &= k^+ \cdot m_2 \cdot s - k^- \cdot c_2 - \beta_{c_2} \cdot c_2,
\end{aligned} \tag{6}$$

where m_2 and c_2 denote the concentrations of the competitor mRNA and the complex it forms with an sRNA molecule, respectively. k^+, k^- and β_{c2} are the binding rate, unbinding rate and the degradation rate of this new complex.

We assume that the competitor is induced at the time t = 150 min. Then, from t = 0 to 150 min, the dynamics is the same as Eq (1). At t = 150 min, Eq (6) are being integrated, which corresponds to the competitor gene being turned on. With zero initial concentrations for all components (at t = 150 min, new variables r = v = 0), an example simulation is shown in Fig 7.

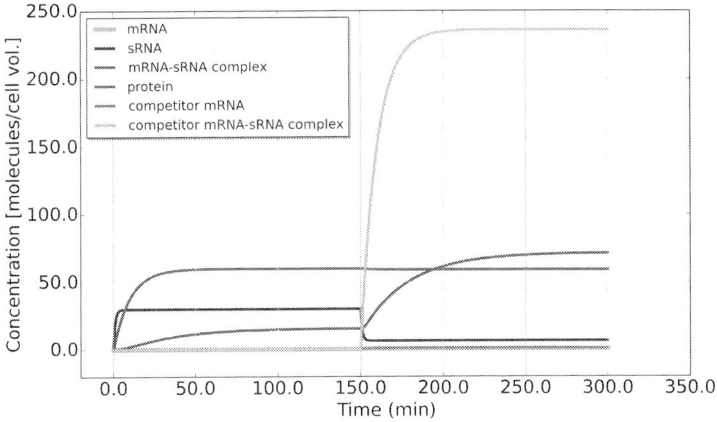

Figure 7. Introduction of a competitor mRNA.

Concentrations of sRNA, mRNA, mRNA-sRNA complex, competitor mRNA, complex formed by competitor mRNA and sRNA and protein as a function of time. Note the non-zero steady state concentrations after the triggering event resulting from the new dynamics. The parameters in Eq (6) are as follows: the first 10 parameters are the same as in Fig 4. Parameter values for competitor mRNA are chosen from within the experimental values of mRNA synthesis rate and life time, with synthesis rate $\alpha_2 = 4.0$ and degradation rate $\beta_2 = 0.6$. The competitor mRNA-sRNA complex binding and unbinding rates are $k_+ = 60$ and $k_- = 1.0$. The competitor RNA complex degradation rate is $\beta_{c2} = 0.1$. Toxin mRNA level is found to be low throughout the simulation.

A transient protein concentration peak can be observed with some specific combinations of parameters, but the majority of the simulations we run show no transient peak, as in Fig 7. However, a non-zero steady state toxin concentration exists after induction of the competitor which can be effective in killing the host cell, since the mRNA encoding the toxic protein is still produced from the plasmid. Thus, in contrast to the case considered before, here the toxin-antitoxin system does not perform its key function in a transient dynamical fashion, but rather in the steady state, with the competitor either induced or not induced. Therefore, we define a new fold-increase parameter $\widetilde{R}$ of the toxin concentration as the ratio between the steady states of the protein concentration before and after the synthesis of the competitor RNA.

Similar to the procedure in Section "Dynamic Equations and Analytical Solution of Steady State Concentrations", the new steady state concentration for the mRNA $\widetilde{m_2^*}$ turns out to be a positive root to the cubic equation:

$$\alpha_2 g\left(1+\frac{k^-}{\beta_{c_2}}\right)m^2 - \frac{\beta_s}{\beta_2 h^+}\left\{\left[\beta_2\left(1+\frac{k^-}{\beta_{c_2}}\right) - \frac{k^+}{h^+}\beta_m\left(1+\frac{h^-}{\beta_c}\right)\right]m + \frac{k^+}{h^+}\alpha_m g\left(1+\frac{h^-}{\beta_c}\right)\right\}$$
$$\times\left\{-\frac{\beta_m}{\beta_s}h^+m^2 - \beta_m\left(1+\frac{h^-}{\beta_c}\right)m - \frac{(\alpha_s-\alpha_m-\alpha_2)gh^+}{\beta_s}m + \alpha_m g\left(1+\frac{h^-}{\beta_c}\right)\right\} = 0 \tag{7}$$

After algebraic manipulations, it can be written in canonical form as follows.

$$Am^3 + Bm^2 + Cm + D = 0 \tag{8}$$

where A, B, C and D are functions of the rates. An analytical formula for the solutions of a cubic equation can always be given explicitly, but the expressions are excessively lengthy and hence not shown here. With a root finding algorithm such as the bisection method one can easily find the three roots to the cubic equation numerically. Using values similar to the experimental values, we found that even in cases where there is more than one positive root, we can still pick out the correct solution because usually one of the two positive roots corresponds to an unrealistically large concentration.

The analytical solution to Eq (7) is then compared to the numerically generated steady state concentrations after integration, and very good agreement (up to floating point error) is found. Once we obtain the steady state mRNA concentration $\widetilde{m_2^*}$, then following $\widetilde{p_2^*} = \widetilde{m_2^*}\cdot\alpha_p/\beta_p$ one can solve for the steady state protein concentration. The ratio $\widetilde{p_2^*}/p^*$, where p* is given by Eq (2), gives an analytical value for the fold-increase $\widetilde{R}$. This is how we could theoretically obtain an analytical formula for $\widetilde{R}$ in the case of competitive RNA binding, which is just as effective in killing the host cell at some triggering event as the original toxin-antitoxin mechanism.

Fig 8 shows the single parameter variation response plots generated with the same method as used in Fig 4. We observe that as soon as the synthesis of the competitor is turned on, i.e. $\alpha_2 > 0$, there is a substantial fold-increase in the toxin concentration, i.e. $\widetilde{R} > 1$. As α_2 becomes large, the increase in 0$\widetilde{R}$ is diminished because all sRNA molecules are bound. Changing the degradation rate of the competitor mRNA β_2 results in small variation in the toxin fold-increase response $\widetilde{R}$, meaning that for the default parameter combination, there are very few free competitor mRNA molecules and most are bound in complexes. The degradation rate of the competitor mRNA-sRNA complex β_{c2} positively affects $\widetilde{R}$, because it increases the destruction of the antitoxin in bound state, allowing more target mRNA to be in free form available for protein synthesis.

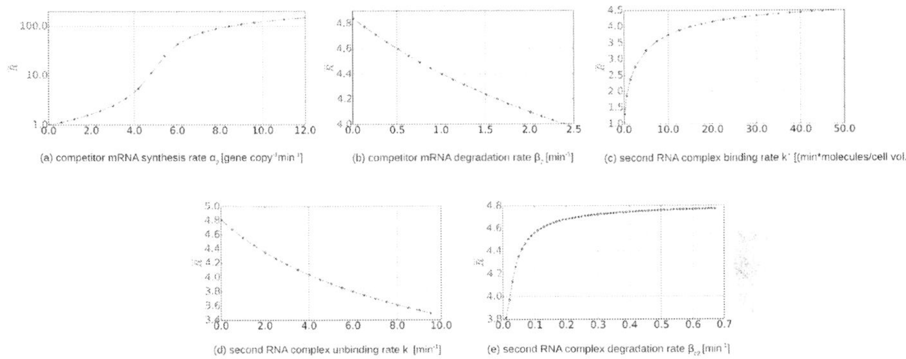

Figure 8: Dependence on the parameters of the competitor.

Each new parameter of the system is increased along the x axis. Toxic protein concentration peak fold-increase $\widetilde{R}$ is plotted on the y axis for each run corresponding to the value the new parameter takes on. The default or fixed values of the parameters are the same as in Fig 7.

The parameter variation in Fig 8 shows that the synthesis rate of the new competitor mRNA typically plays the dominant role compared to other new parameters, a general feature of cross-talk in sRNA regulation [33].

CONCLUSION AND SUMMARY

We have studied the toxin-antitoxin mechanism for plasmid maintenance through post-segregational killing using a theoretical model. The model extends previously studied models for sRNA-dependent regulation that have considered the limiting cases of rapid equilibrium or irreversible binding, but have thereby largely ignored the nonlinear nature of the RNA complex binding and the dynamics of the complex itself. Here we have taken the full dynamics into account and given an analytical solution for the steady state concentrations (where the plasmid copy number remains constant). We simulated the dynamics before and after plasmid loss by numerical integration and showed that the toxin-antitoxin system performs its function by inducing a transient peak in the toxin concentration, which must exceed a threshold for host killing. The formation of this peak depends on the release of the mRNA template of the protein toxin from the mRNA-sRNA complex. The accumulation of the mRNA molecules is only possible when sRNA is degraded at a faster rate than the mRNA.

We notice that the scenario of post-segregational killing of host cells that have lost the plasmid is not the only possible function of a toxin-antitoxin system of a plasmid. Our simulations of sequential loss of plasmid copies show a gradual increase of the toxin concentration without exceeding the threshold for killing. It is possible that an these sublethal toxin concentrations also contribute to plasmid maintenance: If sublethal toxin concentrations slow down growth of the host cell, this would provide additional time for plasmid replication, allowing to restore the plasmid number to the level before the loss of one or more copies. This effect is not considered in our present model, which does not included growth modulation by sublethal toxin concentrations or the control mechanism of plasmid replication, but will be studied in future work.

Using two kinds of parameter variations to study the parameter dependence of the system (individual parameter are varied in isolation, and all parameters are varied simultaneously in a random manner within a given range), we can draw a number of conclusions with regards to which ones of the 10 system parameters have significant effects on the system, and why when they take on certain ranges of values, the system functions more optimally than in other cases. In general we have found that, for an efficient toxin-antitoxin mechanism, the synthesis rate of toxin's mRNA template should be lower than that of the sRNA antitoxin, the mRNA template should be more stable compared to the sRNA antitoxin, and the mRNA-sRNA complex should be more stable than that of the sRNA antitoxin. Analytically approximating the protein peak by allowing all mRNA to be released at once gives us an analytical expression for the peak, which despite overpredicting its height nonetheless gives qualitative insights that are consistent with previous numerical observations.

Finally, we also studied the possibility of inducing the toxic protein with a competitor mRNA, which sequesters the sRNA antitoxin. Such a mechanism has been proposed as an antibacterial strategy [15, 16]. Here the effectiveness in killing the host cell depends on the ratio between the steady states of the toxin concentration before and after introducing the competitor RNA. An analytical solution can be given for this ratio by solving a cubic equation. We performed a single-parameter variation study for this scenario and found that the dominant parameter dependence here is on the synthesis rates. A sufficiently high synthesis rate results in all antitoxin being sequestered and the synthesis of protein toxin. In such a way the cross-talk inherent in sRNA regulation could be utilized towards inducible killing of the cells.

REFERENCES

1. Altuvia S. Identification of bacterial small non-coding RNAs: experimental approaches. Curr Opin Microbiol. 2007;10(3):257–61. pmid:17553733
2. Livny J, Waldor MK. Identification of small RNAs in diverse bacterial species. Curr Opin Microbiol. 2007;10(2):96–101. pmid:17383222
3. Waters LS, Storz G. Regulatory RNAs in bacteria. Cell. 2009;136(4):615–28. pmid:19239884
4. Levine E, Zhang Z, Kuhlman T, Hwa T. Quantitative characteristics of gene regulation by small RNA. PLoS Biol. 2007;5(9):e229. pmid:17713988
5. Summers DK. The Biology of Plasmids. Oxford: Blackwell Science Ltd; 1996.
6. Paulsson J, Ehrenberg M. Noise in a minimal regulatory network: plasmid copy number control. Q Rev Biophys. 2001;34:1–59. pmid:11388089
7. Eguchi Y, Itoh T, Tomizawa J. Antisense RNA. Annu Rev Biochem. 1991;60(1):631–52. pmid:1715680
8. Gerdes K, Wagner EG. RNA antitoxins. Curr Opin Microbiol. 2007;10(2):117–24. pmid:17376733
9. Magnuson RD. Hypothetical functions of toxin-antitoxin systems. J Bacteriol. 2007;189(17):6089–92. pmid:17616596

10. Lewis K. Persister cells, dormancy and infectious disease. Nat Rev Micro. 2007;5(1):48–56.
11. Gerdes K, Bech FW, Jorgensen ST, Løbner-Olesen A, Rasmussen PB, Atlung T, et al. Mechanism of postsegregational killing by the hok gene product of the parB system of plasmid R1 and homology with the relF gene product of the E. coli relB operon. EMBO J. 1986;5(8):2023–29. pmid:3019679
12. Gerdes K, Nielsen A, Thorsted P, Wagner EG. Mechanism of killer gene activation. Antisense RNA-dependent RNase III cleavage ensures rapid turn-over of the stable hok, srnB and pndA effector messenger RNAs. J Mol Biol. 1992;226(3):637–49. pmid:1380562
13. Legewie S, Dienst D, Wilde A, Herzel H, Axmann I. Small RNAs establish delays and temporal thresholds in gene expression. Biophys J. 2008;95(7):3232–8. pmid:18599624
14. Mitarai N, Benjamin JAM, Krishna S, Semsey S, Csiszovszki Z, Massé E, et al. Dynamic features of gene expression control by small regulatory RNAs. Proc Natl Acad Sci USA. 2009;106:10655–59. pmid:19541626
15. Williams JJ, Hergenrother PJ. Artificial activation of toxin-antitoxin systems as an antibacterial strategy. Trends Microbiol. 2012;20(6):291–98. pmid:22445361
16. Faridani OR, Nikravesh A, Pandey DP, Gerdes K, Good L. Competitive inhibition of natural antisense Sok-RNA interactions activates Hok-mediated cell killing in Escherichia coli. Nucleic Acids Res. 2006;34(20):5915–22. pmid:17065468
17. Liang ST, Bipatnath M, Xu YC, Chen SL, Dennis P, Ehrenberg M, et al. Activities of constitutive promoters in Escherichia coli. J Mol Biol. 1999;292(1):19–37. pmid:10493854
18. Lin-Chao S, Bremer H. Activities of the RNAI and RNAII promoters of plasmid pBR322. J Bacteriol. 1987;169(3):1217–22. pmid:2434459
19. Gerdes K, Thisted T, Martinussen J. Mechanism of post-segregational killing by the hoklsok system of plasmid R1: sok antisense RNA regulates formation of a hok mRNA species correlated with killing of plasmid-free cells. Mol Microbiol. 1990;4(11):1807–18. pmid:1707122

20. Bernstein J, Khodursky A, Lin P, Lin-Chao S, Cohen S. Global analysis of mRNA decay and abundance in Escherichia coli at single-gene resolution using two-color fluorescent DNA microarrays. Proc Natl Acad Sci USA. 2002;99(15):9697–702. pmid:12119387
21. Wagner EG, Altuvia S, Romby P. Antisense RNAs in bacteria and their genetic elements. Adv Genet. 2002;46:361–98. pmid:11931231
22. Jasiecki J, Wegrzyn G. Growth-rate dependent RNA polyadenylation in Escherichia coli. EMBO Rep. 2003;4(2):172–7. pmid:12612607
23. Lin-Chao S, Cohen SN. The rate of processing and degradation of antisense RNAI regulates the replication of ColE1-type plasmids in vivo. Cell. 1991;65:1233–42. pmid:1712252
24. Hjalt T, Wagner EG. The effect of loop size in antisense and target RNAs on the efficiency of antisense RNA control. Nucleic Acids Res. 1992;20(24):6723–32. pmid:1282705
25. Nordgren S, Slagter-Jäger JG, Wagner GH. Real time kinetic studies of the interaction between folded antisense and target RNAs using surface plasmon resonance. J Mol Biol. 2001;310(5):1125–34. pmid:11502000
26. Gerdes K, Helin K, Christensen O, Løbner-Olesen A. Translational control and differential RNA decay are key elements regulating postsegregational expression of the killer protein encoded by the parB locus of plasmid R1. J Mol Biol. 1988;203(1):119–29. pmid:2460630
27. Brenner M, Tomizawa T. Quantitation of ColE1-encoded replication elements. Proc Natl Acad Sci USA. 1991;88(2):405–409. pmid:1703297
28. Stougaard P, Molin S, Nordstrom K. RNAs involved in copy-number control and incompatibility of plasmid R1. Proc Natl Acad Sci USA. 1981;78(10):6008–6012. pmid:6171808
29. Lenz DH, Mok KC, Lilley BN, Kulkarni RV, Wingreen NS, Bassler BL. The small RNA chaperone Hfq and multiple small RNAs control quorum sensing in Vibrio harveyi and Vibrio cholerae. Cell. 2004;118(1):69–82. pmid:15242645

30. Weaver KE, Ehli EA, Nelson JS, Patel S. Antisense RNA Regulation by Stable Complex Formation in the Enterococcus faecalis Plasmid pAD1 par Addiction System. J Bacteriol. 2004;186:6400–6408. pmid:15375120
31. Nédélec F. Computer simulations reveal motor properties generating stable antiparallel microtubule interactions. The Journal of Cell Biology. 2002;158(6):1005–1015. pmid:12235120
32. Catozzi S, Di-Bella JP, Ventura AC, Sepulchre JA. Signaling cascades transmit information downstream and upstream but unlikely simultaneously. BMC Systems Biology. 2016;10(1):84. pmid:27561377
33. Levine E, Hwa T. Small RNAs establish gene expression thresholds. Curr Opin Microbiol. 2008;11(6):574–9. pmid:18935980

SECTION III: PLASMID MAINTENANCE AND INHERITANCE

CHAPTER

5

NOVEL ASSAY TO MEASURE THE PLASMID MOBILIZING POTENTIAL OF MIXED MICROBIAL COMMUNITIES

Uli Klümper, Ariadni Droumpali, Arnaud Dechesne and Barth F. Smets

Department of Environmental Engineering, Technical University of Denmark, Kongens Lyngby, Denmark

ABSTRACT

Mobilizable plasmids lack necessary genes for complete conjugation and are therefore non-self-transmissible. Instead, they rely on the conjugation system of conjugal plasmids to be horizontally transferred to new recipients. While community permissiveness, the fraction of a mixed microbial community that can receive self-transmissible conjugal plasmids, has been studied, the intrinsic ability of a community to mobilize plasmids that lack conjugation systems is unexplored. Here, we present a novel framework and experimental method to estimate the mobilization potential of mixed communities. We compare the transfer frequency of a mobilizable plasmid to that of a mobilizing and conjugal plasmid measured for a model strain and for the assayed community.

With *Pseudomonas putida* carrying the *gfp*-tagged mobilizable IncQ plasmid RSF1010 as donor strain, we conducted solid surface mating experiments with either a *P. putida* strain carrying the mobilizing IncP-1α plasmid RP4 or a model bacterial community that was extracted from the inner walls of a domestic shower conduit. Additionally, we estimated the permissiveness of the same community for RP4 using *P. putida* as donor strain. The permissiveness of the model community for RP4 [at 1.16×10^{-4} transconjugants per recipient (T/R)] was similar to that previously measured for soil microbial communities. RSF1010 was mobilized by the model community at a frequency of 1.16×10^{-5} T/R, only one order of magnitude lower than its permissiveness to RP4. This mobilization frequency is unexpectedly high considering that (i) mobilization requires the presence of mobilizing conjugal plasmids within the permissive fraction of the recipients; (ii) in pure culture experiments with *P. putida* retromobilization of RSF1010 through RP4 only took place in approximately half of the donors receiving the conjugal plasmid in the first step. Further work is needed to establish how plasmid mobilization potential varies within and across microbial communities. This method has the potential to provide such insights; in addition it allows for the direct isolation of *in situ* mobilizing plasmids together with their endogenous hosts.

Keywords: Plasmid mobilization, permissiveness, RSF1010, RP4, plasmid transfer, conjugation, horizontal gene transfer

INTRODUCTION

Plasmid transfer is believed to be a main mechanism in rapid bacterial adaption to environmental changes (Sørensen et al., 2005; Grohmann, 2011; Heuer and Smalla, 2012). Plasmids can be classified into two main groups based on the presence of genes associated with the transfer phenotype (Smillie et al., 2010). Conjugal plasmids encode a complete set of transfer genes needed to be self-transmissible. Mobilizable plasmids, on the other hand, lack some of the genes encoded in the transfer operon (*tra*), which encodes most of the functions involved in mating pair formation (MPF; Thomas and Nielsen, 2005).

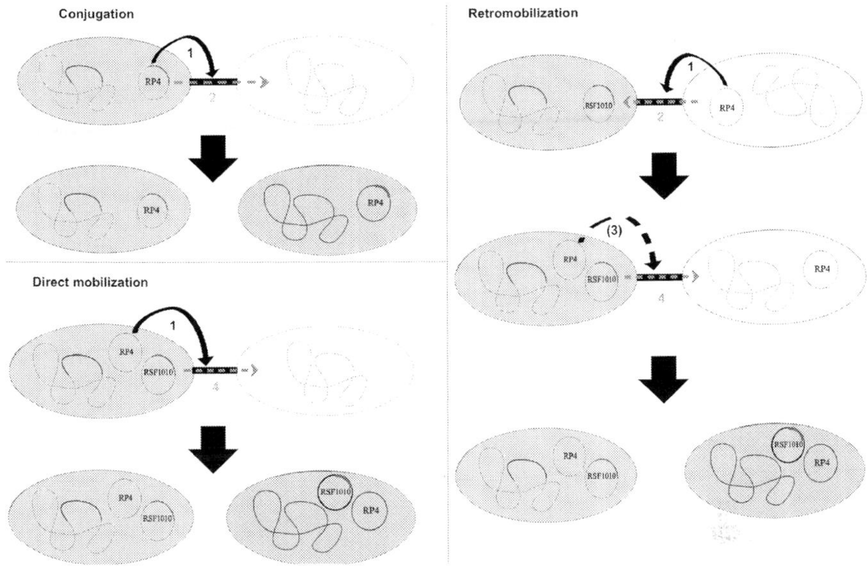

Figure 1: Conjugation, direct mobilization, and retromobilization of the conjugative/mobilizable plasmid couple RP4/RSF1010. In all shown combinations, the donor strains are displayed in red as chromosomally tagged with the red fluorescent protein gene *mCherry* and a *gfp* repressor gene (blue). Recipients transition from being colorless to being green after the *gfp*-tagged plasmid is transferred and thus freed from the chromosomal repression of the donor. **(A)** Conjugal transfer of the self-transmissible IncP-1 plasmid RP4. Step 1 illustrates the establishment of the pilus between donor and recipient as part of the type IV secretion system (T4SS) encoded by the conjugal plasmid. Step 2 displays the transfer of the conjugal plasmid through its own secretion system into the recipient. (B) Direct mobilization of the mobilizable IncQ plasmid RSF1010 from the donor to the recipient by the co-resident conjugal plasmid RP4. The conjugal plasmid establishes the pilus as part of its T4SS and interconnects donor and recipient strain (Step 1). The mobilizable plasmid does not encode for its own T4SS and transfers through the established pilus into the recipient cell (Step 4). The conjugal plasmid might or might not transfer along with the mobilizable plasmid in the direct mobilization process. (C) Retromobilization process of plasmid RSF1010, mobilized by a conjugal plasmid from the recipient cell. In this process, the conjugal plasmid from the recipient establishes the conjugal connection between recipient and donor (Step 1) and transfers from the recipient to the donor (Step 2). The mobilizable plasmid can subsequently transfer through the established connection (Step 4) or through a potential new connection established by the now co-resident conjugal plasmid (Step 3).

Conjugal plasmids possess an origin of transfer (*oriT*), a relaxase, type IV coupling proteins (T4CP) and a type IV secretion system (T4SS). The relaxase is a key protein of the conjugal machinery, common to all conjugal and mobilizable plasmids. Conjugal transfer of self-transmissible plasmids like the IncP-1α plasmid RP4 is based on pilus establishment between donor and recipient cells coded by the T4SS. The plasmid then transfers through the pilus into the recipient (Figure 1). Mobilizable plasmids encode only a *MOB* module (with or without the T4CP) and need the MPF apparatus of a co-resident (i.e., located within the same cell) conjugal plasmid to be transmissible by conjugation (Smillie et al., 2010). To be transferred, they take advantage of a conjugal plasmid that initiates replication through expression of its *rep* genes. These genes are involved in pilus formation and connection of the relaxosome with proteins enabling passage of the DNA across the membranes (Yano et al., 2013). Direct mobilization involves a presently co-resident conjugal plasmid; in retromobilization the donor cells (harboring the mobilizable plasmid) must first receive a mobilizing conjugal plasmid from the recipient, which thereafter mobilizes the mobilizable plasmid toward the recipient (Figure 1).Therefore, microbial communities need a high intrinsic conjugal plasmid content to allow mobilization of mobilizable plasmids with potentially useful genetic content, when no co-resident conjugal plasmids are present in the newly introduced donor strain.

The most well-studied non-self-transmissible, mobilizable plasmids belong to the IncQ group. Compared to the broad host range IncP-1 conjugal plasmids, they are relatively small (5.1–14.2 kb; Loftie-Eaton and Rawlings, 2012). Thanks to their host independent replication system, these plasmids have a broader host range than any other known replicating components in bacteria (Meyer, 2009). They can be conjugally mobilized by a variety of different plasmid encoded type IV transporters (Meyer, 2009) as well as through integrative and conjugative elements (ICEs; Lee et al., 2012), both often at high frequencies (Gregory et al., 2008; Meyer, 2009).

Mobilization by the IncP-1 plasmids has contributed extensively to the dissemination of IncQ plasmids (Meyer, 2009) and the coupling of the transfer machinery of the IncP-1 RP4 plasmid to mobilize the IncQ RSF1010 plasmid has been well studied (Lessl et al., 1993; Haase et al., 1995).

In order to assess a conjugal plasmid's potential contribution to horizontal gene transfer in a microbial community, the permissiveness of the community toward the plasmid is a main parameter. We have defined permissiveness as the fraction of a community able to receive and maintain a target exogenous plasmid (Musovic et al., 2010; Klümper et al., 2014). Different factors such as phylogenetic diversity, cell density, and various environmental stress factors may affect community permissiveness (Musovic et al., 2010; Heuer et al., 2011). While some bacteria are known to exude signal molecules in order to obtain plasmids (Hirt, 2002), permissiveness toward a self-transmissible, conjugal plasmid is probably a passive trait of the bacterial community. The ability of a community to receive genes located on mobilizable non-self-transmissible plasmids, on the other hand, would rely on the community's own content of conjugal plasmids. While the spread and contribution of conjugal plasmids to gene exchange has been intensely studied (Heuer et al., 2012; Shintani et al., 2014; Zhang et al., 2014), the mobilization potential of microbial communities and the contribution of mobilizable plasmids to horizontal gene flow have been comparably poorly studied (Top et al., 1995). Exogenous isolation techniques to capture mobilizing and mobilizable plasmids from natural communities have been developed (Top et al., 1994; van Elsas et al., 1998; Smalla et al., 2000). However, the characterization of communities based on their mobilization potential has mainly been carried out using indirect measures through triparental matings where both donor and terminal recipient were artificially introduced to the communities (Hill et al., 1992). For example, manure addition was shown to increase a soil microbial community's ability to support mobilization of a mobilizable plasmid between two introduced strains through an increased intrinsic plasmid content (Götz and Smalla, 1997). Direct mobilization of mobilizable plasmids into indigenous bacteria of a mixed community has been detected before (Hill et al., 1992; van Elsas et al., 1998), but was never directly quantified.

Here, we present a novel experimental method to estimate the plasmid mobilization potential of a mixed bacterial community, using IncQ RSF1010 as model plasmid. We quantify the mobilization potential of a model community extracted from a domestic shower conduit. We evaluated the transfer frequency by comparing it to the community's permissiveness toward the mobilizing, conjugal plasmid RP4. We finally related the observed transfer frequencies to those measured in transfer between isogenic strains. We additionally aimed to isolate transconjugants that mobilized the RSF1010 plasmid, assuming that retromobilization is the main mobilization process.

MATERIAL AND METHODS

Principle of Plasmid Transfer Detection

The recipient community was challenged with various plasmid combinations introduced through *Pseudomonas putida* in solid surface filter matings (Figure 2). All strains used or constructed for this study can be found in Table 1. The plasmids (Table 2) were marked with a genetic tag encoding a conditionally expressible fluorescent marker. The used entranceposon (Bahl et al., 2009) carries a *lacI^q^*repressible promoter upstream of the *gfpmut3* gene, coding for the green fluorescent protein (*gfp*). The plasmid donor strain was chromosomally tagged with a gene cassette encoding constitutive red fluorescence and constitutive *lacI^q^* production. As a result, there is no *gfp* expression in the donor strain, but upon plasmid transfer to recipient bacteria, *gfp* expression is possible, resulting in green fluorescent cells or microcolonies, which can be detected and quantified by fluorescence microscopy or sorted by fluorescent activated cell sorting (FACS), respectively. *P. putida* KT2440 served as the donor strain in all the experiments, and was tagged through electroporation with plasmid pGRG36-$lacI^q$-Km^R-*Lpp-mCherry*carrying both the transposase genes and the Tn7 $lacI^q$-*Lpp-mCherry*-Km^R region for specific integration of the $lacI^q$-*Lpp-mCherry*-Km^R gene cassette into the chromosomal attTn7 site as described earlier (Bahl et al., 2009).

Conjugal plasmids possess an origin of transfer (*oriT*), a relaxase, type IV coupling proteins (T4CP) and a type IV secretion system (T4SS). The relaxase is a key protein of the conjugal machinery, common to all conjugal and mobilizable plasmids. Conjugal transfer of self-transmissible plasmids like the IncP-1α plasmid RP4 is based on pilus establishment between donor and recipient cells coded by the T4SS. The plasmid then transfers through the pilus into the recipient (Figure 1). Mobilizable plasmids encode only a *MOB* module (with or without the T4CP) and need the MPF apparatus of a co-resident (i.e., located within the same cell) conjugal plasmid to be transmissible by conjugation (Smillie et al.,

2010). To be transferred, they take advantage of a conjugal plasmid that initiates replication through expression of its *rep* genes. These genes are involved in pilus formation and connection of the relaxosome with proteins enabling passage of the DNA across the membranes (Yano et al., 2013). Direct mobilization involves a presently co-resident conjugal plasmid; in retromobilization the donor cells (harboring the mobilizable plasmid) must first receive a mobilizing conjugal plasmid from the recipient, which thereafter mobilizes the mobilizable plasmid toward the recipient (Figure 1).Therefore, microbial communities need a high intrinsic conjugal plasmid content to allow mobilization of mobilizable plasmids with potentially useful genetic content, when no co-resident conjugal plasmids are present in the newly introduced donor strain.

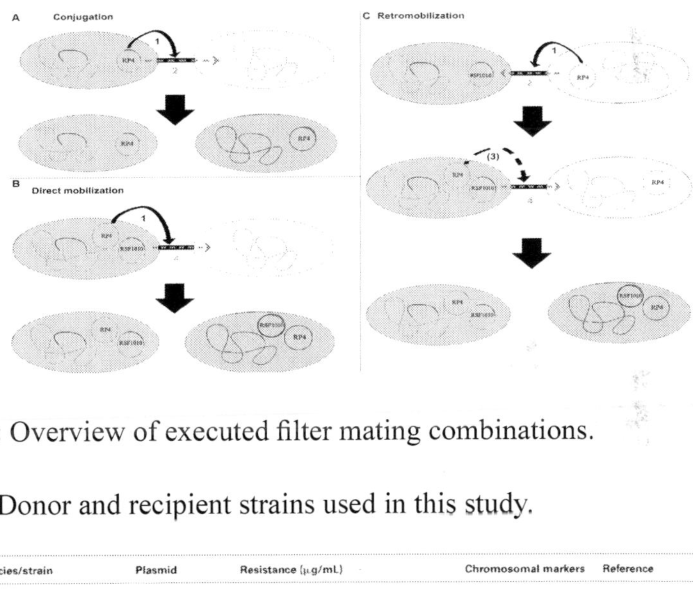

Figure 2: Overview of executed filter mating combinations.

Table 1: Donor and recipient strains used in this study.

Species/strain	Plasmid	Resistance (μg/mL)	Chromosomal markers	Reference
Pseudomonas putida KT2442	RP4::*gfp*	*Km*[R], *Amp*[R], *Tet*[R] (50, 100, 10)	*Rif*[R]	Musovic et al. (2010)
P. putida KT2440	RSF1010::*gfp*	*Strep*[R] (100)	*lacI*[q]-*pLpp*-*mCherry*, *Km*[R]	This study
P. putida KT2440	RSF1010::*gfp*, RP4	*Strep*[R], *Amp*[R], *Tet*[R], *Km*[R] (100, 100, 40, 50)	*lacI*[q]-*pLpp*-*mCherry*, *Km*[R]	This study
P. putida KT2440	–	–	–	Nelson et al. (2002)
P. putida KT2440	RP4	*Tet*[R], *Km*[R], *Amp*[R] (40, 50, 100)	–	This study

Table 2: Plasmids used in this study.

Plasmid	Transfer	Size	Incompatibility	Resistance (μg/mL)	Host range	Degradation pathways	Reference
RP4	Conjugal	60 kb	IncP-1α	*Amp*^{R}, *Km*^{R}, *Tet*^{R} (100, 50, 20)	broad	BP, 4CBP	Barth and Grinter (1977)
RSF1010	Mobilizable	8.7 kb	IncQ	*Strep*^{R} (100)	broad	Arginine, Ornithine	Honda et al. (1991)

The 8.7 kbp IncQ plasmid, RSF1010, originally isolated from *Escherichia coli* (Scholz et al., 1989), harbors streptomycin and sulphonamide resistance determinants and genes for the degradation of arginine and ornithine. For *gfp*-tagging the P_{A10403}-*gfpmut3*-*Km*^{R} section of entranceposon [*Km*^{R}, P_{A10403}-*gfpmut3*] was amplified by PCR, subjected to subsequent enzyme digestion and ligated to the RSF1010 vector cut with the same enzyme. The correct insert location at the enzyme cut site of [*Km*^{R}, P_{A10403}-*gfpmut3*] in plasmid RSF1010 was confirmed by sequencing from the inserted fragment in one direction using primer Seq_Bw_Ent_gfp: 5′-GCCAGAACCGTTATGATGTCGG-3′. The selected *gfpmut3*-tagged RSF1010 (abbreviated as RSF1010::*gfp*) plasmid was finally introduced by transformation into the donor strain, *P. putida* KT2440::*Km*^{R}-*Lpp-mCherry*.

A donor *P. putida* KT2440::*Km*^{R}-*Lpp-mCherry* with both RSF1010::*gfp* and the wild type conjugal plasmid RP4 was also constructed. The previously created donor strain *P. putida* KT2440::*Km*^{R}-*Lpp-mCherry* carrying the RSF1010::*gfp* plasmid was mated with *E. coli* J5 harboring an untagged version of the RP4. Mating was carried out on microfiber filters (GF/C Whatman filter, 24 mm). Cells were detached from the mating filters and *P. putida* donor strains hosting both plasmids were selected for on 10 mM citrate medium supplemented with streptomycin and tetracycline and checked for red and green fluorescence after IPTG induction of *gfp*.

Donor and Recipient Strain Growth and Preparation

The *P. putida* recipient and donor strains were grown overnight on R2A medium supplemented with the plasmid specific antibiotics (Table 2) and harvested by centrifugation at 10,000 × *g* for 10 min. Harvested cells were resuspended and washed twice with sterile 0.9% NaCl solution to remove residual antibiotics and thereafter adjusted to a bacterial density of 3×10^6 bacteria/mL using Thoma chamber counts and sterile 0.9% NaCl solution for dilutions.

Recipient Community Extraction and Preparation

As model recipient microbial communities, we extracted biofilms that colonized the inner walls of a domestic shower PVC hose from a private residence. The shower hose was first drained in a sterile 50 mL Falcon tube. The emptied hose was then incised with a sterilized steel scalpel blade and the biofilm at its inner surface removed by scraping. The removed biofilm was transferred to the same 50 mL Falcon tube. The suspension was centrifuged for 8 min at 10.000 × *g*. The pellet was resuspended in 5 mL TTSP [tetrasodium pyrophosphate (50 mM), Tween 80® (0.05%)], vortexed at maximum speed for 5 min, and sonicated 60 s in a Branson Sonifier 250 (Branson, MO, USA) at 40% power at 200 W to disrupt cell aggregates. The bacterial suspension was then filtered through a sterile 20 μm pore-size filter. This filtrate was used as the recipient community in mating assays after adjusting the bacterial density to ~3×10^6bacteria/mL, as confirmed by Thoma chamber counts.

Solid Surface Filter Mating Assay

The recipient communities were challenged with the plasmids introduced through the constructed donor via solid surface filter matings (Musovic et al., 2010) at a 1:1 initial donor to recipient cell ratio and an initial density of approximately 30,000 bacteria/mm^2 filter surface area, with 10-fold diluted R2A as solid 1.5% agar mating medium. Conjugation was verified by epifluorescence stereomicroscopy after 48 h incubation at room temperature and the transfer events quantified (Musovic et al., 2010). R2A was chosen as filter mating medium as it is presumed optimal for water borne organisms (Reasoner et al., 1979). However, to simulate low nutrient conditions typical of drinking water distribution systems (Boe-Hansen et al., 2002), the R2A medium was diluted to the maximum extent possible, while maintaining high enough bacterial activity for growth of microcolonies, to establish donor to recipient cell contact during the mating, and for expression of the plasmid encoded *gfp*-gene after plasmid transfer. Five different dilutions of R2A (1:5, 1:10, 1:50, 1:100, 1:1000) were tested and the 10-fold diluted R2A was finally chosen, as it was the highest dilution at which transconjugants were still observed for all tested plasmids.

Visualization and Quantification of Transfer Events by Stereomicroscopy and Image Analysis

Successful plasmid transfer was visualized *in situ* by stereomicroscopy and quantified by automated image analysis (Image Pro Plus 7.1; Media Cybernetics, Silver Spring, MD, USA) as previously described (Musovic et al., 2010), using a Leica MZ16 FA fluorescence stereomicroscope equipped with a 10x plan apochromatic objective, a 10× eyepiece (10×/21B), a 40× magnification zoom. Conditions for *gfp*- and *mCherry*-based fluorescence detection were 480/20 nm with emission at 525/40 nm and 580/25 with emission at 650/60 nm, respectively, and images were acquired with a Leica DFC300 fluorescence camera. A representative scanning zone of 7 × 7 fields of 980 × 732 μm each were analyzed per filter. With a total filter area of 270 mm^2, the scanned and quantified area corresponded to approximately 13% of the total filter area. Triplicate filters were analyzed for each donor/recipient combination.

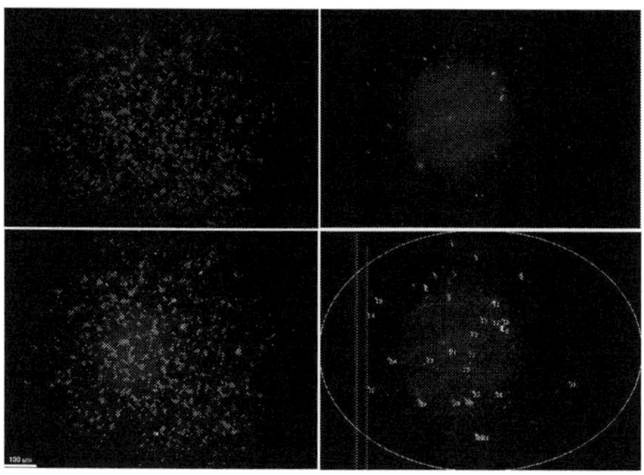

Figure 3: Detection and quantification of transconjugant microcolonies by fluorescent microscopy. Fluorescence based stereomicroscopic images and image analysis of an example filter mating of *Pseudomonas putida* KT2440::*lacI^q-Lpp-mCherry-Km^R* (RSF1010::*gfp*) with the recipient community. (A) corresponds to the red fluorescent channel, displaying donor microcolonies. (B) shows the green fluorescent channel, corresponding to the transconjugal microcolonies that received the plasmid through retromobilization. (C) is a composite image of both channels with increased contrasts. Transconjugal microcolonies can be found in direct proximity to donor colonies. (D) illustrates counting of transcon-

jugal colonies through a macro that increases contrast of the images, subtracts background, eliminates the poorly illuminated corners and counts green fluorescent object larger than 4 μm^2.

Quantification of transfer events was performed with a custom-made macro written in Image Pro Plus 7.1. This macro successively extracts and subtracts the background from the original image, performs a best-fit equalization of the image intensity, before detecting bright objects larger than 4 μm^2 based on automatic segmentation. Analysis of images was limited to the brightly illuminated elliptic central area of the field of view (Figure 3). All images were manually controlled for enumeration errors, and values corrected if deviations were noted. The number of *gfp*-positive colonies (transfer events) detected was scaled up to the total filter area and transfer frequency was calculated by dividing this number by the number of potential recipients originally placed on the filter.

Cell Collection and Fluorescence Activated Cell Sorting of Transconjugants

Cells from the filter mating between *P. putida* (RSF1010::*gfp*) and the model community were removed by vortexing in 2 mL of a 0.9% NaCl-solution for 3 min. Flow cytometric detection of cells and *gfp*-based isolation of transconjugants were carried out using a FACSAria IIIu Flowcytometer (Becton Dickinson Biosciences, San Jose, CA, USA), as previously described (Klümper et al., 2014).

RESULTS

Permissiveness of the Recipient Community for Conjugal IncP-1 Plasmid RP4

We explored the intrinsic ability of an extracted model microbial community to mobilize the broad host range mobilizable plasmid RSF1010 as well as its ability to receive the conjugal broad host range plasmid RP4. Both plasmids were introduced via a red fluorescent-tagged donor *P. putida* in which plasmid encoded *gfp* expression is repressed (Table 1). Microscopic examination and enumeration of the mating events (Figure 3) between the recipient microbial community and *P. putida* (RP4::*gfp*) revealed a transfer frequency of 1.16×10^{-4} transconjugants per potential recipient (T/R; Figure 4). A higher transfer frequency (1.76×10^{-3} T/R) was observed in the mating assay using isogenic *P. putida* donor and recipient strains (Figure 4). In this experiment, all recipients were obviously within the plasmid host range and any incompatibility effect with RP4 could be ruled out because they were all initially plasmid-free. Hence, the observed transfer frequency in these intrastrain experiments was not limited by the recipient permissiveness, but only by donor promiscuity (the fraction of donor cells expressing conjugal genes), successful completion of initiated plasmid transfer events to *P. putida* recipient cells, and the degree of donor–recipient contact saturation.

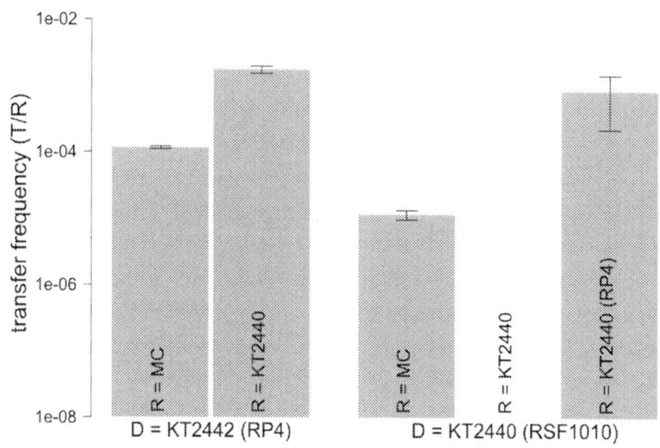

Figure 4: Transfer frequencies of RSF1010 and RP4. Transfer frequencies defined as transconjugant microcolonies per recipient were obtained in solid surface filter matings with the recipient community and in *P. putida* intrastrain matings. Values are shown as mean of triplicates with standard error of mean. The *gfp*-tagged plasmid in the donor strain (D) is shown on the x-axis. RP4 or RSF1010 were each introduced through KT2440 or KT2442 (*P. putida*KT2440/ KT2442::*lacI^q-Lpp-mCherry-Km^R*) into the recipients. Recipients (R) are shown within the bars (MC: model community; KT2440: *P. putida* KT2440). For the combination D= KT2440 (RSF1010) with R= KT2440 no transfer was observed.

We can now express the community's permissiveness against the defined co-culture experiments: The community permissiveness for the conjugal RP4 (1.16×10^{-4} T/R) is divided by the conjugal transfer frequency of plasmid RP4 in intrastrain matings, where all *P. putida* recipients can potentially take up RP4 (1.76×10^{-3} T/R), as a standard.

The resulting community permissiveness for RP4 is 0.066 RP4 intrastrain equivalents.

Mobilizing Potential of the Recipient Community for Plasmid RSF1010

When the model community was challenged with *P. putida* (RSF1010::*gfp*), a transfer frequency of 1.16 × 10^{-5} T/R was measured. This value is one order of magnitude lower than the community's measured permissiveness for the conjugal plasmid RP4 (Figure 4).

In these experiments RSF1010 must have been retromobilized into the recipient community by cells carrying IncQ compatible mobilizing conjugal plasmids (Figure 1). In order to explore the retrotransfer frequency of RSF1010 further, isogenic *P. putida* strains were used to execute two different intrastrain matings, taking advantage of all *P. putida* recipient cells being potential RSF1010 hosts. In the first experiment, a plasmid-free, non-*mCherry*-tagged *P. putida* strain served as recipient. In the second experiment, a non-*mCherry*-tagged *P. putida* strain hosting the untagged wild-type conjugal, mobilizing RP4 plasmid served as recipient. In the first experiment no RSF1010 transfer was observed, consistent with RSF1010's non-self-transmissible nature. In the second experiment with *P. putida* (RP4) as recipient, retrotransfer was observed, with a measured frequency of 8.34 × 10^{-4} T/R. Successful RSF1010 retrotransfer requires initial conjugal plasmid transfer from recipients to RSF1010 donors, before RSF1010 is mobilized and retransferred to the recipients (Top et al., 1992; Figure 1C).

RSF1010 retrotransfer frequency by *P. putida* (RP4) results from a combination of the RP4 transfer process from the recipient to the donor (Figure 1C Steps 1 and 2) and the subsequent mobilization of RSF1010 through the now co-resident RP4 plasmid (Figure 1C Steps 3 and 4). It can be contrasted with the measured RP4 intraspecies transfer frequency of 1.76 × 10^{-3} T/R. RP4 intrastrain transfer corresponds to the first two steps in RSF1010 retrotransfer (Figure 1A). Hence, the probability for a cell that recently acquired RP4 via conjugal transfer to mobilize RSF1010 can be estimated at 47.4% [8.34 × 10^{-4} T/R for *P. putida* (RSF1010::*gfp*) to *P. putida* (RP4) divided by 1.76 × 10^{-3} (T/R) for *P. putida* (RP4::*gfp*) to *P. putida*]. For this specific pair of mobilizing and mobilizable plasmid, retrotransfer is high (Figure 4).

The retrotransfer of RSF1010 to the recipient community occurs at a frequency of 10% compared to its permissiveness for the RP4 plasmid. Still, as shown above, mobilization of RSF1010 is realized only approximately every second time a conjugal plasmid is transferred from the recipient community into the donor strain, based on mobilization through RP4. If all these potential mobilization events were realized, the maximal mobilization potential of the recipient community is reached. The theoretical maximal mobilization potential toward RSF1010 can be quantitatively assessed as 2.45×10^{-5} T/R by dividing its transfer frequency toward the community (1.16×10^{-5} T/R) by the now established 47.4% probability of retrotransfer. When subsequently dividing 2.45×10^{-5} T/R through the community's permissiveness toward RP4 (1.16×10^{-4} T/R) as a standard, this results in 0.211 RP4 permissiveness equivalents as the maximal mobilization potential.

Potential Community Permissiveness Toward Mobilizable Plasmid RSF1010

In a final experiment, we quantified the intrinsic permissiveness of the model community for RSF1010. To do so, we augmented the community's own RSF1010 mobilizing potential by adding an exogenous RSF1010 mobilizing strain. Hence, the recipient community was challenged with *P. putida* hosting both the RSF1010::*gfp* and the wild-type RP4, which can directly mobilize RSF1010 (Figure 1B). The observed transfer frequency of RSF1010 in this mating was 3.14×10^{-3} T/R. This frequency is, surprisingly, higher (~30-fold) than the community's permissiveness for RP4. As expected, this value is also substantially higher (~2 orders of magnitude) than the RSF1010 mobilization frequency (Figure 4) relying on the community's inherent retromobilization potential only.

FACS Based Sorting of RSF1010 Transconjugants

Cell suspensions from matings between the recipient community and *P. putida* (RSF1010::*gfp*) were collected, resuspended and subjected to FACS to isolate green fluorescent transconjugants (Klümper et al., 2014). 200 transconjugants were successfully sorted, despite a sorting time exceeding 24 h, due to the low initial relative abundance of transconjugant cells at less than 1:1,000,000 events sorted.

DISCUSSION

Plasmids of the promiscuous, conjugal IncP-1 group illustrate the enormous potential of horizontal gene transfer among an extremely wide variety of gram-negative and gram-positive bacterial species (Gelder et al., 2005; Klümper et al., 2014; Musovic et al., 2014; Shintani et al., 2014). Studies on conjugal gene flow mainly focused on the passive characteristics of a mixed community to receive self-transmissible plasmids. Former approaches to assess the mobilization potential of mixed communities were using an indirect approach through triparental matings where both donor and terminal recipient were artificially introduced to the communities (Hill et al., 1992; Götz and Smalla, 1997) and even capture the mobilizing (van Elsas et al., 1998) or mobilizable (Smalla et al., 2000) genetic elements from natural communities. This study is the first one to directly quantify the potential of a microbial community to actively mobilize non-self-transmissible, mobilizable plasmids to its indigenous bacteria. It also illustrates how the community's intrinsic plasmid content can contribute to an increased gene uptake potential. To estimate the maximum mobilization potential of a community, we utilized filter matings at maximized cell-to-cell contact of donor and potential recipients (Musovic et al., 2010). The spatial limitations for contact in water distribution systems might be small compared to other environments like the ones reported for soil (Dechesne et al., 2005). However, the initial invasion of the plasmid donor into the biofilm community might be limited to the surface of the biofilm and further reduced at high water flow conditions (Licht et al., 1999; Merkey et al., 2011; Król et al., 2013). Therefore, using our maximum cell-to-cell contact assay instead of natural conditions allows every single recipient cell to establish contact with donor cells and potentially engage in gene transfer. However, using this assay might limit the retransfer potential of the plasmid from new transconjugants to further recipients. Recipients that newly acquired the plasmid might only be surrounded by *P. putida* donor cells and not by other cells from the recipient community and can thus not retransfer the plasmid to other recipients. This retransfer process can especially be crucial for mobilizable plasmids. The first retromobilization transfer event leads to the co-occurrence of the mobilizable plasmid with the mobilizing conjugal plasmid(s) in the same cell. Through this co-occurrence the transconjugant cell significantly increases its transfer frequency of the mobilizable plasmid to the recipient community by switching the mechanism from retromobilization to direct mobilization, thereby omitting the steps involved in transferring the mobilizer to the donor cell. We measured a more than 300-fold increase in plasmid transfer for *P. putida* to the mixed community between retro- to direct mobilization. This large increase in transfer frequency was also reported earlier with a difference of over three orders of magnitude for direct mobilization versus retromobilization for a different mobilizable plasmid among pure strains (Top et al., 1995). Therefore, experiments that assess how this retransfer process influences the mitigation and invasion of a mobilizable plasmid from the initial donor through a mixed and spatially stratified biofilm community might be needed. To conclude, once mobilizable plasmids are in co-occurrence with a promiscuous mobilizing plasmid, they can significantly contribute to horizontal gene transfer in mixed communities.

We show here that the IncQ model plasmid RSF1010 can be easily mobilized by the bacterial community extracted from a household water distribution system. The permissiveness of this microbial community toward the conjugal plasmid RP4 is comparable in magnitude with that measured in diverse soil communities (Musovic et al., 2014). The lower permissiveness toward RP4 measured for mixed recipient communities compared to *P. putida* intraspecies transfer results primarily from the inability of a fraction of the bacterial community to either receive, transiently maintain, or express plasmid encoded genes.

The community's potential to retromobilize and subsequently receive RSF1010 is only one order of magnitude lower than its permissiveness toward RP4. This surprisingly high transfer frequency may result from the fact that IncQ plasmids have a broader host range than any other known replicating component in bacteria (Meyer, 2009) combined with an extremely efficient transfer mechanism (Gregory et al., 2008; Meyer, 2009). The numbers appear even higher taking into account that in pure culture experiments with *P. putida*, only half of the microcolonies that recently received RP4 retromobilized RSF1010. Earlier retrotransfer experiments between two *E. coli* strains (Top et al., 1992) showed T/R ratios within the same orders of magnitude (10^{-3}–10^{-4}) as our intrastrain matings. But, they suggested that retrotransfer of the mobilizable plasmid appears at rates lower than 1% once the first step of acquiring a conjugal plasmid is realized. In that work transfer was quantified based on single cells and after 2.33 h. Our far higher numbers (~50%) might therefore result from quantifying transfer on a microcolony basis after 48 h. Only one retrotransfer event within a microcolony is needed for quantification as successful transfer event and due to increased incubation time retrotransfer can happen not only through the initial, but also through newly established conjugal pili. Nonetheless, the observed retromobilization requires the presence of mobilizing, conjugal plasmids within the permissive fraction of the recipients. Other mobilization possibilities involve conjugation-independent transfer of plasmids through the formation of nanotubes from members of the complex community toward the donor cells (Dubey and Ben-Yehuda, 2011), but are only realized if nanotubes from the recipient to the *P. putida* donor are established. Therefore, a high intrinsic conjugal plasmid content of the model recipient community in combination with RSF1010's efficient transfer mechanism is the most likely reason for the observed high mobilization potential.

IncP type IV secretion systems can conjugally connect a large variety of organisms (Grahn et al., 2000; Thomas and Nielsen, 2005; Klümper et al., 2014). But like the plasmids encoding them, they are evolutionary adapted to connect their mainly Gram-negative hosts. These self-transmissible plasmids might easily reach dead ends after being transferred, if the secretion system is not encoded efficiently for retransfer in the new host. Contrarily, mobilizable plasmids might less frequently reach dead ends once acquired, since they can utilize the conjugal connections build through adapted resident plasmids in their new host (Meyer, 2009) or through ICEs (Lee et al., 2012). Additionally, mobilizable plasmids are relatively stable, as their high copy number (Meyer, 2009) increases retention in a host until new transfer becomes possible. These two facts in connection with their strictly host-independent initiation of replication helps them to sustain in a very broad host range, including *Pseudomonas sp.*, related species in the Proteobacteria, as well as phylogenetically distant species within the Firmicutes, Actinomycetes and even Cyanobacteria (Meyer, 2009) or plants (Buchanan-Wollaston et al., 1987). Consequently, RSF1010, as a mobilizable plasmid, has a far higher replication host range than RP4.

RSF1010 can even spread to a mixed community at a more than 30-fold higher transfer frequency when directly mobilized through co-occurring plasmid RP4 in the same donor cell compared to RP4 itself. Therefore, mobilizable plasmids might contribute to long term gene spread and acquisition to a so far underestimated extent, especially in environments with high intrinsic mobilizing plasmid content. In our current experiment, we use a simplified system and are able to deliver insights into the mobilization potential of a community at the first acquisition event of a newly introduced mobilizable plasmid. The wide variety of mobilization systems possibly involved might not resemble the one encoded by RP4 in efficiency. Still, equivalents based on the community's permissiveness toward RP4 can be used here, since long term maintenance and retransfer are not taken into account. For more complex natural systems and experiments that allow extensive retransfer we recommend assessing the intrinsic mobilization potential of microbial communities based on absolute transfer frequencies, as the transfer and maintenance processes of RSF1010 and RP4 differ too much in the long term.

Apart from quantification of the mobilization potential, the method presented here provides several possibilities to study plasmid ecology and mobilization mechanisms. FACS based sorting of RSF1010 carrying transconjugants from the recipient community was possible. Studying the diversity of transconjugants might provide insights into the enormous host range of mobilizable plasmids, compared to those of broad host range conjugal plasmids (Klümper et al., 2014). But the high amount of sorting time prohibits intensive studies at this point. However, taking advantage of FACS sorting, even at low speed, new possibilities for plasmid isolation emerge. The mobilizing, conjugal plasmid can, now, after retromobilization, co-occurring with RSF1010 in the transconjugant, be subsequently isolated within its original environmental host. Compared to common exogenous plasmid isolation techniques our method has the potential to also capture plasmids that are only transiently hosted and therefore quickly lost in the introduced capturing strains. Since these plasmids remain stable in their original hosts, we gain the ability to isolate them with our method. Isolated plasmids need therefore only stable maintenance in their natural hosts rather than in an artificially introduced strain. This increases the range of obtainable plasmids and immediately supplies information on where they naturally occur. This method reverses the exogenous isolation technique for mobilizable plasmids (Top et al., 1994) and is cultivation independent. Additionally using the tools presented here in combination with FACS sorting, single cell observations to better understand the exact mechanisms proposed for retromobilization (Top et al., 1992, 1995) might become possible.

In conclusion, this method is the first one to assess the plasmid mobilization potential of a microbial community on a quantitative level by estimating transfer frequencies through fluorescent microscopy. Using the new method, we discovered that a mixed microbial community has the potential to easily mobilize a newly introduced mobilizable plasmid at high rates compared to a conjugal plasmid. We also showed that the mobilizable plasmid is spread at far increased frequencies once directly mobilized by a co-occurring conjugal plasmid from within the same cell.

Conflict of Interest Statement

The authors declare that the research was conducted in the absence of any commercial or financial relationships that could be construed as a potential conflict of interest.

ACKNOWLEDGMENTS

We thank L. Riber and S. J. Sørensen for access to the tagged RSF1010 plasmid, L. K. Jensen for technical assistance in the laboratory and S. M. Milani for assistance in FACS sorting. This work was funded by the Villum Kann Rasmussen Foundation Center of Excellence CREAM (Center for Environmental and Agricultural Microbiology).

REFERENCES

1. Bahl, M. I., Oregaard, G., Sørensen, S. J., and Hansen, L. H. (2009). Construction and use of flow cytometry optimized plasmid-sensor strains. *Methods Mol. Biol.* 532, 257–268. doi: 10.1007/978-1-60327-853-9_15
2. Barth, P. T., and Grinter, N. J. (1977). Map of plasmid RP4 derived by insertion of transposon C. *J. Mol. Biol.* 113, 455–474. doi: 10.1016/0022-2836(77)90233-9
3. Boe-Hansen, R., Albrechtsen, H.-J., Arvin, E., and Jørgensen, C. (2002). Bulk water phase and biofilm growth in drinking water at

low nutrient conditions. *Water Res.* 36, 4477–4486. doi: 10.1016/S0043-1354(02)00191-4

4. Buchanan-Wollaston, V., Passiatore, J. E., and Cannon, F. (1987). The mob and oriT mobilization functions of a bacterial plasmid promote its transfer to plants. *Nature* 328, 172–175. doi: 10.1038/328172a0
5. Dechesne, A., Pallud, C. C., Bertolla, F., Grundmann, G. L., and Grundmann, L. G. (2005). Impact of the microscale distribution of a Pseudomonas strain introduced into soil on potential contacts with indigenous bacteria. *Appl. Environ. Microbiol.* 71, 8123–8131. doi: 10.1128/AEM.71.12.8123-8131.2005
6. Dubey, G. P., and Ben-Yehuda, S. (2011). Intercellular nanotubes mediate bacterial communication. *Cell* 144, 590–600. doi: 10.1016/j.cell.2011.01.015
7. Gelder, L., De Vandecasteele, F. P. J., Celeste, J., Forney, L. J., Top, E. M., Brown, C. J.,et al. (2005). Plasmid donor affects host range of promiscuous IncP-1beta plasmid pB10 in an activated-sludge microbial community. *Appl. Environ. Microbiol.* 71, 5309–5317. doi: 10.1128/AEM.71.9.5309-5317.2005
8. Grahn, A. M., Haase, J., Bamford, D. H., and Lanka, E. (2000). Components of the RP4 conjugative transfer apparatus form an envelope structure bridging inner and outer membranes of donor cells: implications for related macromolecule transport systems. *J. Bacteriol.*182, 1564–1574. doi: 10.1128/JB.182.6.1564-1574.2000
9. Gregory, R., Saunders, J. R., and Saunders, V. (2008). Rule-based modelling of conjugative plasmid transfer and incompatibility. *Biosystems* 91, 201–215. doi: 10.1016/j.biosystems.2007.09.003
10. Grohmann, E. (2011). "Horizontal gene transfer between bacteria under natural conditions," in *Microbes and Microbial Technology: Agricultural and Environmental Applications,* eds I. Ahmad, F. Ahmad, and J. Pichtel (New York, NY: Springer), 163–187. doi: 10.1007/978-1-4419-7931-5_7
11. Götz, A., and Smalla, K. (1997). Manure enhances plasmid mobilization and survival of *Pseudomonas putida* introduced into field soil. *Appl. Environ. Microbiol.* 63, 1980–1986.
12. Haase, J., Lurz, R., Grahn, A. M., Bamford, D. H., and Lanka, E.

(1995). Bacterial conjugation mediated by plasmid RP4:RSF1010 mobilization, donor-specific phage propagation, and pilus production require the same Tra2 core components of a proposed DNA transport complex. *J. Bacteriol.* 177, 4779–4791.

13. Heuer, H., Binh, C. T. T., Jechalke, S., Kopmann, C., Zimmerling, U., Krögerrecklenfort, E.,et al. (2012). IncP-1ε plasmids are important vectors of antibiotic resistance genes in agricultural systems: diversification driven by Class 1 integron gene cassettes. *Front. Microbiol.*3:2. doi: 10.3389/fmicb.2012.00002
14. Heuer, H., Schmitt, H., and Smalla, K. (2011). Antibiotic resistance gene spread due to manure application on agricultural fields. *Curr. Opin. Microbiol.* 14, 236–243. doi: 10.1016/j.mib.2011.04.009
15. Heuer, H., and Smalla, K. (2012). Plasmids foster diversification and adaptation of bacterial populations in soil. *FEMS Microbiol. Rev.*36, 1083–1104. doi: 10.1111/j.1574-6976.2012.00337.x
16. Hill, K. E., Weightman, A. J., and Fry, J. C. (1992). Isolation and screening of plasmids from the epilithon which mobilize recombinant plasmid pD10. *Appl. Environ. Microbiol.* 58, 1292–1300.
17. Hirt, H. (2002). In vivo induction of virulence and antibiotic resistance transfer in *Enterococcus faecalis* mediated by the sex pheromone-sensing system of pCF10. *Infect. Immun.* 70, 716–723. doi: 10.1128/IAI.70.2.716-723.2002
18. Honda, Y., Sakai, H., Hiasa, H., Tanaka, K., Komano, T., and Bagdasarian, M. (1991). Functional division and reconstruction of a plasmid replication origin: molecular dissection of the oriV of the broad-host-range plasmid RSF1010. *Proc. Natl. Acad. Sci. U.S.A.* 88, 179–183. doi: 10.1073/pnas.88.1.179
19. Klümper, U., Riber, L., Dechesne, A., Sannazzarro, A., Hansen, L. H., Sørensen, S. J.,et al. (2014). Broad host range plasmids can invade an unexpectedly diverse fraction of a soil bacterial community. *ISME J.* doi: 10.1038/ismej.2014.191 [Epub ahead of print].
20. Król, J. E., Wojtowicz, A. J., Rogers, L. M., Heuer, H., Smalla, K., Krone, S. M.,et al. (2013). Invasion of *E. coli* biofilms by antibiotic resistance plasmids. *Plasmid* 70, 110–119. doi: 10.1016/j.plasmid.2013.03.003

21. Lee, C. A., Thomas, J., and Grossman, A. D. (2012). The Bacillus subtilis conjugative transposon ICEBs1 mobilizes plasmids lacking dedicated mobilization functions. *J. Bacteriol.* 194, 3165–3172. doi: 10.1128/JB.00301-12
22. Lessl, M., Balzer, D., Weyrauch, K., and Lanka, E. (1993). The mating pair formation system of plasmid RP4 defined by RSF1010 mobilization and donor-specific phage propagation. *J. Bacteriol.* 175, 6415–6425.
23. Licht, T. R., Christensen, B. B., Krogfelt, K. A., and Molin, S. (1999). Plasmid transfer in the animal intestine and other dynamic bacterial populations: the role of community structure and environment. *Microbiology* 145, 2615–2622.
24. Loftie-Eaton, W., and Rawlings, D. E. (2012). Diversity, biology and evolution of IncQ-family plasmids. *Plasmid* 67, 15–34. doi: 10.1016/j.plasmid.2011.10.001
25. Merkey, B. V., Lardon, L. A., Seoane, J. M., Kreft, J.-U., and Smets, B. F. (2011). Growth dependence of conjugation explains limited plasmid invasion in biofilms: an individual-based modelling study. *Environ. Microbiol.* 13, 2435–2452. doi: 10.1111/j.1462-2920.2011.02535.x
26. Meyer, R. (2009). Replication and conjugative mobilization of broad host-range IncQ plasmids. *Plasmid* 62, 57–70. doi: 10.1016/j.plasmid.2009.05.001
27. Musovic, S., Dechesne, A., Sørensen, J., and Smets, B. F. (2010). Novel assay to assess permissiveness of a soil microbial community toward receipt of mobile genetic elements. *Appl. Environ. Microbiol.* 76, 4813–4818. doi: 10.1128/AEM.02713-09
28. Musovic, S., Klümper, U., Dechesne, A., Magid, J., and Smets, B. F. (2014). Long-term manure exposure increases soil bacterial community potential for plasmid uptake. *Environ. Microbiol. Rep.* 6, 125–130. doi: 10.1111/1758-2229.12138
29. Nelson, K. E., Weinel, C., Paulsen, I. T., Dodson, R. J., Hilbert, H., Martins dos Santos, V. A. P.,et al. (2002). Complete genome sequence and comparative analysis of the metabolically versatile *Pseudomonas putida* KT2440. *Environ. Microbiol.* 4, 799–808. doi: 10.1046/j.1462-2920.2002.00366.x

30. Reasoner, D. J., Blannon, J. C., and Geldreich, E. E. (1979). Rapid seven-hour fecal coliform test. *Appl. Environ. Microbiol.* 38, 229–236.

31. Scholz, P., Haring, V., Wittmann-Liebold, B., Ashman, K., Bagdasarian, M., and Scherzinger, E. (1989). Complete nucleotide sequence and gene organization of the broad-host-range plasmid RSF1010. *Gene* 75, 271–288. doi: 10.1016/0378-1119(89)90273-4

32. Shintani, M., Matsui, K., Inoue, J.-I., Hosoyama, A., Ohji, S., Yamazoe, A.,et al. (2014). Single-cell analyses revealed transfer ranges of IncP-1, IncP-7, and IncP-9 plasmids in a soil bacterial community. *Appl. Environ. Microbiol.* 80, 138–145. doi: 10.1128/AEM.02571-13

33. Smalla, K., Heuer, H., Götz, A., Niemeyer, D., Krogerrecklenfort, E., and Tietze, E. (2000). Exogenous isolation of antibiotic resistance plasmids from piggery manure slurries reveals a high prevalence and diversity of IncQ-like plasmids. *Appl. Environ. Microbiol.* 66, 4854–4862. doi: 10.1128/AEM.66.11.4854-4862.2000

34. Smillie, C., Garcillán-Barcia, M. P., Francia, M. V., Rocha, E. P. C., and de la Cruz, F. (2010). Mobility of plasmids. *Microbiol. Mol. Biol. Rev.* 74, 434–452. doi: 10.1128/MMBR.00020-10

35. Sørensen, S. J., Bailey, M., Hansen, L. H., Kroer, N., and Wuertz, S. (2005). Studying plasmid horizontal transfer in situ: a critical review. *Nat. Rev. Microbiol.* 3, 700–710. doi: 10.1038/nrmicro1232

36. Thomas, C. M., and Nielsen, K. M. (2005). Mechanisms of, and barriers to, horizontal gene transfer between bacteria. *Nat. Rev. Microbiol.* 3, 711–721. doi: 10.1038/nrmicro1234

37. Top, E., De Smet, I., Verstraete, W., Dijkmans, R., and Mergeay, M. (1994). Exogenous isolation of mobilizing plasmids from polluted soils and sludges. *Appl. Environ. Microbiol.* 60, 831–839.

38. Top, E., Rore, H., Collard, J.-M. J., Gellens, V., Slobodkina, G., Verstraete, W.,et al. (1995). Retromobilization of heavy metal resistance genes in unpolluted and heavy metal polluted soil. *FEMS Microbiol. Ecol.* 18, 191–203. doi: 10.1111/j.1574-6941.1995.tb00176.x

39. Top, E., Vanrolleghem, P., Mergeay, M., and Verstraete, W. (1992).

Determination of the mechanism of retrotransfer by mechanistic mathematical modeling. *J. Bacteriol.* 174, 5953–5960.

40. van Elsas, J. D., Gardener, B. B., Wolters, A. C., and Smit, E. (1998). Isolation, characterization, and transfer of cryptic gene-mobilizing plasmids in the wheat rhizosphere. *Appl. Environ. Microbiol.* 64, 880–889.
41. Yano, H., Rogers, L. M., Knox, M. G., Heuer, H., Smalla, K., Brown, C. J.,et al. (2013). Host range diversification within the IncP-1 plasmid group. *Microbiology* 159, 2303–2315. doi: 10.1099/mic.0.068387-0
42. Zhang, M., Visser, S., Pereira E Silva, M. C., and van Elsas, J. D. (2014). IncP-1 and PromA group plasmids are major providers of horizontal gene transfer capacities across bacteria in the mycosphere of different doil fungi. *Microb. Ecol.* doi: 10.1007/s00248-014-0482-6

CHAPTER

6

PARB PARTITION PROTEINS: COMPLEX FORMATION AND SPREADING AT BACTERIAL AND PLASMID CENTROMERES

Barbara E. Funnell

Department of Molecular Genetics, University of Toronto, Toronto, ON, Canada

ABSTRACT

In bacteria, active partition systems contribute to the faithful segregation of both chromosomes and low-copy-number plasmids. Each system depends on a site-specific DNA binding protein to recognize and assemble a partition complex at a centromere-like site, commonly called *parS*. Many plasmid, and all chromosomal centromere-binding proteins are dimeric helix-turn-helix DNA binding proteins, which are commonly named ParB. Although the overall sequence conservation among ParBs is not high, the proteins share similar domain and functional organization, and they assemble into similar higher-order complexes. *In vivo*, ParBs "spread," that is, DNA binding extends away from the *parS* site into the surrounding non-specific DNA, a feature that reflects higher-order complex assembly. ParBs bridge and pair DNA at *parS* and non-specific DNA sites. ParB dimers interact with each other via flexible conformations of an N-terminal region. This review will focus on the properties of the HTH centromere-binding protein, in light of recent experimental evidence and models that are adding to our understanding of how these proteins assemble into large and dynamic partition complexes at and around their specific DNA sites.

Keywords: Chromosome dynamics, segregation, ParABS, DNA-binding, bridging

In bacteria, the segregation, or partition, of low-copy-number plasmids and cellular chromosomes depends on the activity of site-specific DNA binding proteins to recognize one or more copies of a centromere-like DNA site. These "centromere-binding proteins" generally work in concert with an ATPase or GTPase, resulting in dynamic movement and positioning of plasmids or chromosomal domains during the cell cycle (reviewed in Wang et al., 2013; Baxter and Funnell, 2014; Bouet et al., 2014). Centromere-binding proteins fall into one of two structural classes, as helix-turn-helix (HTH), or ribbon-helix-helix site-specific DNA binding proteins. In bacteria, the proteins of all chromosomal, and many plasmid partition systems are members of the HTH class. They share similar properties *in vivo* and *in vitro*, including similar domain organization and DNA-binding properties, although there are also interesting differences. They all form large partition complexes *in vivo* that can be visualized as foci using fluorescence approaches. This review will focus on the properties of the HTH centromere-binding proteins, and in particular, how they assemble into large partition complexes. I will discuss the contribution of protein domains, the "spreading" phenomenon that has been reported as a general property of HTH centromere-binding proteins, and how flexibility in the protein allows multiple conformations and binding modes in complex assembly.

The components of partition systems are commonly named ParA (the partition ATPase), ParB (the centromere-binding protein), and *parS* (the centromere or partition site). Plasmids typically contain one *parS* located near the *parA* and *parB* genes, although some have multiple *parS* sites. Bacterial chromosomes contain several *parS* sites, which are primarily located in the chromosomal domain that contains the replication origin. Bacterial ParA and ParB are also called Soj and Spo0J, respectively, because their genes were first defined by roles in sporulation of *Bacillus subtilis* (Ireton et al., 1994). For simplicity, I will use the ParABS nomenclature, with specific names for some of the discussion to be consistent with published literature.

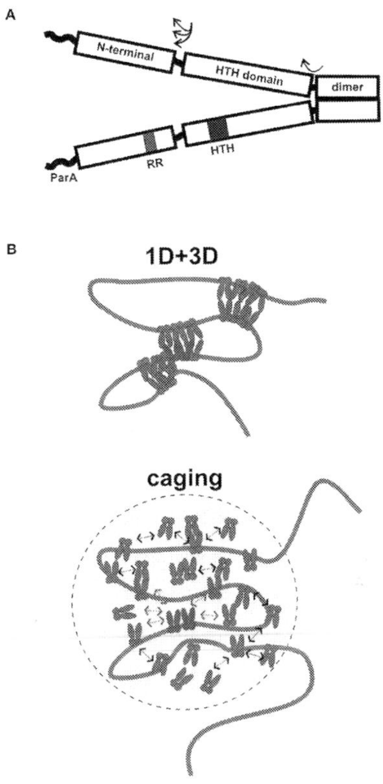

Figure 1: Assembly of ParB partition complexes. (A) Cartoon of the conserved domain structure of HTH ParBs, shown as a dimer. The black rectangles represent the regions of the protein for which there is some structural information: the C-terminal dimer domain, the HTH DNA binding domain, and the N-terminal domain. The three regions are connected by flexible linker sequences (arrows). The linker length here is represented as short, as in the *Hp*Spo0J and P1 ParB published structures (Schumacher and Funnell, 2005; Chen et al., 2015), but may be longer in other ParBs. The wavy line represents the region that interacts with ParA in many, although not all, ParBs. The position of the HTH motif (blue) and the conserved arginine patch motif (RR, red) are indicated in one monomer. **(B)** Diagrams of 1D + 3D and caging models for higher-order

ParB binding and partition complex assembly. ParB dimers bound to *parS* (in red) nucleate complex assembly and interact with other ParBs in green. Arrows in the caging architecture illustrate that dynamic associations maintain the cluster of ParB.

HTH ParBs share a similar domain organization although the primary sequence conservation is not high among members of this family. In general, the protein is divided into three regions: A central HTH DNA binding domain is flanked by a C-terminal dimer domain and an N-terminal region necessary for protein oligomerization (Figure 1A). Flexible linkers connect the domains, and flexibility in domain organization, orientation, and folding have been observed in biochemical experiments and crystal structures. The most highly conserved sequence among plasmid and chromosomal ParBs is a short arginine-rich motif in the N-terminus, which is often called an arginine patch (Yamaichi and Niki, 2000). ParA interactions are often specified by residues near the N-terminus of ParB (Radnedge et al., 1998; Figge et al., 2003; Leonard et al., 2005; Ah-Seng et al., 2009), although there are exceptions. For example, the ParA interactions for RK2 KorB and *Pseudomonas aeruginosa* ParB map to the center and dimer domain, respectively (Lukaszewicz et al., 2002; Bartosik et al., 2004). There are also added complexities to the general arrangement. For example, P1 ParB and its relatives contain an additional site-specific DNA binding activity within the dimer domain. These ParBs recognize both an inverted repeat and a second DNA motif in their *parS* sites, via their HTH domain and dimer domains, respectively (Schumacher and Funnell, 2005).

Partition Complex Assembly, Spreading, and Bridging

Partition complex assembly begins with the recognition of *parS* by a dimer of ParB, followed by loading of multiple ParB dimers to form a very large protein-DNA complex (Baxter and Funnell, 2014). These higher-order complexes are necessary as both the substrates and the activators of the mechanisms of partition. The number of ParB protein foci observed inside cells is usually lower than the number of *parS* sites, leading to the idea that inter and intra-molecular pairing of *parS* sites occurs in plasmids and chromosomal domains.

Spreading is an unusual feature for site-specific DNA binding proteins that is common to HTH ParBs, and it reflects how ParB assembles into higher-order complexes (Rodionov et al., 1999; Murray et al., 2006; Breier and Grossman, 2007; Sanchez et al., 2015). Measured by ChIP approaches, *in vivo* ParB binding extends beyond *parS* into the surrounding non-specific DNA, often many kb away from the site. Binding is maximal at *parS*, and diminishes non-linearly as a function of distance from *parS*. Spreading can be impeded by "roadblocks," which are strong binding sites for other proteins (Rodionov et al., 1999; Murray et al., 2006). Spreading has also been inferred from the ability of some ParBs, especially when overexpressed, to silence expression of nearby genes (Lynch and Wang, 1995; Rodionov et al., 1999; Bartosik et al., 2004; Bingle et al., 2005; Kusiak et al., 2011). Silencing is likely a consequence of protein overexpression and is not necessary for partition (Rodionov and Yarmolinsky, 2004). Spreading ability is required however, as ParB mutants that do not spread are defective in partition (Rodionov et al., 1999; Autret et al., 2001; Breier and Grossman, 2007; Kusiak et al., 2011; Graham et al., 2014). In particular, the arginine patch in ParB is essential for spreading, focus formation, and partition activity *in vivo*.

The first and simplest model described spreading as lateral protein-protein association along the DNA as a one-dimensional filament (Rodionov et al., 1999). However some properties of ParBs lead investigators to question this idea. Studies with plasmid KorB and with *B. subtilis* Spo0J (*Bs*Spo0J) argued that the intracellular concentration of ParB was insufficient to account for the amount of spreading observed *in vivo* if arranged as a one-dimensional filament (Bingle et al., 2005; Graham et al., 2014). It was also difficult to demonstrate biochemically that ParB binding to *parS* increased the affinity of ParB for adjacent non-specific DNA.

Two other models have emerged recently, based on sophisticated microscope and ChIP-seq technologies as well as computer modeling and traditional biochemistry (Broedersz et al., 2014; Graham et al., 2014; Sanchez et al., 2015). The first proposes that limited lateral ParB-ParB interactions (1D) in combination with inter and intra-molecular looping and bridging (3D) act to build large complexes and coalesce many ParB molecules into foci (Broedersz et al., 2014; Graham et al., 2014; Figure 1B). Computer modeling was used to argue that neither 1D nor 3D interactions alone could generate ParB foci; that the 1D + 3D arrangement allows focus formation because it creates a surface tension on the ParB cluster that counteracts the tendency for entropy to disperse the protein on DNA (Broedersz et al., 2014). Elegant *in vitro* TIRF microscopy experiments provided experimental support for the bridging activity: flow-stretched DNA was condensed by *Bs*Spo0J in a manner that is most consistent with ParB bridging across loops within the same or across different DNA molecules (Graham et al., 2014). The experiments were however unable to demonstrate any sequence-specificity for *parS*, leading to the suggestion that experiments on flow-stretched DNA were not recapitulating an undefined aspect of critical nucleation properties of ParB bound to *parS*. For example, one factor missing from these experiments is DNA supercoiling, which affects chromosome compaction and may strongly influence the DNA binding properties of ParB in higher-order complexes.

Mutations in conserved arginine residues of the arginine patch motif eliminated bridging in the TIRF assay, consistent with the requirement for this motif for spreading and partition *in vivo* (Graham et al., 2014). Interestingly, *Bs*Spo0J mutated at one residue in the arginine patch (G77S), which is unable to form foci or spread *in vivo*, was still able to bridge DNA, and with slightly higher stability than that of wild-type *Bs*Spo0J. Therefore the bridging activity of ParB is necessary but not sufficient for complex formation. These observations lead to the suggestion that the G77S mutation may promote inappropriate bridging and/or alter the dynamics of bridging necessary for proper complex assembly *in vivo*. Modeling the spreading/bridging behavior also predicted that roadblocks would decrease the probability of loops forming in their vicinity, interfering with complex assembly beyond the roadblock (Broedersz et al., 2014).

In contrast, a second model proposes that a network of stochastic binding of ParB explains the clustering of ParB molecules around *parS* (Sanchez et al., 2015). In essence, the nucleation of ParB by *parS* creates, and maintains a very high localized concentration of ParB in a "cage" by many weak but dynamic interactions with itself (dimer-dimer interactions via the N-terminal domains), as well as with non-specific DNA around *parS* (Figure 1B). In caging, these interactions do not need to occur simultaneously or to bridge DNA (Figure 1B). Computer modeling of the patterns of ParB occupancy around *parS* measured by ChIP was used to argue that they are not consistent with either 1D lateral spreading or a combination of 1D spreading and 3D bridging. Biochemical examination showed no evidence that binding of one ParB could stabilize binding of an adjacent ParB. The caging model neither requires nor excludes bridging, although bridging interactions are intuitively attractive as part of the dynamic glue. The model does depend on other properties of the DNA chromosome, such as topology or organization by other nucleoid-binding proteins *in vivo* to help restrict the DNA within the cage. Roadblocks could alter local DNA organization and reduce the proximity of *parS* to the rest of the DNA in three-dimensional space; that is, place this DNA outside the cage.

Both models agree that ParB binding to *parS* must nucleate the formation of higher-order complexes to explain ParB clustering and foci *in vivo* (Broedersz et al., 2014; Sanchez et al., 2015). This is most simple to envision in plasmid systems such as P1, in which ParB's affinity for *parS* is at least 10,000-fold higher than that for non-specific DNA (Funnell, 1991). However this affinity difference is small for some ParBs (Broedersz et al., 2014; Taylor et al., 2015), and it was suggested that a conformational change is induced in ParB by *parS*-specific binding to effectively anchor the focus at the site. Both models agree that multiple ParB-ParB interaction interfaces must be involved in assembly of higher-order partition complexes, and that the dynamics of these interactions are critical for proper assembly and function in partition. How bridging activities detected *in vitro* contribute to ParB activity *in vivo* remains to be resolved. Further refinement of these models will depend on the ability to completely reconstitute the *parS*-dependent complex assembly *in vitro*, which will in turn depend on identifying the other factors necessary for caging or bridging, and on the nature of ParB-ParB and ParB-DNA interactions at the molecular level. The influence of ParA on complex architecture and dynamics has also yet to be defined (see below).

ParB-ParB and ParB-DNA Interactions in Higher-Order-Complex Assembly

Site-specific DNA binding of ParBs to cognate *parS* sites has been examined directly and in detail in many different partition systems, but the *parS*-dependent formation of higher order, large partition complexes has been difficult to reconstitute *in vitro*. However, there are insights arising from structural biology of several plasmid and chromosomal

ParBs, which are leading to a preliminary, albeit incomplete, picture of partition complex assembly.

Although there are no structures of a full length ParB, those of individual domains or combinations of domains, with and without *parS* DNA, have provided clues concerning the three dimensional organization of the protein with respect to DNA and to itself. There are structures of the HTH domains of three plasmid ParBs (P1 ParB, F SopB, and RP4 KorB) in complex with their specific DNA sites, and of their dimer domains (SopB and KorB dimer domains solved separately from the HTH; Delbrück et al., 2002; Khare et al., 2004; Schumacher and Funnell, 2005; Schumacher et al., 2010). Structures of two chromosomal ParB fragments, each containing the N-terminal region and adjacent HTH domain, have visualized the oligomerization interactions of the proteins. These ParBs are *Thermus thermophilus* Spo0J (*Tt*Spo0J) and *Helicobacter pylori* Spo0J (*Hp*Spo0J); structure of the latter was solved bound to *parS* (Leonard et al., 2004; Chen et al., 2015).

One of the first themes to highlight is that of flexibility. Taken together, the structures indicate that these three regions of the protein are connected by flexible linkers, and that their orientation with respect to each other can vary. The conformation of the N-terminus is particularly flexible (Chen et al., 2015).

Second, ParBs are bona-fide HTH site-specific DNA binding proteins, but with a twist. As expected, the HTH domains contact inverted repeat sequences within *parS* via helix insertion into the major groove of DNA. However, unexpected features emerged from the structures. First, residues outside of the recognition helix also contribute to specificity for *parS* in SopB, KorB, and *Hp*Spo0J (Khare et al., 2004; Schumacher et al., 2010; Sanchez et al., 2013; Chen et al., 2015). Second, both the P1 ParB and *Hp*Spo0J structures demonstrated bridging across *parS* sites mediated by the dimer and N-terminal domains, respectively (Schumacher and Funnell, 2005; Chen et al., 2015). Each monomer of a P1 ParB dimer interacts with a half-site on a different DNA molecule, effectively pairing two *parS* sites. *Hp*Spo0J-*parS*bridging interactions are mediated by the N-terminal oligomerization regions of the protein (Figure 2). Four monomers of *Hp*Spo0J (monomeric because it lacks the C-terminal dimer domain) interact with two *parS* oligos and with each other in a cross-bridge arrangement across the DNA molecules (molecules A–D in Figure 2). The monomers share a common HTH domain, but show different conformations in the extended N-terminal regions as well as different interactions with each other. There are two adjacent (AB and CD) and one transverse (AC) sets of protein-protein interactions, which are distinct. For example, the conserved arginine patch motifs are close to each other at the AC interface, but not at the AB or CD interfaces (Figure 2). In the structure, there is no BD interaction, leaving these surfaces available, perhaps for interactions with different conformations of ParB or with different partners.

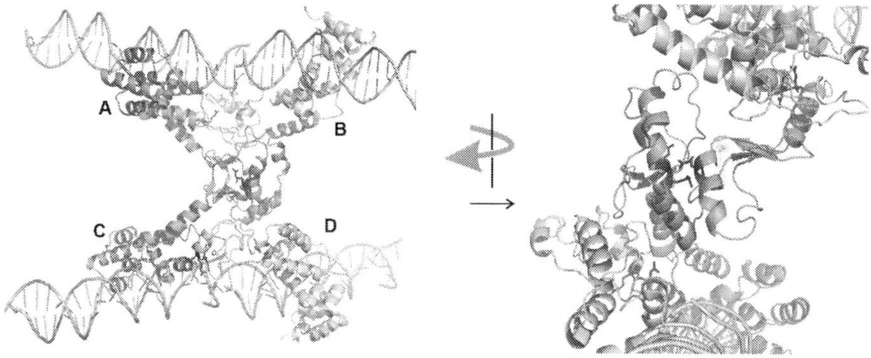

Figure 2: Structure of *Hp*Spo0J monomers (lacking C-terminal dimer domain) bound to and across *parS* DNA (PDB 4UMK, **Chen et al., 2015)**. Four monomers (A to D) make adjacent (AB and CD) and transverse (AC) interactions. The arginine patch motif is illustrated by two red arginines from each monomer. The arrangement on the left is rotated approximately 90° and magnified on the right to illustrate the environments of these arginines in the different interactions. The images were generated using the PyMOL Molecular Graphics System, Version 1.8.2.1 Schrödinger, LLC.

The overall fold of *Hp*Spo0J is similar to that of *Tt*Spo0J, except for a bend in the linker between the N-terminal and HTH domains (Leonard et al., 2004; Chen et al., 2015). It was suggested that the *Tt*Spo0J structure may represent a closed conformation that opens up following DNA binding to the *Hp*Spo0J architecture.

How do these structures inform us of higher-order complex assembly, particularly when ParB binds, bridges, and spreads on non-specific DNA adjacent to and away from *parS*? The simplest model is that the HTH is responsible for both specific and non-specific DNA interactions, which is supported by the observation that a triple substitution in the HTH domain of F SopB impairs both DNA binding activities (Ah-Seng et al., 2009). In this case the *Hp*Spo0J-DNA structure may represent ParB-ParB and ParB-DNA interactions during spreading at any DNA site. The requirement for the N-terminus in higher-order complex formation *in vivo* is also consistent with this picture. The flexibility of and variation in the cross-monomer interactions and interfaces seen in the *Hp*Spo0J structure make it an attractive model for the ability of ParB to make multiple and flexible interactions during complex assembly and maintenance. However, one recent study suggests that the specific and non-specific DNA binding activities may be distinct (Taylor et al., 2015). The specific and non-specific DNA binding activities of *Bs*Spo0J showed different properties *in vitro*, including different abilities to protect the HTH domain from proteolysis. The results lead to the proposal that the N-terminus contains a DNA binding region that is distinct from the HTH. These observations may also reflect differences between plasmid and chromosomal ParBs. We must await identification of the protein-DNA contacts at non-specific sites before we can confirm the organization of ParB during higher-order partition complex assembly.

The Role of ParA in ParB-DNA Complexes

During partition, ParAs form patterns on the surface of the bacterial nucleoid due to dynamic interactions with ParB bound to *parS* (Hatano et al., 2007; Ringgaard et al., 2009; Hatano and Niki, 2010; Ah-Seng et al., 2013). The patterning is necessary for the segregation of plasmids and chromosomal domains, and the molecular mechanisms involved are still being defined. ParA is not necessary to form the large ParB-DNA complexes seen *in vivo* and *in vitro*, but ParA can influence or modulate these complexes. For example, the behavior of several *parA* and *parB* mutants supports a proposal that ParA is necessary to separate pairs or groups of plasmids during segregation (Fung et al., 2001; Ah-Seng et al., 2013). For the ParBs that interact with ParA via N-terminal regions that are adjacent to the flexible ParB-ParB oligomerization interface, one attractive idea is that ParA-ParB interactions at the N-terminus may influence the available conformations for ParB-ParB interactions next door. For example, specific interference with the transverse ParB-

ParB interactions in the *Hp*Spo0J structure might favor intramolecular associations over intermolecular ones (Figure 2).

Flexibility and Order in Partition Complex Assembly

Taken together, the biochemistry and structural biology of the N-terminal regions of ParBs imply that the folding and structures are flexible, dynamic, and fluid. Recent experiments using magnetic tweezers support the idea that the complexes are not highly ordered (Taylor et al., 2015). ParB-ParB (dimer-dimer) interactions via the N-terminal domain must contribute to higher-order complex assembly and function. It is attractive to consider that the flexibility of the N-terminus, in folding and conformation, is important for the dynamics and architecture of the large higher-order partition complex *in vivo*. This conformational flexibility resembles the properties of so-called "intrinsically disordered" proteins and domains whose unstructured properties allow proteins to sample and bind to multiple targets and in multiple ways (Wright and Dyson, 1999; Uversky, 2016). The same region in ParB could be involved, directly or indirectly, in different binding scenarios with different partners, including ParA, and potentially non-specific DNA. Why are flexibility, dynamics, and size important for these partition complexes? ParBs are engaging in multiple and constantly changing interactions during partition. Clustering creates a condensed, organized DNA substrate and provides a high density of ParBs available to ParA. Defining the molecular nature of these interactions continues to be an essential step toward the understanding of these intriguing DNA binding proteins.

REFERENCES

1. Ah-Seng, Y., Lopez, F., Pasta, F., Lane, D., and Bouet, J.-Y. (2009). Dual role of DNA in regulating ATP hydrolysis by the SopA partition protein. *J. Biol. Chem.* 284, 30067–30075. doi: 10.1074/jbc.M109.044800
2. Ah-Seng, Y., Rech, J., Lane, D., and Bouet, J. Y. (2013). Defining the role of ATP hydrolysis in mitotic segregation of bacterial plasmids. *PLoS Genet* 9:e1003956. doi: 10.1371/journal.pgen.1003956
3. Autret, S., Nair, R., and Errington, J. (2001). Genetic analysis of

the chromosome segregation protein Spo0J of *Bacillus subtilis*: evidence for separate domains involved in DNA binding and interactions with Soj protein. *Mol. Microbiol.* 41, 743–755. doi: 10.1046/j.1365-2958.2001.02551.x

4. Bartosik, A. A., Lasocki, K., Mierzejewska, J., Thomas, C. M., and Jagura-Burdzy, G. (2004). ParB of *Pseudomonas aeruginosa*: interactions with its partner ParA and its target *parS* and specific effects on bacterial growth. *J. Bacteriol.* 186, 6983–6998. doi: 10.1128/jb.186.20.6983-6998.2004
5. Baxter, J. C., and Funnell, B. E. (2014). Plasmid partition mechanisms. *Microbiol. Spectr.* 2, 6. doi: 10.1128/microbiolspec.PLAS-0023-2014
6. Bingle, L. E. H., Macartney, D. P., Fantozzi, A., Manzoor, S. E., and Thomas, C. M. (2005). Flexibility in repression and cooperativity by KorB of broad host range IncP-1 plasmid RK2. *J. Mol. Biol.* 349, 302–316. doi: 10.1016/j.jmb.2005.03.062
7. Bouet, J. Y., Stouf, M., Lebailly, E., and Cornet, F. (2014). Mechanisms for chromosome segregation. *Curr. Opin. Microbiol.* 22, 60–65. doi: 10.1016/j.mib.2014.09.013
8. Breier, A. M., and Grossman, A. D. (2007). Whole-genome analysis of the chromosome partitioning and sporulation protein Spo0J (ParB) reveals spreading and origin-distal sites on the *Bacillus subtilis* chromosome. *Mol. Microbiol.* 64, 703–718. doi: 10.1111/j.1365-2958.2007.05690.x
9. Broedersz, C. P., Wang, X. D., Meir, Y., Loparo, J. J., Rudner, D. Z., and Wingreen, N. S. (2014). Condensation and localization of the partitioning protein ParB on the bacterial chromosome. *Proc. Natl. Acad. Sci. U.S.A.* 111, 8809–8814. doi: 10.1073/pnas.1402529111
10. Chen, B.-W., Lin, M.-H., Chu, C.-H., Hsu, C.-E., and Sun, Y.-J. (2015). Insights into ParB spreading from the complex structure of Spo0J and *parS*. *Proc. Natl. Acad. Sci. U.S.A.* 112, 6613–6618. doi: 10.1073/pnas.1421927112
11. Delbrück, H., Ziegelin, G., Lanka, E., and Heinemann, U. (2002). An Src Homology 3-like domain is responsible for dimerization of the repressor protein KorB encoded by the promiscuous IncP plasmid RP4. *J. Biol. Chem.* 277, 4191–4198. doi: 10.1074/jbc.

M110103200

12. Figge, R. M., Easter, J., and Gober, J. W. (2003). Productive interaction between the chromosome partitioning proteins, ParA and ParB, is required for the progression of the cell cycle in *Caulobacter crescentus*. *Mol. Microbiol.* 47, 1225–1237. doi: 10.1046/j.1365-2958.2003.03367.x
13. Fung, E., Bouet, J.-Y., and Funnell, B. E. (2001). Probing the ATP-binding site of P1 ParA: partition and repression have different requirements for ATP binding and hydrolysis. *EMBO J.* 20, 4901–4911. doi: 10.1093/emboj/20.17.4901
14. Funnell, B. E. (1991). The P1 partition complex at *parS*: the influence of *Escherichia coli* integration host factor and of substrate topology. *J. Biol. Chem.* 266, 14328–14337.
15. Graham, T. G. W., Wang, X., Song, D., Etson, C. M., van Oijen, A. M., Rudner, D. Z., et al. (2014). ParB spreading requires DNA bridging. *Genes Dev.* 28, 1228–1238. doi: 10.1101/gad.242206.114
16. Hatano, T., and Niki, H. (2010). Partitioning of P1 plasmids by gradual distribution of the ATPase ParA. *Mol. Microbiol.* 78, 1182–1198. doi: 10.1111/j.1365-2958.2010.07398.x
17. Hatano, T., Yamaichi, Y., and Niki, H. (2007). Oscillating focus of SopA associated with filamentous structure guides partitioning of F plasmid. *Mol. Microbiol.* 64, 1198–1213. doi: 10.1111/j.1365-2958.2007.05728.x
18. Ireton, K., Gunther, N. W., and Grossman, A. D. (1994). *spo0J* is required for normal chromosome segregation as well as the initiation of sporulation in *Bacillus subtilis*. *J. Bacteriol.* 176, 5320–5329.
19. Khare, D., Ziegelin, G., Lanka, E., and Heinemann, U. (2004). Sequence-specific DNA binding determined by contacts outside the helix-turn-helix motif of the ParB homolog KorB. *Nat. Struct. Mol. Biol.* 11, 656–663. doi: 10.1038/nsmb773
20. Kusiak, M., Gapczyńska, A., Płochocka, D., Thomas, C. M., and Jagura-Burdzy, G. (2011). Binding and spreading of ParB on DNA determine its biological function in *Pseudomonas aeruginosa*. *J. Bacteriol.* 193, 3342–3355. doi: 10.1128/JB.00328-11
21. Leonard, T. A., Butler, P. J. G., and Löwe, J. (2004). Structural analysis of the chromosome segregation protein Spo0J from *Thermus*

thermophilus. *Mol. Microbiol.* 53, 419–432. doi: 10.1111/j.1365-2958.2004.04133.x

22. Leonard, T. A., Butler, P. J., and Löwe, J. (2005). Bacterial chromosome segregation: structure and DNA binding of the Soj dimer – a conserved biological switch. *EMBO J.* 24, 270–282. doi: 10.1038/sj.emboj.7600530
23. Lukaszewicz, M., Kostelidou, K., Bartosik, A. A., Cooke, G. D., Thomas, C. M., and JaguraBurdzy, G. (2002). Functional dissection of the ParB homologue (KorB) from IncP-1 plasmid RK2. *Nucl. Acids Res.* 30, 1046–1055. doi: 10.1093/nar/30.4.1046
24. Lynch, A. S., and Wang, J. C. (1995). SopB protein-meditated silencing of genes linked to the *sopC* locus of *Escherichia coli* F plasmid. *Proc. Natl. Acad. Sci. U.S.A.* 92, 1896–1900.
25. Murray, H., Ferreira, H., and Errington, J. (2006). The bacterial chromosome segregation protein Spo0J spreads along DNA from *parS*nucleation sites. *Mol. Microbiol.* 61, 1352–1361. doi: 10.1111/j.1365-2958.2006.05316.x
26. Radnedge, L., Youngren, B., Davis, M., and Austin, S. (1998). Probing the structure of complex macromolecular interactions by homolog specificity scanning: the P1 and P7 plasmid partition systems. *EMBO J.* 17, 6076–6085.
27. Ringgaard, S., van Zon, J., Howard, M., and Gerdes, K. (2009). Movement and equipositioning of plasmids by ParA filament disassembly. *Proc. Natl. Acad. Sci. U.S.A.* 106, 19369–19374. doi: 10.1073/pnas.0908347106
28. Rodionov, O., Lobocka, M., and Yarmolinsky, M. (1999). Silencing of genes flanking the P1 plasmid centromere. *Science* 283, 546–549.
29. Rodionov, O., and Yarmolinsky, M. (2004). Plasmid partitioning and the spreading of P1 partition protein ParB. *Mol. Microbiol.* 52, 1215–1223. doi: 10.1111/j.1365-2958.2004.04055.x
30. Sanchez, A., Cattoni, D. I., Walter, J.-C., Rech, J., Parmeggiani, A., Nollmann, M., et al. (2015). Stochastic self-assembly of ParB proteins builds the bacterial DNA segregation apparatus. *Cell Syst.* 1, 163–173. doi: 10.1016/j.cels.2015.07.013
31. Sanchez, A., Rech, J., Gasc, C., and Bouet, J. Y. (2013). Insight into centromere-binding properties of ParB proteins: a secondary

binding motif is essential for bacterial genome maintenance. *Nucl. Acids Res.* 41, 3094–3103. doi: 10.1093/nar/gkt018

32. Schumacher, M. A., and Funnell, B. E. (2005). ParB-DNA structures reveal DNA-binding mechanism of partition complex formation. *Nature* 438, 516–519. doi: 10.1038/nature04149
33. Schumacher, M. A., Piro, K. M., and Xu, W. J. (2010). Insight into F plasmid DNA segregation revealed by structures of SopB and SopB-DNA complexes. *Nucl. Acids Res.* 38, 4514–4526. doi: 10.1093/nar/gkq161
34. Taylor, J. A., Pastrana, C. L., Butterer, A., Pernstich, C., Gwynn, E. J., Sobott, F., et al. (2015). Specific and non-specific interactions of ParB with DNA: implications for chromosome segregation. *Nucl. Acids Res.* 43, 719–731. doi: 10.1093/nar/gku1295
35. Uversky, V. N. (2016). Dancing protein clouds: the strange biology and chaotic physics of intrinsically disordered proteins. *J. Biol. Chem.*291, 6681–6688. doi: 10.1074/jbc.R115.685859
36. Wang, X. D., Llopis, P. M., and Rudner, D. Z. (2013). Organization and segregation of bacterial chromosomes. *Nat. Rev. Genet.* 14, 191–203. doi: 10.1038/nrg3375
37. Wright, P. E., and Dyson, H. J. (1999). Intrinsically unstructured proteins: re-assessing the protein structure-function paradigm. *J. Mol. Biol.* 293, 321–331. doi: 10.1006/jmbi.1999.3110
38. Yamaichi, Y., and Niki, H. (2000). Active segregation by the *Bacillus subtilis* partitioning system in *Escherichia coli*. *Proc. Natl. Acad. Sci. U.S.A.* 97, 14656–14661. doi: 10.1073/pnas.97.26.14656

SECTION IV:
SPECIFIC PLASMID SYSTEMS

CHAPTER

7

ESTABLISHMENT OF A REPLICATING PLASMID IN RICKETTSIA PROWAZEKII

David O. Wood[1], Andria Hines[1], Aimee M. Tucker[1], Andrew Woodard[1] , Lonnie O. Driskell[1], Nicole Y. Burkhardt[2], Timothy J. Kurtti[2], Gerald D. Baldridge[2], Ulrike G. Munderloh[2]

[1]Department of Microbiology and Immunology, Laboratory of Molecular Biology, University of South Alabama College of Medicine, Mobile, Alabama, United States of America,

[2]Department of Entomology, University of Minnesota, St. Paul, Minnesota, United States of America

ABSTRACT

Rickettsia prowazekii, the causative agent of epidemic typhus, grows only within the cytosol of eukaryotic host cells. This obligate intracellular lifestyle has restricted the genetic analysis of this pathogen and critical tools, such as replicating plasmid vectors, have not been developed for this species. Although replicating plasmids have not been reported in R. prowazekii, the existence of well-characterized plasmids in several less pathogenic rickettsial species provides an opportunity to expand the genetic systems available for the study of this human pathogen. Competent R. prowazekii were transformed with pRAM18dRGA, a 10.3 kb vector derived from pRAM18 of R. amblyommii. A plasmid-containing population of R. prowazekii was obtained following growth under antibiotic selection, and the rickettsial plasmid was maintained extrachromosomally throughout multiple passages. The transformant population exhibited a generation time comparable to that of the wild type strain with a copy number of approximately 1 plasmid per rickettsia. These results demonstrate for the first time that a plasmid can be maintained in R. prowazekii, providing an important genetic tool for the study of this obligate intracellular pathogen.

INTRODUCTION

Rickettsia prowazekii, the causative agent of epidemic typhus and a Category B Select Agent, is an obligate intracellular bacterium that grows directly within the cytosol of eukaryotic host cells. The obligate intracellular nature of *R. prowazekii* growth places considerable restrictions on the genetic manipulation of this pathogen. Foremost, it is essential that manipulations do not prevent the rickettsiae from infecting host cells, a requirement for rickettsial survival and growth. Since *R. prowazekii* cannot be grown as colonies on an agar surface using axenic media, standard bacterial cloning protocols are unavailable. In addition, plaquing techniques currently used for cloning spotted fever group rickettsiae that can polymerize actin for intracellular movement are problematic for *R. prowazekii*, which is deficient in actin polymerization [1], [2], [3]. Thus, clonal populations of *R. prowazekii* mutants must be isolated using labor-intensive and time-consuming techniques such as limiting dilution [4], [5], [6], [7]. This inability to form colonies or efficiently form plaques prohibits the precise determination of *R. prowazekii*transformation frequencies [8]. However, despite these barriers, advances in the genetic manipulation of this intractable organism have been made. *For example*, identification of antibiotics suitable for the selection of rickettsial transformants [4], [5], the use of fluorescent proteins as reporter genes [9], the adaptation of transposon systems for generating *R. prowazekii* insertional mutants [5], [6], and the directed knockout of a rickettsial gene [7] have now been reported. Although complementation of an *R. rickettsii* gene mutation using the

Himar1 transposon system was recently achieved [8], the *R. prowazekii* genetic toolbox still lacks a replicating plasmid for extrachromosomal gene expression studies that would not result in chromosomal disruption. Fortunatcly, the demonstration that some rickettsial species harbor plasmids has added another genetic component to the rickettsial gene repertoire.

Originally, the first rickettsial genome sequencing projects targeting rickettsial pathogens failed to find plasmids, supporting the hypothesis that rickettsial species did not contain extrachromosomal elements. However, beginning with the identification of plasmids in *R. felis*[10], followed by plasmid identifications in rickettsiae ranging from other pathogens (e.g. *R. akari*) to arthropod endosymbionts (e.g. *R. amblyommii*, *R. bellii*, *R rhipicephali*, and the rickettsial endosymbionts of *Ixodes scapularis*, REIS) [11], [12], [13], it became evident that plasmids are not uncommon in rickettsial species. However, after multiple sequencing projects examining different strains, plasmids still have not been identified in *R. prowazekii*. To evaluate whether *R. prowazekii* can maintain a plasmid and to generate an additional tool for the genetic analysis of this pathogen, we introduced a recombinant plasmid derived from one of the natural plasmids of *R. amblyommii* into *R. prowazekii* and characterized its maintenance and its effect on rickettsial growth. To our knowledge this is the first plasmid shown to stably replicate in *R. prowazekii*, opening the door to genetic analyses requiring an extrachromosomal platform.

MATERIALS AND METHODS

Bacterial Strains and Culture Conditions

R. prowazekii Madrid E strain rickettsiae (Passage 283) were propagated and purified from hen egg yolk sacs [14] and L929 mouse fibroblasts (American Type Culture Collection, Manassas, VA, ATCC Number CCL-1) as described previously [5]. Purified rickettsiae were stored frozen in a sucrose-phosphate-glutamate-magnesium solution (0.218 M sucrose, 3.76 mM KH_2PO_4, 7.1 mM K_2HPO_4, 4.9 mM potassium glutamate, and 10 mM $MgCl_2$). Rickettsiae-infected L929 cells were grown in modified Eagle medium (Mediatech, Inc., Herndon, VA) supplemented with 10% newborn calf serum (Hyclone, Logan, UT) and 2 mM L-glutamine (Mediatech) in an atmosphere of 5% CO_2 at 34°C. *Escherichia coli* strain

XL1-Blue (Stratagene, La Jolla, CA) was used as a recipient for shuttle vector pRAM18dRGA [15] and for preparation of plasmid DNA used in rickettsial transformation. XL1-Blue was cultured in Luria-Bertani (LB) medium at 37°C. For selection of *E. coli* transformants, rifampin was added to a final concentration of 50 μg/ml.

R. prowazekii transformation

Purified rickettsiae were made competent for transformation and electroporated, as previously described [5], [16], in the presence of 14 μg of pRAM18dRGA plasmid DNA. Twenty-four hours following electroporation and infection of mouse fibroblast L929 cells, rifampin was added to a final concentration of 200 ng/ml and rifampin selection was maintained throughout the experiment. The introduction of a gene conferring rifampin resistance into *R. prowazekii* has been approved by both the University of South Alabama Institutional Biosafety Committee and the Centers for Disease Control, Division of Select Agents and Toxins. Rickettsial infection and growth was monitored by microscopic examination of Gimenez-stained [17] infected cells on cover slips. For infection levels and calculations of rickettsiae per cell, 100 cells were analyzed at each time point. Fluorescent images were acquired using a Nikon Eclipse T2000-U fluorescent microscope and images overlaid using MetaMorph Imaging System software (Universal Imaging Corporation).

Plasmid Recovery

Total DNA from the rifampin-resistant rickettsial population (designated ME-pRAM18dRGA) grown in L929 cells was extracted using the DNEasy Blood & Tissue Kit (Qiagen, Valencia, CA). Following total DNA extraction, plasmid DNA was isolated using the Qiagen Plasmid Mini kit. Plasmid DNA (200 ng) was electroporated into XL-1 Blue electrocompetent *E. coli* and transformants selected on LB medium agar plates containing 50 μg/ml rifampin. Resistant colonies were amplified, plasmid DNA extracted and subsequently sequenced by primer walking at the Iowa State University DNA Facility.

Rickettsial Growth Analyses

To compare growth characteristics of the ME-pRAM18dRGA to that of the parent Madrid E strain, L929 cells were infected in suspension for 1 hour with either ME-pRAM18dRGA or the wild type Madrid E strain at similar multiplicities of infection. The infected cells were seeded in 60 mm dishes. Samples from each infection were harvested approximately every 24 hours. DNA was extracted from approximately 1×10^6 infected cells using the Archive Pure DNA Cell/Tissue Kit (5 Prime Inc., Gaithersburg, MD). At sampling times, the medium was removed and adherent cells were gently rinsed with phosphate buffered saline (PBS). Cells were immediately lysed in the dish using 800 µl of Archive Pure lysis solution and total DNA (L929 and rickettsiae) was extracted. DNA concentrations were determined using the NanoDrop 1000 spectrophotometer (Thermo Scientific, Wilmington, DE) and aliquots were diluted to a final working concentration of 10 ng/µl in RT-PCR water (Applied Biosystems, Austin, TX). Samples were prepared from master mixes, so that each reaction contained 10 ng of DNA, and analyzed by quantitative PCR (QPCR). Bacterial, host, and plasmid genome equivalents were determined by targeting the single-copy *R. prowazekii rho* (RP521) chromosomal gene, the host β-actin gene, and the gfp_{uv} gene of pRAM18dRGA. Primer pairs specific for each gene are listed in Table 1. Prior to QPCR analyses, primer pair specificity was validated for each target at the working dilution using either rickettsial genomic DNA, L929 cell genomic DNA, or the pRAM18dRGA plasmid. Assays were performed using the LightCycler® DNA Master SYBR Green I QPCR master mix (Roche, Mannheim, Germany), 500 nM concentration of each primer, and a Cepheid Smart Cycler® according to the manufacturer›s protocol (Cepheid, Sunnyvale, CA). Cycle parameters were 1 cycle at 95°C for 2 min, followed by 40 cycles at 95°C for 15 s, 52.3°C (*rho*), 54°C (actin) or 58°C (gfp_{uv}) for 15 s, and 72°C for 15 s. Amplification specificity was confirmed by melting-curve analysis. Data acquisition and analysis were performed using the Cepheid Smart Cycler Version 2.0c software. Genome equivalents were determined by comparison to standard curves. Three independent biological samples were analyzed in duplicate for each time point.

Table 1: Primers pairs.

Gene	Strand	Sequence	Size (bp)	Efficiency[a]
Actin	F	CCCTACAGTGCTGTGGGTTT	178	93%
	R	GACATGCAAGGAGTGCAAGA		
Rho	F	CCTGCAAGTAGACATGTGC	179	95%
	R	AGTGCATTAGCATCAACACC		
GFPuv	F	CTTTTCGTTGGGATCTTTCG	119	104%
	R	ACATCACGGCAGACAAACAA		
GFP probe	F	ATGAGTAAAGGAGAAGAACTTTTCAC	716	
	R	TTATTTGTAGAGCTCATCCATGCC		

[a]Efficiency = 10^(−1/slope of the standard curve)−1)*100.
doi:10.1371/journal.pone.0034715.t001

Southern blot analyses

Total DNA was isolated from rickettsiae grown in L929 cells using the DNeasy Blood and Tissue Kit (Qiagen). DNA (2 μg) was digested with SpeI, HindIII, or XhoI and the resulting DNA fragments separated by agarose gel electrophoresis. DNA was subsequently transferred to Nytran® SuPerCharge membranes (Schleicher & Schuell, Keene, NH). A gfp_{uv}-specific probe used in Southern hybridizations [18] was PCR amplified (Table 1) and labeled using [α-^{32}P]dATP (MP Biomedicals, Inc., Philadelphia, PA) and the Multiprime DNA labeling system (Amersham, Piscataway, NJ). Hybridized fragments were visualized using a Cyclone Plus phosphoimager (Perkin Elmer, Waltham, MA).

RESULTS

Plasmid maintenance in *R. prowazekii*

Plasmid pRAM18dRGA was introduced into competent *R. prowazekii* Madrid E rickettsiae via electroporation. Selection with rifampin was initiated 24 hours following electroporation. On day 11, a sample of the rickettsiae-infected cells was harvested, DNA extracted and examined by PCR for the presence of a rickettsial chromosomal gene (*sdh*) and for two genes ($arr2_{RP}$ and gfp_{uv}) contained on the plasmid. A predicted PCR product was obtained for each gene (data not shown), indicating the presence of plasmid DNA within the rifampin resistant population. Continued expansion of the infected host cell population generated a slowly increasing population of rifampin-resistant rickettsiae that exhibited fluorescence when examined microscopically (Fig. 1). Rickettsiae were isolated from this population and used to infect L929 cells at a high multiplicity of infection to increase the percentage of infected cells. The resulting rifampin-resistant rickettsial population (designated ME-pRAM18dRGA) was subsequently analyzed for the presence of plasmid DNA by Southern hybridization (Fig. 2) using a probe that spans the coding region of *gfp*$_{uv}$. The three physical forms of plasmid DNA (linear, covalently closed circular, and open circular) can be observed in the Uncut lane of Figure 2, demonstrating the extrachromosomal nature of pRAM18dRGA in the ME-pRAM18dRGA strain. This is supported by the presence of predicted restriction patterns for SpeI (linearizes the 10 kb plasmid), HindIII (generates a 2854 bp fragment containing the *gfp*$_{uv}$ gene) and XhoI (cuts within the *gfp*$_{uv}$ gene generating two labeled fragments of 4263 bp and 5985 bp). The two faint bands that appear below the linear fragment in the SpeI digested DNA are likely the result of DNA degradation. No hybridization to *R. prowazekii* chromosomal DNA or to L929 cell DNA was observed (data not shown). The absence of additional bands on the Southern blot supports the data that the plasmid has remained extrachromosomal and has not incorporated into the rickettsial chromosome at a detectable level.

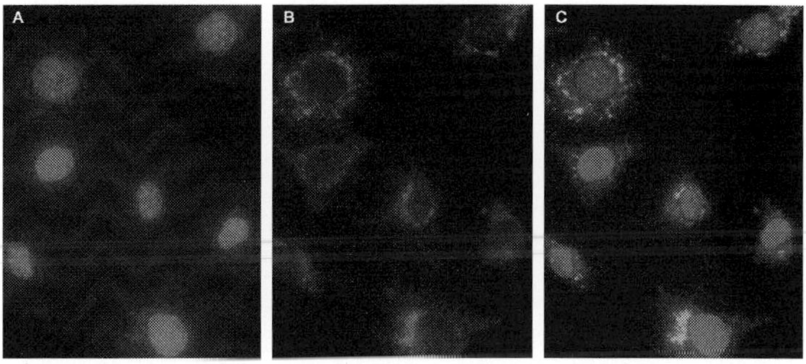

Figure 1: Fluorescent imaging of ME-pRAM18dRGA growing in L929 fibroblasts.

L929 fibroblasts infected with ME-pRAM18dRGA were analyzed for the expression of GFP$_{uv}$. (A) Eukaryotic nuclei and rickettsiae are visualized using DNA counter-stain DAPI. (B) Fluorescent rickettsiae expressing GFP$_{UV}$. (C) A single plane overlay of images in Panels A and B showing that each infected cell contains fluorescent transformed rickettsiae.

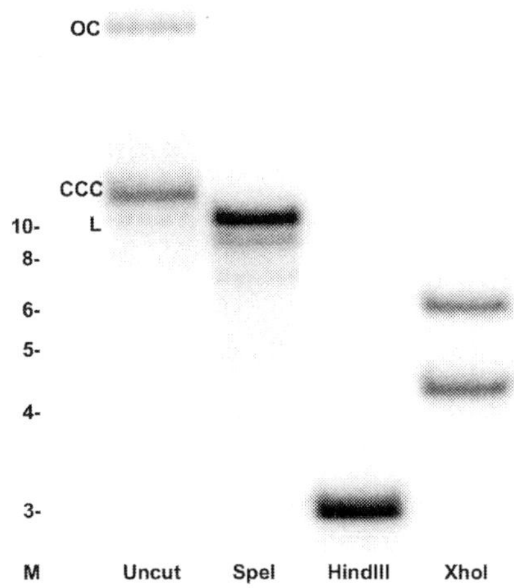

Figure 2: Southern blot analysis of DNA extracted from ME-pRAM-18dRGA using a probe specific for *gfp*$_{UV}$.

Lanes are labeled as uncut or by the enzyme used in the digestion. Locations of molecular size markers (in kB) are indicated on the left. Plasmid forms in the uncut lane are L- linear, CCC- closed circular, and OC- open circular.

The ME-pRAM18dRGA population was serially passaged more than 10 times for a period of greater than 30 days, under selection, without the loss of fluorescence. For serial passages, confluent cell monolayers were trypsin-treated and cells harvested and seeded into new flasks at a 1:3 dilution. Cells reached confluence after approximately three days of growth. After extensive passaging, an extrachromosomal plasmid, with the predicted restriction pattern, could still be isolated from the rickettsiae. In addition, pRAM18dRGA could still be isolated from rickettsiae harvested from L929 cells that were serially passaged in the absence of selective pressure (rifampin) for more than two weeks.

The effect of plasmid maintenance on rickettsial growth

Growth of ME-pRAM18dRGA was evaluated by determination of rickettsial genome equivalents per host cell using QPCR and primers (Table 1) specific for the *R. prowazekii rho* gene and the host cell actin gene. Infections were initiated with rickettsiae harvested from L929 cells, and for each infection, the percent of infected L929 cells was greater than 90%. Evaluating the increasing number of rickettsiae per adherent cell permitted a comparison of the two strains (Madrid E and ME-pRAM18dRGA) (Fig. 3). Growth of the two rickettsial strains was comparable with each exhibiting a generation time of approximately 13 and 12 hours respectively.

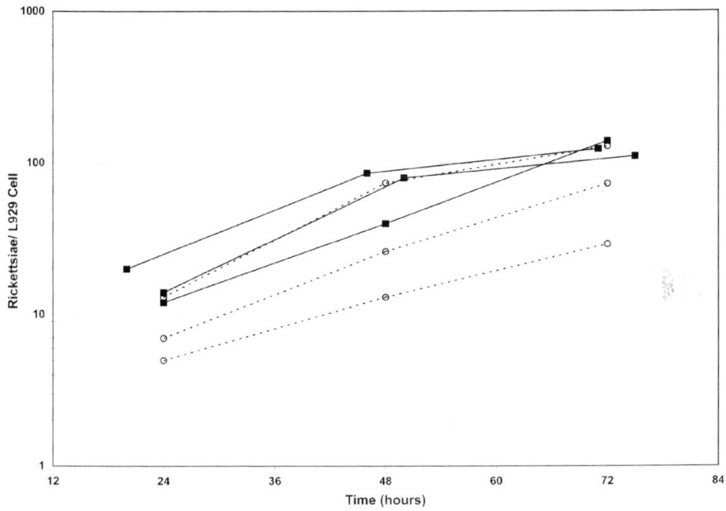

Figure 3: Growth curves for Madrid E (- - - -) and ME-pRAM18dRGA(—).

Rickettsial growth in cultures of adherent L929 cells was followed by determining genome equivalents of an *R. prowazekii* chromosomal gene (*rho*) and the host cell actin gene. Growth is expressed as the number of rickettsiae per host cell. Each line represents an independent experiment.

While the Southern blot and QPCR data cannot eliminate the possibility that spontaneous rifampin-resistant rickettsiae lacking a plasmid exist in the population, plasmid copy number (see next section) did not change appreciably over the time course of the growth curve suggesting that such a background population did not significantly affect the growth analysis. In addition, examination of cells infected with a population of plasmid transformed *R. prowazekii*revealed that every infected cell contained fluorescent rickettsiae (Fig. 1). Absolute confirmation will require the isolation and expansion of a single rickettsia by limiting dilution.

Plasmid Copy Number

Plasmid copy number was determined using QPCR. The relative ratio of the plasmid *gfp*$_{uv}$ gene to the single-copy *R. prowazekii rho* chromosomal gene was determined using gene-specific primers (Table 1). Copy number was evaluated over two growth curves at daily intervals for five days (10 independent determinations). These experiments revealed (Fig. 4) that pRAM18dRGA maintained a low copy number per rickettsia of approximately 1 (0.86+/−0.3, mean+/−S.D.) which falls at the lower range of rickettsial plasmid copy numbers (2.4–9.2) established for naturally-occurring rickettsial plasmids [11].

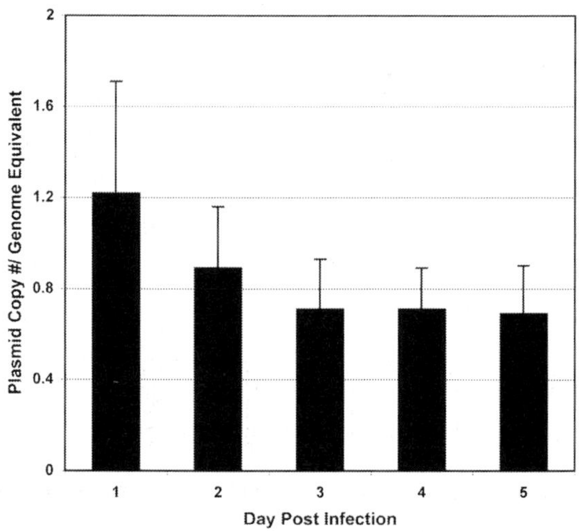

Figure 4. Plasmid copy number determination.

Plasmid copy number/rickettsial genome equivalent was determined by comparing genome equivalents of the rickettsial chromosomal gene, *rho*, and the plasmid-borne *gfp*$_{uv}$ gene. Genome equivalents were determined by comparison to standard curves. Two independent biological samples were analyzed in duplicate.

DISCUSSION

This study documents for the first time that *R. prowazekii* can support plasmid replication. Characterization of rickettsial plasmids identified in non-pathogenic rickettsial species led to the construction of the vector pRAM18dRGA [15]. This plasmid contains a 7.3 kb fragment from the replicative pRAM18 plasmid of *R. amblyommii* [11] ligated to a pGEM vector, containing the *rpsL*^{P}*-Rparr-2/ompA*^{P}*-gfp*$_{uv}$ selection/detection cassette [6], for replication in *E. coli*. This vector represents the smallest pRAM18 derived plasmid tested that was capable of replicating in a non-*R. amblyommii* rickettsial species and encodes proteins with homology to the DnaA-like replication initiator and the ParA plasmid partitioning protein [15].

pRAM18dRGA, was introduced into pathogenic *R. prowazekii* via electroporation and transformants isolated using rifampin selection. The observation of GFP$_{uv}$ expression, under the control of a rickettsial promoter, also demonstrates that plasmid-borne genes can be expressed and detected, providing a promising first step in complementation assays. Although pRAM18dRGA was maintained at a low copy number, this is characteristic of known rickettsial plasmids and suggests that the machinery for maintaining plasmids in this pathogenic species is functional. Interestingly, when this plasmid was transformed into several other rickettsial species, the copy number of pRAM18dRGA was noticeably higher, ranging from 5.5+/--0.65 to 28.1+/−1.89 [15]. The low copy number of pRAM18dRGA in *R. prowazekii* supports its future use in expression studies, alleviating concerns of over-expression of multi-copy plasmid-borne genes in complementation assays.

The *R. amblyommii* fragment in pRAM18dRGA encodes only four proteins; a DnaA-like protein, a ParA partitioning protein, a TPR repeat-containing protein, and a homolog to Xre, a putative repressor protein. The DnaA-like protein was initially described during the annotation of the pRF plasmid of *R. felis* [10]. Interestingly, the annotation is based on homology of the carboxy-terminal 75 amino acids of the plasmid-expressed protein with the first 65 amino acids of the archetypical DnaA protein of *E. coli* [19]. The N-terminal region, or Domain I, of the DnaA protein is responsible for the loading of the helicase, DnaB [20]. However, the remaining 700 amino acids of the DnaA-like protein show no homology to DnaA and display no conserved domains, despite their conservation among the rickettsial plasmid sequences. In contrast to the highly conserved rickettsial chromosomal *parA* genes, the *parA* genes from several rickettsial plasmids were found to be highly diverse and clustered with *parA* genes found on plasmids from other bacterial genera [11], [21]. In addition to these DnaA-like and ParA proteins, the pRAM fragment encodes a protein with a tetratricopeptide repeat (TPR) motif, originally identified in yeast as a protein-protein interaction module, and a protein with homology to a helix-turn-helix transcription regulator, Xre. Interestingly, the Xre homolog in *Bacillus subtilis* is described as a probable repressor necessary for the maintenance of the lysogenic state of the defective prophage pbsX [22].

Fortunately, for future studies that might employ pRAM18dRGA, the presence of a plasmid appeared to have minimal effects on *R. prowazekii* growth. Growth of the plasmid-containing strain was comparable to that of wild-type Madrid E strain. In fact the plasmid-containing strain exhibited a slightly shorter generation time than the wild-type control. However, in the growth experiments presented here, the control Madrid E bacteria was isolated from hen egg yolk sacs and only passaged in L929 cells for two days prior to initiation of growth curve assays. A recent report demonstrated that the Madrid E strain grows slower in cell culture without adaptation to the specific host cell environment [23]. However, the 13-hour replication time is similar to the published 8–12 hour generation time [24], [25], [26], [27].

The demonstration of plasmid replication in *R. prowazekii* provides an important genetic tool and model genetic system for the study of this obligate intracellular pathogen. The absence of an extrachromosomal platform for the genetic analysis of *R. prowazekii* has prevented the evaluation of gene function by classical genetic complementation techniques. While it is possible to use transposon systems to evaluate gene complementation [8], an extrachromosomal location may be preferred for some experiments, since the plasmid would not disrupt the bacterial chromosome potentially contributing to an observed phenotype. The maintenance of the pRAM18dRGA shuttle plasmid suggests that other rickettsial plasmids may be maintained as well expanding the genetic tools available for the study of this rickettsial pathogen.

ACKNOWLEDGMENTS

We thank Amanda Beaster for her critical review of the manuscript.

AUTHOR CONTRIBUTIONS

Conceived and designed the experiments: DOW AMT AW NYB TJK GB UGM. Performed the experiments: AH AMT AW LOD. Analyzed the data: DOW AH AMT AW LOD. Contributed reagents/materials/analysis tools: NYB TJK GB UGM. Wrote the paper: DOW AMT AW.

REFERENCES

1. Heinzen RA, Grieshaber SS, Van Kirk LS, Devin CJ (1999) Dynamics of actin-based movement by Rickettsia rickettsii in vero cells. Infect Immun 67: 4201–4207.RA HeinzenSS GrieshaberLS Van KirkCJ Devin1999Dynamics of actin-based movement by Rickettsia rickettsii in vero cells.Infect Immun6742014207
2. Heinzen RA, Hayes SF, Peacock MG, Hackstadt T (1993) Directional actin polymerization associated with spotted fever group Rickettsia infection of Vero cells. Infect Immun 61: 1926–1935. RA HeinzenSF HayesMG PeacockT. Hackstadt1993Directional actin polymerization associated with spotted fever group Rickettsia infection of Vero cells.Infect Immun6119261935
3. Van Kirk LS, Hayes SF, Heinzen RA (2000) Ultrastructure of Rickettsia rickettsii actin tails and localization of cytoskeletal proteins. Infect Immun 68: 4706–4713.LS Van KirkSF HayesRA Heinzen2000Ultrastructure of Rickettsia rickettsii actin tails and localization of cytoskeletal proteins.Infect Immun6847064713
4. Rachek LI, Hines A, Tucker AM, Winkler HH, Wood DO (2000) Transformation of Rickettsia prowazekii to erythromycin resistance encoded by the Escherichia coli ereB gene. J Bacteriol 182: 3289–3291.LI RachekA. HinesAM TuckerHH WinklerDO Wood2000Transformation of Rickettsia prowazekii to erythromycin resistance encoded by the Escherichia coli ereB gene.J Bacteriol18232893291
5. Qin A, Tucker AM, Hines A, Wood DO (2004) Transposon mutagenesis of the obligate intracellular pathogen Rickettsia prowazekii. Appl Environ Microbiol 70: 2816–2822.A. QinAM TuckerA. HinesDO Wood2004Transposon mutagenesis of the obligate intracellular pathogen Rickettsia prowazekii.Appl Environ Microbiol7028162822

6. Liu Z-M, Tucker AM, Driskell LO, Wood DO (2007) mariner-based transposon mutagenesis of Rickettsia prowazekii. Appl Environ Microbiol 73: 6644–6649.Z-M LiuAM TuckerLO DriskellDO Wood2007mariner-based transposon mutagenesis of Rickettsia prowazekii.Appl Environ Microbiol7366446649
7. Driskell LO, Yu XJ, Zhang L, Liu Y, Popov VL, et al. (2009) Directed mutagenesis of the Rickettsia prowazekii pld gene encoding phospholipase D. Infect Immun 77: 3244–3248.LO DriskellXJ YuL. ZhangY. LiuVL Popov2009Directed mutagenesis of the Rickettsia prowazekii pld gene encoding phospholipase D.Infect Immun7732443248
8. Clark TR, Ellison DW, Kleba B, Hackstadt T (2011) Complementation of Rickettsia rickettsii RelA/SpoT restores a nonlytic plaque phenotype. Infect Immun 79: 1631–1637.TR ClarkDW EllisonB. KlebaT. Hackstadt2011Complementation of Rickettsia rickettsii RelA/SpoT restores a nonlytic plaque phenotype.Infect Immun7916311637
9. Baldridge GD, Burkhardt NY, Herron MJ, Kurtti TJ, Munderloh UG (2005) Analysis of fluorescent protein expression in transformants of Rickettsia monacensis, an obligate intracellular tick symbiont. Appl Environ Microbiol 71: 2095–2105.GD BaldridgeNY BurkhardtMJ HerronTJ KurttiUG Munderloh2005Analysis of fluorescent protein expression in transformants of Rickettsia monacensis, an obligate intracellular tick symbiont.Appl Environ Microbiol7120952105
10. Ogata H, Renesto P, Audic S, Robert C, Blanc G, et al. (2005) The genome sequence of Rickettsia felis identifies the first putative conjugative plasmid in an obligate intracellular parasite. PLoS Biol 3: e248.H. OgataP. RenestoS. AudicC. RobertG. Blanc2005The genome sequence of Rickettsia felis identifies the first putative conjugative plasmid in an obligate intracellular parasite.PLoS Biol3e248
11. Baldridge GD, Burkhardt NY, Labruna MB, Pacheco RC, Paddock CD, et al. (2010) Wide dispersal and possible multiple origins of low-copy-number plasmids in Rickettsia species associated with blood-feeding arthropods. Appl Environ Microbiol 76: 1718–1731.GD BaldridgeNY BurkhardtMB LabrunaRC PachecoCD Paddock2010Wide dispersal and possible multiple origins of low-

copy-number plasmids in Rickettsia species associated with blood-feeding arthropods.Appl Environ Microbiol7617181731

12. Baldridge GD, Burkhardt NY, Felsheim RF, Kurtti TJ, Munderloh UG (2007) Transposon insertion reveals pRM, a plasmid of Rickettsia monacensis. Appl Environ Microbiol 73: 4984–4995.GD BaldridgeNY BurkhardtRF FelsheimTJ KurttiUG Munderloh2007Transposon insertion reveals pRM, a plasmid of Rickettsia monacensis.Appl Environ Microbiol7349844995
13. Baldridge GD, Burkhardt NY, Felsheim RF, Kurtti TJ, Munderloh UG (2008) Plasmids of the pRM/pRF family occur in diverse Rickettsia species. Appl Environ Microbiol 74: 645–652.GD BaldridgeNY BurkhardtRF FelsheimTJ KurttiUG Munderloh2008Plasmids of the pRM/pRF family occur in diverse Rickettsia species.Appl Environ Microbiol74645652
14. Winkler HH (1976) Rickettsial permeability: an ADP-ATP transport system. J Biol Chem 251: 389–396.HH Winkler1976Rickettsial permeability: an ADP-ATP transport system.J Biol Chem251389396
15. Burkhardt NY, Baldridge GD, Williamson PC, Billingsley PM, Felsheim RF, et al. (2011) Development of a shuttle vector for transformation of diverse Rickettsia species. PLoS One 6: e29511. NY BurkhardtGD BaldridgePC WilliamsonPM BillingsleyRF Felsheim2011Development of a shuttle vector for transformation of diverse Rickettsia species.PLoS One6e29511
16. Rachek LI, Tucker AM, Winkler HH, Wood DO (1998) Transformation of Rickettsia prowazekii to rifampin resistance. J Bacteriol 180: 2118–2124.LI RachekAM TuckerHH WinklerDO Wood1998Transformation of Rickettsia prowazekii to rifampin resistance.J Bacteriol18021182124
17. Gimenez DF (1964) Staining rickettsiae in yolk-sac cultures. Stain Technol 39: 135–140.DF Gimenez1964Staining rickettsiae in yolk-sac cultures.Stain Technol39135140
18. Southern EM (1975) Detection of specific sequences among DNA fragments separated by gel electrophoresis. J Mol Biol 98: 503. EM Southern1975Detection of specific sequences among DNA fragments separated by gel electrophoresis.J Mol Biol98503

19. Sutton MD, Carr KM, Vicente M, Kaguni JM (1998) Escherichia coli DnaA protein: The N-terminal domain and loading of DnaB helicase at the E. coli chromosomal origin. J Biol Chem 273: 34255–34262. MD SuttonKM CarrM. VicenteJM Kaguni1998Escherichia coli DnaA protein: The N-terminal domain and loading of DnaB helicase at the E. coli chromosomal origin.J Biol Chem2733425534262
20. Seitz H, Weigel C, Messer W (2000) The interaction domains of the DnaA and DnaB replication proteins of Escherichia coli. Mol Micro 37: 1270–1279.H. SeitzC. WeigelW. Messer2000The interaction domains of the DnaA and DnaB replication proteins of Escherichia coli.Mol Micro3712701279
21. Felsheim RF, Kurtti TJ, Munderloh UG (2009) Genome sequence of endosymbiont Rickettsia peacockii and comparison with virulent Rickettsia rickettsii: identification of virulence factors. PLoS ONE 4: e8361.RF FelsheimTJ KurttiUG Munderloh2009Genome sequence of endosymbiont Rickettsia peacockii and comparison with virulent Rickettsia rickettsii: identification of virulence factors. PLoS ONE4e8361
22. Wood HE, Devine KM, McConnell DJ (1990) Characterisation of a repressor gene (xre) and a temperature-sensitive allele from the Bacillus subtilis prohage, PBSX. Gene 96: 83–88.HE WoodKM DevineDJ McConnell1990Characterisation of a repressor gene (xre) and a temperature-sensitive allele from the Bacillus subtilis prohage, PBSX.Gene968388
23. Bechah Y, El Karkouri K, Mediannikov O, Leroy Q, Pelletier N, et al. (2010) Genomic, proteomic, and transcriptomic analysis of virulent and avirulent Rickettsia prowazekii reveals its adaptive mutation capabilities. Genome Res 20: 655–663.Y. BechahK. El KarkouriO. MediannikovQ. LeroyN. Pelletier2010Genomic, proteomic, and transcriptomic analysis of virulent and avirulent Rickettsia prowazekii reveals its adaptive mutation capabilities. Genome Res20655663
24. Weiss E (1982) The biology of rickettsiae. Ann Rev Microbiol 36: 345–370.E. Weiss1982The biology of rickettsiae.Ann Rev Microbiol36345370
25. Weiss E (1973) Growth and physiology of rickettsiae. Bacteriol Rev

37: 259–283.E. Weiss1973Growth and physiology of rickettsiae. Bacteriol Rev37259283

26. Wisseman CL Jr, Waddell AD (1975) In vitro studies on rickettsia-host cell interactions: intracellular growth cycle of virulent and attenuated Rickettsia prowazekii in chicken embryo cells in slide chamber cultures. Infect Immun 11: 1391–1404.CL Wisseman JrAD Waddell1975In vitro studies on rickettsia-host cell interactions: intracellular growth cycle of virulent and attenuated Rickettsia prowazekii in chicken embryo cells in slide chamber cultures.Infect Immun1113911404
27. Silverman DJ, Wisseman CL Jr, Waddell A (1980) In vitro studies of Rickettsia-host cell interactions: ultrastructural study of Rickettsia prowazekii-infected chicken embryo fibroblasts. Infect Immun 29: 778–790.DJ SilvermanCL Wisseman JrA. Waddell1980In vitro studies of Rickettsia-host cell interactions: ultrastructural study of Rickettsia prowazekii-infected chicken embryo fibroblasts.Infect Immun29778790

CHAPTER

8

GENOME ANALYSIS OF A NOVEL BRADYRHIZOBIUM SP. DOA9 CARRYING A SYMBIOTIC PLASMID

Shin Okazaki[1], Rujirek Noisangiam[2], Takashi Okubo[3], Takakazu Kaneko[4], Kenshiro Oshima[5], Masahira Hattori[5] , Kamonluck Teamtisong[2] , Pongpan Songwattana[2] , Panlada Tittabutr[2] , Nantakorn Boonkerd[2] , Kazuhiko Saeki[6], Shusei Sato[3], Toshiki Uchiumi[7], Kiwamu Minamisawa[3] , Neung Teaumroong[2]

[1] Graduate School of Agriculture, Tokyo University of Agriculture and Technology, Tokyo, Japan,

[2] School of Biotechnology, Institute of Agricultural Technology, Suranaree University of Technology, Nakhon Ratchasima, Thailand,

[3] Graduate School of Life Science, Tohoku University, Sendai, Japan,

[4] Faculty of Life Sciences, Kyoto Sangyo University, Motoyama, Kamigamo, Kita-Ku, Kyoto 603-8555, Japan,

[5] Center of Omics and Bioinformatics, Graduate School of Frontier Sciences, University of Tokyo, Tokyo, Japan,

[6] Department of Biological Sciences, Faculty of Science, Kyousei Science Center for Life and Nature, Nara Women's University, Kitauoya Nishimachi, Nara 630-8506, Japan,

[7] Graduate School of Science and Engineering, Kagoshima University, Kagoshima, Japan

ABSTRACT

Bradyrhizobium sp. DOA9 isolated from the legume *Aeschynomene americana* exhibited a broad host range and divergent nodulation (*nod*) genes compared with other members of the *Bradyrhizobiaceae.* Genome analysis of DOA9 revealed that its genome comprised a single chromosome of 7.1 Mbp and a plasmid of 0.7 Mbp. The chromosome showed highest similarity with that of the *nod* gene-harboring soybean symbiont *B. japonicum* USDA110, whereas the plasmid showed highest similarity with pBBta01 of the *nod* gene-lacking photosynthetic strain BTAi1, which nodulates *Aeschynomene* species. Unlike in other bradyrhizobia, the plasmid of DOA9 encodes genes related to symbiotic functions including nodulation, nitrogen fixation, and type III/IV protein secretion systems. The plasmid has also a lower GC content (60.1%) than the chromosome (64.4%). These features suggest that the plasmid could be the origin of the symbiosis island that is found in the genome of other bradyrhizobia. The *nod* genes of DOA9 exhibited low similarity with those of other strains. The *nif* gene cluster of DOA9 showed greatest similarity to those of photosynthetic bradyrhizobia. The type III/IV protein secretion systems of DOA9 are similar to those of *nod* gene-harboring *B. elkanii* and photosynthetic BTAi1. The DOA9 genome exhibited intermediate characteristics between *nod* gene-harboring bradyrhizobia and *nod* gene-lacking photosynthetic bradyrhizobia, thus providing the evidence for the evolution of the *Bradyrhizobiaceae* during ecological adaptation. *Bradyrhizobium* sp. DOA9 isolated from the legume *Aeschynomene americana* exhibited a broad host range and divergent nodulation (*nod*) genes compared with other members of the *Bradyrhizobiaceae.* Genome analysis of DOA9 revealed that its genome comprised a single chromosome of 7.1 Mbp and a plasmid of 0.7 Mbp. The chromosome showed highest similarity with that of the *nod*gene-harboring soybean symbiont *B. japonicum* USDA110, whereas the plasmid showed highest similarity with pBBta01 of the *nod* gene-lacking photosynthetic strain BTAi1, which nodulates *Aeschynomene* species. Unlike in other bradyrhizobia, the plasmid of DOA9 encodes genes related to symbiotic functions including nodulation, nitrogen fixation, and type III/IV protein secretion systems. The plasmid has also a lower GC content (60.1%) than the chromosome (64.4%). These features suggest that the plasmid could be the origin of the symbiosis island that is found in the genome of other bradyrhizobia. The *nod* genes of DOA9 exhibited low similarity with those of other strains. The *nif* gene cluster of DOA9 showed greatest similarity to those of photosynthetic bradyrhizobia. The type III/IV protein secretion systems of DOA9 are similar to those of *nod* gene-harboring *B. elkanii* and photosynthetic BTAi1. The DOA9 genome exhibited intermediate characteristics between *nod* gene-harboring bradyrhizobia and *nod* gene-lacking photosynthetic bradyrhizobia, thus providing the evidence for the evolution of the *Bradyrhizobiaceae* during ecological adaptation.

INTRODUCTION

The interaction between leguminous plants and rhizobia results in the formation of root nodules. Within legume nodules rhizobia fix atmospheric nitrogen to ammonia, which can support the growth of host plants. Rhizobia possess a set of modulation (*nod*) genes that control the synthesis of lipochitooligosaccharides called nod factors (NFs) [1], which induce signal transduction cascades in host plants, ultimately leading to nodule formation [1]. The rhizobial *nod* genes were previously thought to be essential for nodule formation; however, the recent discovery of *Bradyrhizobium* spp. ORS278 and BTAi1, which lack *nod* genes, but are able to nodulate some species of the semi-aquatic legume genus *Aeschynomene*, contradicts this long-held dogma [2].

Aeschynomene species can be divided into two groups according to the mechanism of nodule initiation. One group, including *A. afraspera* and *A. americana*, initiates nodulation through NFs, whereas the other group, including *A. indica* and *A. sensitiva*, initiates nodulation without NFs. Although the genome of some *Aeschynomene* symbionts including ORS278 and BTAi1 were sequenced [2], the genetic basis for NF-independent nodulation and the evolutionary relationship between NF-independent and NF-dependent nodulation remains mostly unclear [2,3].

A recent interest in the origin and evolution of *Bradyrhizobiaceae* rhizobia has developed because the genomes of several strains have been sequenced. The genome analyses of *B. japonicum* USDA110 [4] and *B. japonicum* USDA6 ^{T} [5] revealed that genes for nodule symbiosis, such as *nod*, for nitrogen fixation, and for secretion systems are located on integrated genomic islands (symbiosis islands), suggesting that the symbiotic genes were acquired by lateral gene transfer from an unknown donor. This evolution scenario is supported by the detailed genome comparison between symbiotic (*B. japonicum* USDA110) and nonsymbiotic (*Bradyrhizobium* sp. S23321) bradyrhizobia [6]. However, the origin and transmission mechanism of symbiotic genes remains poorly understood.

We recently isolated several strains of bradyrhizobia from the root nodules of *A. americana* [7]. One isolate, *Bradyrhizobium* sp. DOA9, was found to nodulate a broad range of leguminous hosts, including dalbergioid, millettioid, and robinoid trives and was also found to be an endophyte of rice, although DOA9 did not form nodules on NF-independent groups of *Aeschynomene* [8]. DOA9 possesses highly divergent *nod* genes that could not be amplified by general primer sets designed on other bradyrhizobia [7]. Furthermore, Southern blot hybridization suggested that some of the *nod* genes and nitrogen fixation (*nif*) genes of DOA9 are localized on the plasmid unlike in other bradyrhizobia [8]. These results suggested that DOA9 possesses a novel type of genome with *nod* genes divergent from previously reported bradyrhizobia.

Here we analyzed the genome of DOA9 and compared it with the genomes of other bradyrhizobia. Several symbiosis-related genes, including those involved in nodulation, nitrogen fixation, and secretion systems, were investigated in order to highlight the genetic basis of the broad host range of DOA9 and the evolutionary relationships between *nod* gene-harboring soybean bradyrhizobia and *nod* gene-lacking *Aeschynomene* bradyrhizobia.

MATERIALS AND METHODS

Bacterial strains and DNA preparation

Bradyrhizobium sp. DOA9 was cultured for 4 d at 28°C in arabinose–gluconate medium [9]. Genomic DNA was prepared as described by Wilson [10].

Sequencing and annotation

The genome sequence of DOA9 was determined by 454 pyrosequencing analysis using a GS FLX Titanium system (Roche Diagnostics Co., Indianapolis, IN, USA). Genomic DNA (5 μg) was sheared using nebulization to obtain fragments ranging from 300 to 800 bp. Template DNA was prepared according to the supplier's protocol. The pyrosequencing data were assembled using the MIRA assembler ver. 3 [11], and were curated by comparison with the Newbler ver. 2.8 (Roche Diagnostics Co., Indianapolis, IN, USA) assembly or with some close reference strains. Gap closing and resequencing of low-quality regions of the assembled data were performed by PCR and Sanger sequencing. Regions encoding structural RNAs, rRNAs, tRNAs, tmRNAs, noncoding RNAs, and proteins were predicted using the Genaris Annotation System (Genaris, Inc., Kanagawa, Japan).

Genome Comparisons and Ortholog Analysis

A circular genome map showing the GC skew and the GC content was created using the CGview server [12] with default parameters. Similarity was compared between DOA9 and other bacteria using GenomeMatcher [13]. Putative orthologous genes among *Bradyrhizobium*strains DOA9, USDA110, and ORS278 were identified using bidirectional BLASTN comparisons with an e-value cut-off of 10^{-20}. Orthologous relationships were depicted in a Venn diagram. Phylogenetic analysis was performed by comparing the sequences aligned using the CLUSTALW program [14]. Neighbor-joining trees were constructed using MEGA version 5.02 [15] and 1,000 bootstrap replicates were used to generate a consensus tree.

Pulse-field Gel Electrophoresis

DNA plugs for pulse-field gel electrophoresis were prepared using CHEF Genomic DNA Plug Kits (Bio-Rad Laboratories Inc., Hercules, CA, USA) as follows. DOA9 was cultured for 6 d at 28°C in peptone salts yeast extract medium [16]. Bacterial cells were collected by centrifugation, washed twice with 0.85% NaCl, and resuspended with 0.85% NaCl to an OD_{600} of 5. The cell suspension (0.5 mL) was centrifuged at 8,000 × g for 5 min at 10°C, then resuspended with 0.5 mL of cell suspension buffer and mixed thoroughly with 0.5 mL of 2% CleanCut agarose (Bio-Rad Laboratories Inc., Hercules, CA, USA). The mixture was transferred to plug molds and solidified at 4°C for 30 min. The solidified agarose plugs were treated with lysozyme (1 mg/mL) at 37°C for 4 h and with proteinase K (30 U/mL) at 50°C overnight. Fragments of 225–6,000 kb and 225–2,200 kb were separated on 0.8% certified megabase agarose (Bio-Rad Laboratories Inc., Hercules, CA, USA) in TAE buffer or 1% certified megabase agarose in 0.5xTBE buffer, respectively. Contour-clamped homogeneous electric field (CHEF) electrophoresis was conducted at 14°C in a temperature-controlled cooling unit using the autoalgorithm mode in the CHEF Mapper gel electrophoresis system (Bio-Rad Laboratories Inc., Hercules, CA, USA). The gel was stained with 0.5 mg mL^{-1} ethidium bromide for 1 h and then destained in H_2O for 1 h.

RESULTS

Phylogeny of *Bradyrhizobium* sp. DOA9

To examine the relationships between *Bradyrhizobium* sp. DOA9 and other members of the Bradyrhizobiaceae, phylogenetic trees were constructed based on 16S rRNA and internal transcribed spacer sequences (Fig. 1). Using both approaches, DOA9 sequences were clustered within a group of **nonphotosyntetic** *nod* gene-containing bradyrhizobia, including *B. japonicum*, *B. liaoningense*, and *B. yuamingense* and were separated from those of photosynthetic *nod* gene-lacking strains, such as *Bradyrhizobium* spp. ORS278 and BTAi1. These results suggest that the DOA9 genome resembles more closely those of nonphotosynthetic *nod* gene-containing soybean bradyrhizobia than those of photosynthetic *nod* gene-lacking bradyrhizobia.

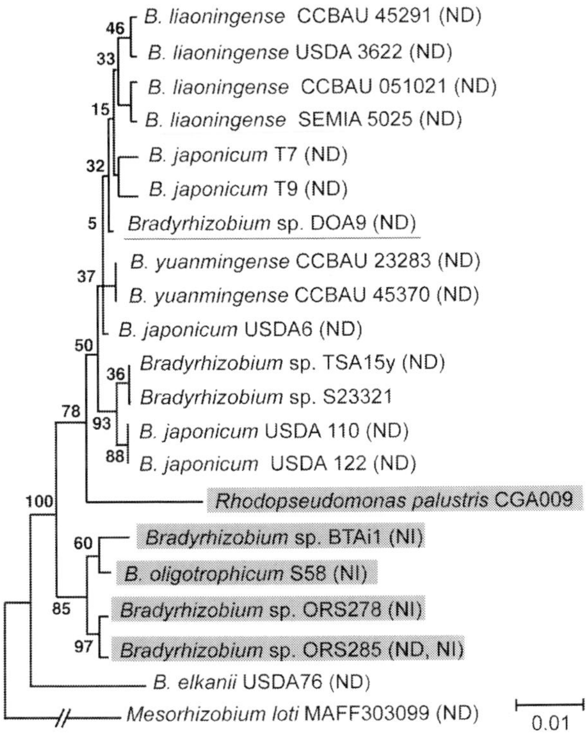

Figure 1: Phylogenetic relationships of *Bradyrhizobium* sp. DOA9 and related bacteria based on 16S rRNA gene sequences.

Bootstrap values are expressed as percentages of 1,000 replications. Evolutionary distances were computed using the Kimura two-parameter method. The bar represents one estimated substitution per 100-nucleotide positions. Strains capable of Nod factor-dependent and -independent nodulation are marked with (ND) and (NI), respectively. Photosynthetic strains are highlighted in gray.

Sequencing

The nucleotide sequence of the DOA9 genome was deduced by assembling 454 pyrosequencing data and Sanger sequencing data. In total, 1,853,745 reads from 454 pyrosequencing data were assembled using the MIRA assembler (version 4.0), and 591 contigs were generated. After curation of the MIRA software assembly by comparison with the Newbler software assembly or with the genomes of close reference strains, 48 contigs remained. Gap closing and resequencing of low-quality regions in the assembled data were performed by PCR and Sanger sequencing.

The final draft genome consisted of two circular scaffolds of about 7.1 Mbp and 0.7 Mbp (Fig. 2). The 7.1-Mb scaffold comprised five contigs and was considered to be the main chromosome of DOA9 (Fig. 2A). Due to the presence of small but highly repetitive elements, five gaps of 1–6 kb in the scaffold could not be closed. The smaller scaffold consisted of one contig of 736,085 bp (Fig. 2B) and was considered to be the plasmid.

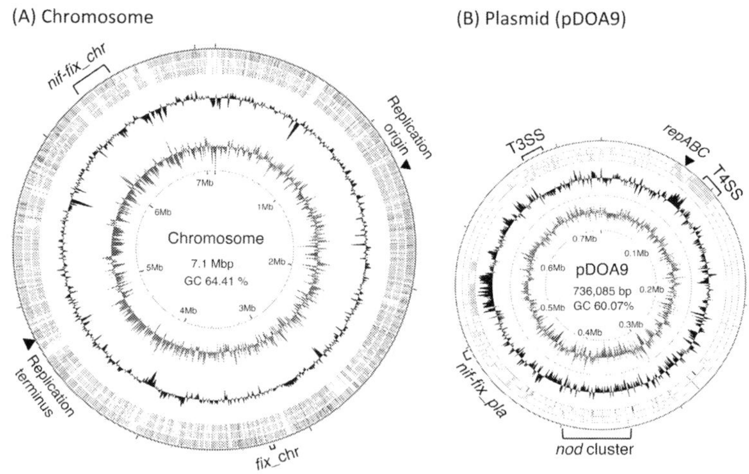

Figure 2: The genome structure of *Bradyrhizobium* sp. DOA9.

(A) Circular representation of the chromosome of *Bradyrhizobium* sp. DOA9. The outermost, second, and third circles represent BLASTN comparisons with *B. japonicum*USDA110, *Bradyrhizobium* sp. ORS278, and *Bradyrhizobium* sp. ORS28, respectively (e-value < 10^{-10}). The innermost and second-innermost circles show the GC skew and the GC content, respectively. The GC content circle shows the deviation from the average GC content of the entire sequence (higher than average GC content is represented in green, and lower than average content is represented in purple). The markings inside the innermost circle represent genome positions (in Mb). The positions of the putative replication origin, putative replication terminus, and nitrogen-fixation genes are shown outside of the outermost circle. (B) Circular representation of the plasmid (pDOA9) of DOA9. The outermost, second, and third circles represent BLASTN comparisons with the plasmid pBBta01 of *Bradyrhizobium* sp. BTAi1, the draft genome of *Bradyrhizobium elkanii* 587 (GenBank accession number AJJK00000000), and *B. japonicum* USDA110, respectively (e-value < 10^{-10}). The innermost and second-innermost circles show the GC skew and the GC content, respectively. The GC content circle shows the deviation from the average GC content of the entire sequence (higher than average GC content in represented in green, and lower than average is represented in purple). The markings inside the innermost circle represent genome positions (in Mb). The positions of the *repABC* operon, T3/T4SS, *nod* genes, *nif-fix* gene clusters, and *hup* cluster are shown outside of the outermost circle.

In order to confirm the replicon structure of DOA9, genomic DNA was analyzed by pulse-field gel electrophoresis. A larger fragment (more than 5 Mbp) and a smaller fragment (between 680 and 750 kb) were detected (Fig. 3), similar in size to the larger scaffold (7.1 Mb) and the smaller scaffold (736,085 bp) of the sequencing results, respectively. The canonical *repABC* operon, which is required for the replication of bacterial plasmids, was found in the smaller scaffold (BDOA9_0200660 to BDOA9_0200680, see below). Therefore, we concluded that the smaller scaffold was the plasmid of DOA9 and designated it as pDOA9. The sequences of the chromosome and plasmid pDOA9 are available in the DDBJ/GenBank/EMBL database (accession numbers DF820425 and DF820426, respectively).

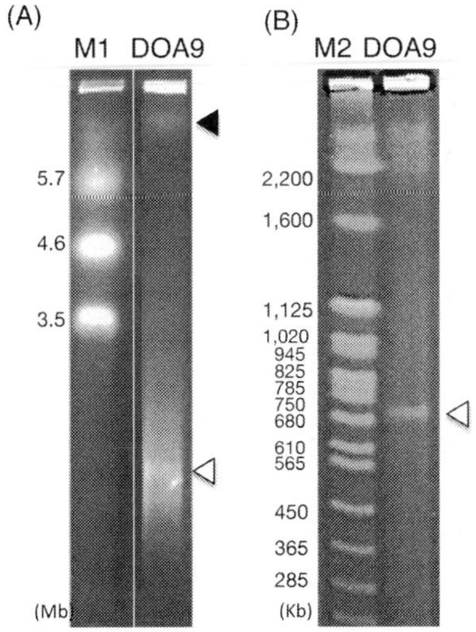

Figure3: Pulse-field gel electrophoresis of *Bradyrhizobium* sp. DOA9 genomic DNA.

DOA9 cells were digested in 1% pulse field grade (PFG) agarose plugs with proteinase K, as described in the Materials and Methods, and run 0.8% certified megabase agarose in TAE buffer to separate fragments of 225–6,000 kb (A), or 1% certified megabase agarose in 0.5 x TBE buffer (B) to separate fragments of 225–2,200 kb, respectively. Closed arrowheads and open arrowheads indicate the putative chromosome and the plasmid, respectively. Lane M1: PFGE marker, 3.5–5.7 Mb, *Saccharomyces pombe*chromosomal DNA. Lane M2: low-range (2.03–194 kb) PFG marker DNA ladder. DOA9: DOA9 genomic DNA.

General genome description

The average GC contents of the chromosome and the plasmid were 64.41% and 60.07%, respectively (Table 1). A GC skew analysis was performed to predict the locations of the putative replication origin and terminus of the chromosome. Two shifts of the GC skew were observed in the chromosome at coordinates 1.3 Mb and 4.86 Mb, respectively (Fig. 2A, innermost circle). The putative replication origin determined by comparison with the genome of *B. japonicum* USDA110 was found at 1.3 Mb. In *B. japonicum* USDA110, the putative replication origin was positioned between two conserved hypothetical proteins bll0636 and blr0637 [4]. The orthologs of these two genes, BDOA9_0112090 and BDOA9_0112080, were found at coordinates 1,313,579 to 1,313,070 and 1,312,684 to 1,311,839, respectively. The *dif*sequence, which is required to convert a dimer chromosome to monomers after aberrant DNA duplication, was found at coordinates 4,862,982 to 4,863,009 (Fig. 2A). The DNA replication may terminate in this region.

Table 1: General genome features of DOA9 and related bradyrhizobiaceae

Strain	*B. japonicum*			*Bradyrhizobium* sp.		
	USDA110	USDA6	DOA9[a]	ORS285	ORS278	BTAi1
Genome size (bp)	9,105,828	9,207,384	7,850,677[b] (736,085)	7,632,258	7,456,587	8,493,515
G + C content (%)	64.06	63.67	64.41 (60.07)	65.23	65.51	64.80
tRNA coding genes	50	51	50 (1)	51	51	51
rRNA genes	3	6	3 (0)	4	6	6
Genes	8,374	8,882	7273 (676)	6,848	6,748	7,723
Photosynthetic	-	-	-	-	+	+
CI group	1	1	1	2	3	3
Nod-dependent (ND) or Nod-independent (NI)	ND	ND	ND	ND and NI	NI	ND
Original host plant	Soybean	Soybean	Aeschynomene	Aeschynomene	Aeschynomene	Aeschynomene

[a] The values of plasmid is shown in parentheses.
[b] The estimated values with gaps.

doi:10.1371/journal.pone.0117392.t001

DNA dot plot analysis was performed to compare the genome structures of Bradyrhizobiaceae. The genome of DOA9 showed extensive similarity with the chromosome of the nonphotosynthetic *B. japonicum* spp. USDA6 and USDA110 and the photosynthetic non-nodulating *Bradyrhizobium* sp. S23321. Conversely, the chromosome of DOA9 showed low similarity with those of the photosynthetic strains, including *Bradyrhizobium* spp. BTAi1, ORS278, and S58. These results are in agreement with the phylogenetic analysis based on 16S rRNA gene sequences (Fig. 1).

RNA- and protein-encoding genes

The DOA9 genome contains one copy of the rRNA gene cluster at coordinates 1,939,820 to 1,939,935 of the chromosome. In total, 50 tRNA genes, which correspond to all 20 of the standard amino acids, were scattered throughout the main chromosome. A total of 7,326 (chromosome: 6,650; plasmid: 676) opening reading frames were predicted and annotated.

A BiBlast comparison was conducted among bradyrhizobial strains DOA9, USDA110, and BTAi1 (Fig. 4). Overall, 3,896 genes (30.8%) are conserved among all three strains (Fig. 4) representing 53.6% (3,896/7,273) of the total number of DOA9 genes. The number of genes unique to DOA9 (1,337 genes; 10.6%) is lower than that of both *B. japonicum* sp. USDA110 (2,268 genes; 17.9%) and *Bradyrhizobium* sp. BTAi1 (2,453 genes; 19.4%). DOA9 and *B. japonicum* sp. USDA110 share 1,437 genes (11.3%) that were not found in *Bradyrhizobium* sp. BTAi1—a markedly higher number than the 556 genes (4.4%) shared by DOA9 and BTAi1. These results indicate that the DOA9 genome is similar to the genome of *B. japonicum* USDA110 in terms of gene content.

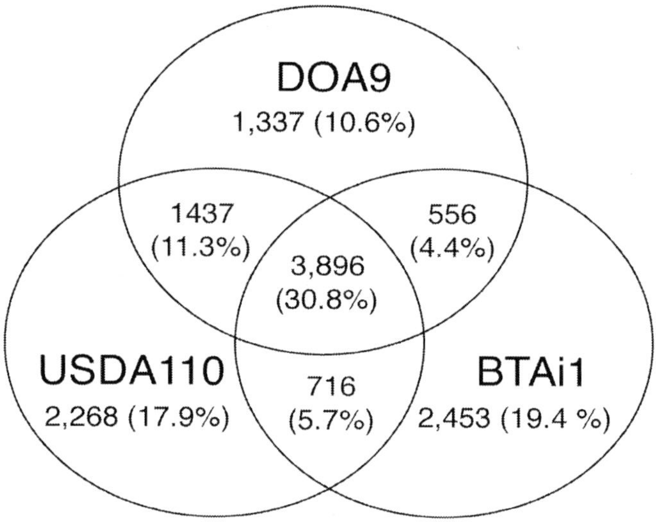

Figure 4: Comparative genomic analysis among *Bradyrhizobium* sp. DOA9, *B. japonicum* USDA110, and *Bradyrhizobium* sp. BTAi1.

Each genome is represented by a circle, and the numbers of shared and unique genes are shown by the overlapping and nonoverlapping regions. The proportion of total genes represented by each area of the diagram is shown in parentheses. The total number of genes in each genome is shown in square brackets.

Properties of pDOA9

The BLASTN comparison indicated that pDOA9 showed similarity to the plasmids of *Rhizobium*and *Agrobacterium*. The dendrogram based on genomic dissimilarity revealed that pDOA9 is grouped with pBBTa01of *Bradyrhizobium* sp. BTAi1 (Fig. 5). There was no evidence of extensive synteny between pDOA9 and pBBTa01 or other symbiotic plasmids, except in gene regions with symbiosis-related functions such as nodulation, nitrogen fixation, type III/IV secretion systems, and hydrogenase (Fig. 2B). The GC content of pDOA9 (60.07%) was lower than that of the chromosome (64.95%); this characteristic is similar to *Bradyrhizobium* sp. BTAi1, where the GC content is 60.69% for pBBTa01 and 64.92% for the chromosome.

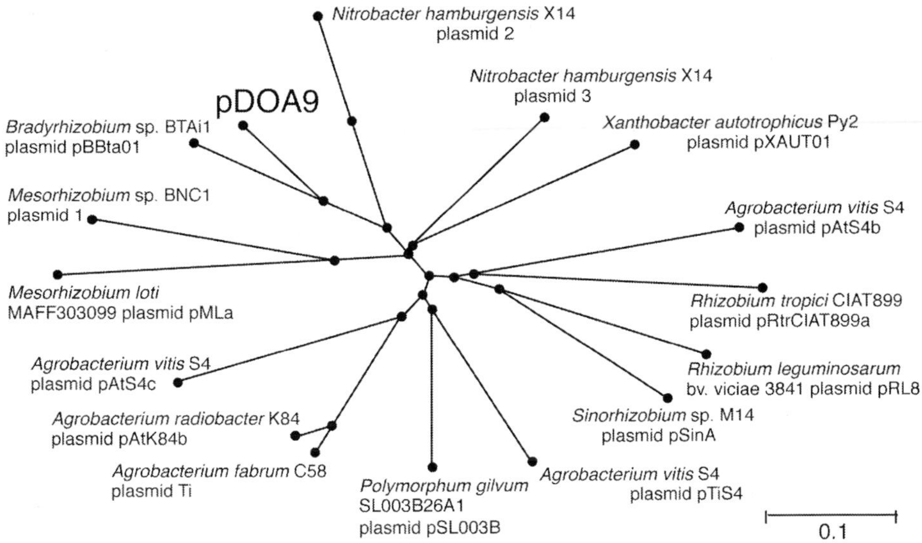

Figure 5: Phylogenetic relationships of pDOA9 and related plasmids.

Bootstrap values are expressed as percentages of 1,000 replications. Evolutionary distances were computed using the Kimura two-parameter method. The bar represents one estimated substitution per 100-nucleotide positions.

Nodulation genes

Nodulation genes encode proteins involved in the biosynthesis and transport of NFs, which induce nodule organogenesis in most rhizobia. A large *nod* gene cluster was found on pDOA9 (Fig. 2B). The organization and repertoire of *nod* genes are similar to those of *B. japonicum*, although losses, insertions, and inversions were found (Fig. 6 and Table 2). For example, *B. japonicum* USDA110 possesses *nodYABCSUIJnolMNO* arranged in a single operon, whereas DOA9 lacks *nodY*, *nodJ*, *nolM*, and *nolN*. In addition, *B. japonicum* has two *nodD* genes (*nodD1* and *nodD2*) within the cluster, whereas DOA9 has only one (*nodD1*, locus tag: BDOA9_0203500) in the cluster and the other (*nodD2*, BDOA9_0205810) located 0.27 Mb away. Furthermore, *nodQ*, *nodP*, *noeE*, *noeL*, and *nolK*

(BDOA9_0203370–0203390, 0203580–0203590) are incorporated into the *nod* gene cluster of DOA9 but are located far from the cluster in *B. japonicum*. Unlike *B. japonicum* USDA110, DOA9 possesses two copies of *nodA*(*nodA1*, BDOA9_0206740 and *nodA2*, BDOA9_0203720), one in the *nodABCSUI* operon and the other in the *nodA2IJ* operon. *nodA1* shares only 32–36% identity with the copies in other *Bradyrhizobium* strains, whereas *nodA2* shares 63–69% identity (Table 2). *nodB* and *nodC*share more than 60% identity with the corresponding genes in *B. japonicum* (Table 2). Phylogenetic analyses of the common nodulation genes *nodA*, *nodB* and *nodC* showed that those of DOA9 were placed on unclassified branches separated from the known *nod* gene-containing rhizobia. Although most of the *nod* genes involved in NF biosynthesis, such as *nodABC*, are located on pDOA9, some are also found on the chromosome (Table 2): *nolG*, *nodV*, *nodW*, *nodQ*, *nodN*, and *nodT*. Eight *nod* box consensus sequences [17], where NodD (the transcriptional activator that responds to flavonoids) binds to the DNA, were identified in pDOA9. Most of the *nod* boxes precede genes involved in Nod factor biosynthesis and regulation, as in other rhizobia. No *nod* boxes were identified in the chromosome using our search criteria.

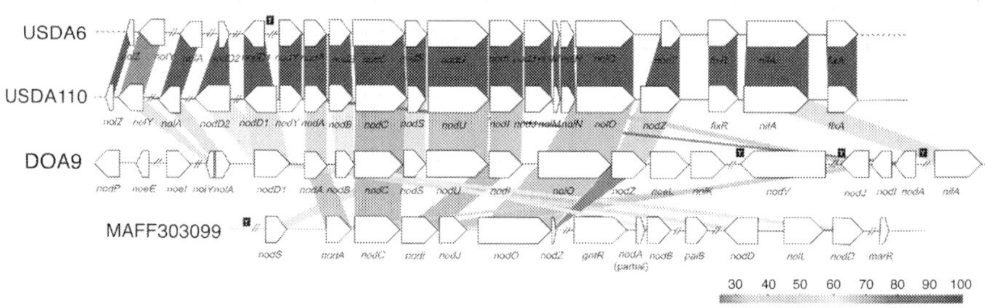

Figure 6: Comparison of nodulation gene clusters in *Bradyrhizobium* sp. DOA9 and related bacteria.

Double slash marks represent DNA regions that are not shown. Colored strips represent the conserved gene regions between the compared strains, and the color indicates the percentage similarity, as detailed in the key. T: region where the transposase genes were located.

Table 2: Nodulation genes detected in the genome of *Bradyrhizobium* sp. DOA9 and higher BLASTP similarity with other microorganisms.

Gene	Locus tag	Percentage of identity in BLASTP
Chromosome		
nolG	BDOA9_0103430	*Bradyrhizobium* sp. BTAi1 (91%), *B. japonicum* USDA110 (90%), *Bradyrhizobium* sp. ORS278 (76%), *R. leguminosarum* bv. *viciae* 3841 (74%)
nolG	BDOA9_0108870	*B. japonicum* USDA110 (95%), *Bradyrhizobium* sp. ORS278 (83%), *Bradyrhizobium* sp. BTAi1 (83%), *A. caulinodans* ORS571 (56%)
nodV	BDOA9_0120640	*B. japonicum* WSM2793 (40%), *Methylobacterium* sp. 10 (31%), *Rhizobium* sp. PDO1-076 (31%)
nodQ	BDOA9_0121760	*B. japonicum* USDA110 (96%), *Bradyrhizobium* sp. ORS278 (80%), *Bradyrhizobium* sp. BTAi1 (80%), *A. caulinodans* ORS571 (60%)
nodW	BDOA9_0136710	*B. japonicum* USDA110 (76%), *S. meliloti* 1021 (58%)
nodW	BDOA9_0136720	*B. japonicum* USDA110 (66%), *R. leguminosarum* bv. viciae 3841 (40%)
nodV	BDOA9_0136850	*B. japonicum* USDA110 (67%), *S. meliloti* 1021 (35%)
nodN	BDOA9_0136940	*B. japonicum* USDA110 (92%), *C. taiwanensis* LMG 19424 (43%)
nodN	BDOA9_0139450	*B. japonicum* USDA110 (96%), *Bradyrhizobium* sp. BTAi1 (83%), *Bradyrhizobium* sp. ORS278 (83%), *M. loti* MAFF303099 (46%)
nolG	BDOA9_0140510	*B. japonicum* USDA110 (97%), *Bradyrhizobium* sp. BTAi1 (90%), *Bradyrhizobium* sp. ORS278 (90%), *A. caulinodans* ORS571 (76%)
nodV	BDOA9_0147240	*B. japonicum* USDA110 (61%), *R. leguminosarum* bv. *viciae* 3841 (32%)
nodW	BDOA9_0147250	*B. japonicum* USDA110 (78%), *A. caulinodans* ORS571 (62%)
nodW	BDOA9_0147260	*B. japonicum* USDA110 (67%), *A. caulinodans* ORS571 (45%)
nodV	BDOA9_0151030	*B. japonicum* USDA110 (88%), *A. caulinodans* ORS571 (34%)
nodW	BDOA9_0151040	*B. japonicum* USDA110 (94%), *A. caulinodans* ORS571 (67%)
nodW	BDOA9_0151050	*B. japonicum* USDA110 (89%), *A. caulinodans* ORS571 (44%)
nolG	BDOA9_0151210	*Bradyrhizobium* sp. S23321 (98%), *B. japonicum* USDA110 (96%)
nodW	BDOA9_0151770	*B. japonicum* USDA110 (89%), *R. etli* CFN42 (50%), *C. taiwanensis* LMG 19424 (47%)
nodV	BDOA9_0153480	*B. japonicum* USDA110 (85%), *Bradyrhizobium* sp. BTAi1 (62%), *Bradyrhizobium* sp. ORS278 (61%)
nolG	BDOA9_0156320	*B. japonicum* USDA110 (91%), *Bradyrhizobium* sp. BTAi1 (87%), *Bradyrhizobium* sp. ORS278 (86%)
nolO	BDOA9_0158070	*B. japonicum* USDA6 (97%), *R. etli* CFN42 (35%), *M. loti* MAFF303099 (35%)
nodT	BDOA9_0162260	*Bradyrhizobium* sp. ORS278 (56%), *Bradyrhizobium* sp. BTAi1 (55%), *S. fredii* NGR234 (54%)
nolG	BDOA9_0165280	*B. japonicum* USDA110 (97%), *Bradyrhizobium* sp. BTAi1 (86%), *Azospirillum* sp. B510 (52%)
Plasmid (pDOA9)		
nodW	BDOA9_0203300	*B. japonicum* USDA110 (65%), *Azoarcus* sp. BH72 (42%)
nodW	BDOA9_0203310	*B. japonicum* USDA110 (67%), *Azoarcus* sp. BH72 (57%)
nodQ	BDOA9_0203370	*Bradyrhizobium* sp. ORS278 (53%), *Mesorhizobium loti* MAFF303099 (48%), *Sinorhizobium fredii* NGR234 (48%)
nodP	BDOA9_0203380	*Bradyrhizobium* sp. ORS278 (73%), *Azospirillum* sp. B510 (71%), *Bradyrhizobium* sp. BTAi1 (70%)
noeE	BDOA9_0203390	*S. fredii* NGR234 (55%)
nodW	BDOA9_0203420	*Mesorhizobium* sp. WSM4349 (33%), *S. terangae* WSM1721 (31%)
noeI	BDOA9_0203430	*B. japonicum* USDA110 (73%), *S. fredii* NGR234 (71%)
nolA	BDOA9_0203490	*Bradyrhizobium* sp. NC92 (62%), *Mesorhizobium* sp. WSM4349 (60%), *Bradyrhizobium* sp. ORS 285 (47%)
nodD1	BDOA9_0203500	*Rhizobium etli* CFN42 (63%), *B. japonicum* USDA110 (62%), *S. meliloti* 1021 (61%), *S. fredii* NGR234 (61%)
nodA1	BDOA9_0206740	*Bradyrhizobium* sp. WSM1417 (36%), *Rhizobium etli* CFN42 (34%), *B. japonicum* USDA110 (32%)
nodB	BDOA9_0203510	*S. fredii* NGR234 (65%), *S. fredii* HH103 (65%), *B. japonicum* USDA110 (62%), *Azorhizobium caulinodans* ORS571 (42%), *Azospirillum* sp. B510 (35%)
nodC	BDOA9_0203520	*B. japonicum* USDA110 (60%), *M. loti* R7A (68%), *R. etli* CFN42 (66%)
nodS	BDOA9_0203530	*S. fredii* NGR234 (59%), *R. etli* CFN42 (55%), *M. loti* MAFF303099 (55%), *B. japonicum* USDA6 (52%)
nodU	BDOA9_0203540	*B. japonicum* USDA110 (69%), *Cupriavidus taiwanensis* LMG 19424 (64%), *A. caulinodans* ORS571 (54%)
nodI	BDOA9_0203550	*M. loti* MAFF303099 (70%), *S. meliloti* 1021 (70%), *R. etli* CFN42 (68%)
nolO	BDOA9_0203560	*B. japonicum* USDA110 (68%), *M. loti* MAFF303099 (67%), *Rhizobium* sp. NGR234 (67%), *R. etli* CFN42 (65%)
nodZ	BDOA9_0203570	*R. etli* CFN42 (66%), *M. loti* MAFF303099 (66%), *S. fredii* NGR234 (64%)
noeL	BDOA9_0203580	*B. japonicum* USDA110 (91%), *Bradyrhizobium* sp. BTAi1 (84%)
nolK	BDOA9_0203590	*B. japonicum* USDA110 (89%), *Bradyrhizobium* sp. BTAi1 (80%), *Bradyrhizobium* sp. ORS278 (80%)
nodV	BDOA9_0203630	*B. japonicum* USDA110 (50%)
nodJ	BDOA9_0203700	*R. etli* CFN42 (66%), *M. loti* MAFF303099 (64%), *S. fredii* NGR234 (64%), *R. leguminosarum* bv. *viciae* 3841 (64%)
nodI	BDOA9_0203710	*B. japonicum* USDA110 (47%), *M. loti* MAFF303099 (38%)
nodA2	BDOA9_0203720	*B. japonicum* USDA110 (71%), *S. fredii* NGR234 (70%), *M. loti* R7A (69%), *C. taiwanensis* LMG 19424 (69%)
nolY	BDOA9_0203790	*B. japonicum* USDA110 (42%), *S. fredii* NGR234 (33%)
nodV	BDOA9_0205070	*B. japonicum* USDA110 (85%), *Bradyrhizobium* sp. BTAi1 (79%), *Bradyrhizobium* sp. ORS278 (79%)
nodD2	BDOA9_0205810	*B. japonicum* USDA110 (69%), *S. meliloti* 1021 (66%), *R. leguminosarum* bv. *viciae* 3841 (63%)

doi:10.1371/journal.pone.0117392.t002

Nitrogen fixation genes

In DOA9 a single cluster of *fix* genes (*fix_chr*) was found on the chromosome and a *nif-fix*cluster was found on both the chromosome (*nif-fix_chr*) and the plasmid (*nif-fix_pla*) (Fig. 2). The *fix_chr* cluster includ *fixK-fixLJ*, which encodes a factor involved in the regulation of symbiotic nitrogen fixation [18], *fixNOQP*, which encodes a microaerobically induced cytochrome oxidase complex [19], and *fixGHIS*, which encodes a membrane-bound complex including a cation pump involved in nitrogen fixation [19]. The comparative analysis showed that the organization of the *fix_chr* cluster is conserved among bradyrhizobial strains.

Comparative analysis showed that the organization of the *nif-fix_chr* cluster is almost identical to that in photosynthetic bradyrhizobial strains, including *Aeschynomene* symbionts (*Bradyrhizobium* spp. ORS278, BTAi1, S58, and ORS285) and the non-nodulating strain *Bradyrhizobium* sp. S23321. An exception was found for the *nifH* gene: photosynthetic bradyrhizobial strains possess two copies of *nifH*, which are located in the *nifHQ* and *nifHDK*operons; by contrast, DOA9 possesses only the *nifHQ* operon, and another copy of *nifH* might have been deleted from the *nifHDK* operon. Notably, a copy of *nifV*, which encodes the homocitrate synthase [20], was found in the cluster, similar to the photosynthetic strains *Bradyrhizobium* spp. ORS285, S58, BTAi1, ORS278, and S23321.

Another *nif-fix* cluster (*nif-fix_pla*) was also found in the plasmid of DOA9. The *nif-fix_pla* is more divergent than the *nif-fix_chr* especially *nifQ* (BDOA9_0204570), which is involved in the incorporation of molybdenum into nitrogenase, and *nifW* (BDOA9_0204580), which is a nitrogenase stabilizing and protective protein. The *nifH* gene in the chromosome (BDOA9_0160870) and the plasmid (BDOA9_0204560) show 96% identity. The phylogenetic analysis showed that *nifH* of DOA9 is placed on a separate branch, distinct from those of nonphotosynthetic species, such as *B. japonicum*, *B. elkanii*, and *B. yuanmingense*. *nifH* on the chromosome of DOA9 shares higher similarity with the same gene of photosynthetic strains than that of nonphotosynthetic strains.

Gene clusters of type III /IV secretion systems

A cluster of genes encoding the type III secretion system was found in pDAO9 spanning a 53-kb region. The organization of genes encoding the type III secretion apparatus (*rhcC*, *rhcJ*, *rhcN*, and *rhcQRSTU*) is well conserved among bradyrhizobial strains. The organization of the *tts* gene cluster of DOA9 is similar to that of *B. elkanii* USDA61, although losses, insertions, and inversions of some genes are present. Phylogenetic analysis using the conserved type III apparatus *rhcTVRUC* genes showed that the *tts* genes of DOA9 are grouped with those of other *Bradyrhizobium* strains. The DOA9 genome contains 10 *tts* boxes, which are conserved motifs located in the promoter region of genes encoding structural components and secrete proteins of the rhizobial *tts* cluster [21]. Most of the *tts* boxes precede genes that encode nodulation outer proteins (*nopB* and *nopX*) or structural components of the type III secretion machinery (*y4yQ* and *rhcT*). *tts* boxes were also identified in the upstream regions of the genes encoding ParA-like ATPase, heat shock protein, cupin, and hypothetical proteins. Homologues of these genes have not been reported so far as type III secreted proteins or related components.

The plasmid pDOA9 also contains genes encoding a VirB/D4 type IV secretion system (T4SS). The *vir* cluster includes *virB1* to *virB11*, which encode the transmembrane complex and pili required for transfer proteins, and *virD4*, which encodes a protein that links the transferred substrates to the transfer apparatus (Ding *et al.* 2003). The *vir* genes of DOA9 are similar to that of BTAi1. The phylogenetic analysis of two representative genes, *virB2* and *virB9*, supports this finding.

DISCUSSION

Bradyrhizobium sp. DOA9 is of particular biological interest because it possesses a broader host range and divergent *nod* genes compared with other bradyrhizobia [7]. The most prominent characteristic of the DOA9 genome is the presence of symbiosis-related genes on the plasmid rather than on the chromosome. A similar trait is seen in fast-growing rhizobia, whose symbiosis-related genes are often clustered on a large plasmid, such as pNGR234a (536,165 bp) of *Sinorhizobium fredii* NGR234 [22], p42d of *Rhizobium etli* CFN42 (371,254 bp) [23], and pRL10 of *Rhizobium leguminosarum* biovar 3841 (488,135 bp) [24].

Although plasmids were found in *Bradyrhizobium* sp. BTAi1 and some soybean bradyrhizobia, they did not contain any *nod* genes [25]. In *nod* gene-carrying bradyrhizobial strains such as *B. japonicum* USDA110, the symbiosis-related genes are clustered in regions of the chromosome with a low GC content, known as symbiosis islands [4,5]. For instance, in *B. japonicum*USDA110 the symbiosis island is a 680-kb region with a relatively low GC content (59.4%) compared with that of the entire genome (64.1%). Notably, in DOA9 the GC content of pDOA9 (60.07%) was lower than that of the chromosome (64.41%). Symbiosis islands have been found to integrate genes of the recipients into the tRNA. Sullivan *et al.* reported [26] that the symbiotic island of *Mesorhizobium loti* was transferred to nonsymbiotic strains present in the soil. It is possible that pDOA9 can be transferred to other nonsymbiotic bradyrhizobia by conjugal transfer and integrated into their chromosomes through a process mediated by an integrase, as reported for many other elements, including the pathogenicity islands of pathogenic bacteria [27]. However, the details of the molecular mechanisms and host range for plasmid transfer remain to be elucidated; the latter is particularly interesting because the plasmid might enhance the fitness of bacteria.

The *nod* genes of DOA9 are highly divergent compared with those of other rhizobia in terms of both gene organization and homology. DOA9 is phylogenetically close to *B. japonicum* (Fig. 1), and their chromosomes share a high degree of similarity. However, their host legume specificities are different and DOA9 is unable to nodulate soybean [7]. In *B. japonicum*, the common *nodABC* genes are cotranscribed with *nodY* [28–30]. The *nod* gene cluster in DOA9 consists of the *nodABCSU* genes but lacks *nodY* and *nolMN* (Fig. 6). *nodY* was reported to be *Bradyrhizobium* specific and to be induced by soybean seed extract and selected isoflavones, primarily genistein and daidzein, but not by flavones [31]. This might account for the nodulation deficiency of DOA9 on soybean. It is possible that soybean bradyrhizobia acquired *nodY* during coevolution with soybean from *Aeschynomene* bradyrhizobia.

Unlike *B. japonicum*, DOA9 harbors additional host-specific genes, *nodQ*, *nodP*, and *noeE*, in the *nod* gene cluster (Fig. 6 and Table 2). These genes exhibit no similarity with those of *B. japonicum*, suggesting that DOA9 acquired them independently from other rhizobia. The roles of these genes in nodulation and host specificity remain to be elucidated.

Moreover, DOA9 harbors two copies of *nodA* in the *nod* gene cluster: *nodA1*, which is divergent, is incorporated into the *nodA1BCSUI* operon, whereas *nodA2* is incorporated into the *nodA2IJ* operon. This structure is similar to the structure of *Rhizobium tropici*, which carries three copies of *nodA* on the symbiotic plasmid and whose *nodA2* and *nodA3* genes have no close homologues [32]. The NF acyl group attached by NodA might contribute to the determination of host range [33]. Additional and divergent *nodA* genes are likely to expand the diversity of NF acyl chains and might broaden the host range of the bacteria [33].

DOA9 harbors several copies of *nodVW* on both the chromosome and the plasmid: seven copies of *nodV* (five on the chromosome and two on the plasmid) and 10 copies of *nodW*(seven on the chromosome and three on the plasmid) (Table 2). NodVW is a member of the classical two-component regulatory family and is essential for the nodulation of cowpea, siratro, and mungbean [34]. In *B. japonicum*, NodVW is considered to recognize plant flavonoids and could therefore contribute to the nodulation of a broader range of host plants by increasing NF synthesis in combination with NodD [20,35]. Although most *nodVW* genes showed high similarity with those of *B. japonicum*, some—BDOA9_0120640, BDOA9_0203420, and BDOA9_0203630—exhibited low similarity (Table 2). These NodVW can recognize additional plant and environmental signals and hence activate symbiosis-related genes. Further study will be necessary to elucidate the involvement of *nodVW* genes in broad host specificity of DOA9.

The genome of DOA9 contains a full set of nitrogen-fixing genes (*nif-fix*) on the chromosome and an incomplete cluster of these genes on pDOA9. The *nif-fix* cluster on the chromosome (*nif-fix_chr*) is highly similar to that of photosynthetic bradyrhizobia as well as of the non-nodulating strain *Bradyrhizobium* sp. S23321. The *nif-fix* cluster on pDOA9 (*nif-fix_pla*) lacks most of the *nif* genes and those present are highly similar to those of photosynthetic bradyrhizobia. As in photosynthetic bradyrhizobia, *nif-fix_chr* cluster in DOA9 carries *nifV*, which encodes homocitrate synthase that is essential for nitrogenase activity in the free-living diazotrophs [20]. *nifV* is absent from most strains of rhizobia that perform nitrogen fixation only in symbiotic states, whereas it is presents in strains that also fix nitrogen in *ex planta* states such as *Azorhizobium caulinodans* [20], *Bradyrhizobium* spp. BTAi1 and ORS278 [36]. The presence of *nifV* in DOA9 suggests that DOA9 has retained the nitrogen-fixation system evolved from the ancestral photosynthetic bradyrhizobia and produces NifV to perform efficient nitrogen fixation during symbiosis.

Clusters of genes encoding elements of the type III secretion system (T3SS) were found in the genomes of various rhizobia including the nonphotosynthetic strains *B. japonicum* USDA6 and USDA110 and *B. elkanii* USDA61, and the photosynthetic strains *Bradyrhizobium* spp. ORS285; however, the cluster was not found in the genome of the most nod-independent bradyrhizobial strains BTAi1, ORS278, and S23321. The phylogenetic analysis using *rhcTVRUC* genes, which encode conserved proteins of the type III secretion apparatus, indicated that the genes encoding T3SS in DOA9 and other bradyrhizobial strains shared the same evolutionary origin. Among these, the components of the type III secretion apparatus of DOA9 showed high identities with those of the soybean symbionts *B. elkanii*USDA61 and *B. japonicum* USDA110. The T3SS genes of these strains were expressed in the presence of the soybean flavonoid genistein, confirming that it functions in the infection of host roots. The induction of T3SS genes involves two regulators: NodD, which is activated by genistein; and TtsI, which is induced by the NodD protein. DOA9 possesses both of these regulators, and they show high identity with the soybean symbionts, despite the fact that DOA9 is unable to nodulate soybean. Expression analysis of T3SS genes and gene-deletion analysis could elucidate the regulation and symbiotic roles of T3SS in DOA9.

The VirB/D4 type IV secretion system (T4SS) has been identified in some rhizobia, including *M. loti* R7A[37] and *R. etli* CFN42[38] but is absent from bradyrhizobia with the exception of *Bradyrhizobium* sp. BTAi1[25]. *M. loti* R7A possesses a gene cluster (*vir*) encoding T4SS, similar to that of *Agrobacterium tumefaciens*. In *M. loti* R7A T4SS is transcriptionally regulated by a VirA/VirG two-component regulatory system [39]. In *M. loti* R7A the *virA* gene is preceded by a *nod* box, and T4SS is activated in a symbiosis-specific manner [39]. The DOA9 genome lack**s** homologues of *M. loti virA and virG*. Similarly, *virA* and *virG* homologues are absent from BTAi1, whereas *R. etli* CFN42 possess v***irA*** and v***irG*** that show 43% and 76% identities, respectively, with those of *M. loti* R7A. These findings suggest that the T4SS of

*Bradyrhizobium*spp. DOA9 and BTAi1 share an origin that differs from that of *M. loti* R7A and other rhizobia.

In conclusion, the genome of DOA9 exhibits a mosaic structure with a unique repertoire of symbiotic genes with various origins. During adaptation to *A. americana* and other legumes, the ancestral strain of DOA9 acquired divergent nodulation genes and other symbiosis-related genes such as the type III and IV secretion systems. The present study also discovered the first symbiotic plasmid among *Bradyrhizobiaceae*. This plasmid will be a valuable tool for future studies on the genetics, physiology, and ecology of these species. For example, it will be interesting to test whether pDOA9 can transform non-symbiotic strains, such as S23321, into symbiotic strains or broaden the host range of BTAi1. Our efforts are now directed toward functional analysis of the divergent *nod* genes in pDOA9, to improve our understanding of the evolution of bradyrhizobia and of their host plants mediated by the symbiotic plasmid.

AUTHOR CONTRIBUTIONS

Conceived and designed the experiments: SO NB KS SS TU KM NT. Performed the experiments: SO RN TO TK KO MH KT PS PT. Analyzed the data: SO RN TO. Contributed reagents/materials/analysis tools: SO NB NT. Wrote the paper: SO RN TO KM NT.

REFERENCES

1. Kouchi H, Imaizumi-Anraku H, Hayashi M, Hakoyama T, Nakagawa T, ct al. (2010) IIow many peas in a pod? Legume genes responsible for mutualistic symbioses underground. Plant and Cell Physiology 51: 1381–1397. pmid:20660226
2. Giraud E, Moulin L, Vallenet D, Barbe V, Cytryn E, et al. (2007) Legumes symbioses: absence of Nod genes in photosynthetic bradyrhizobia. Science 316: 1307–1312. pmid:17540897
3. Chaintreuil C, Arrighi JF, Giraud E, Miché L, Moulin L, et al. (2013) Evolution of symbiosis in the legume genus Aeschynomene. New Phytologist 200: 1247–1259. pmid:23879229
4. Kaneko T, Nakamura Y, Sato S, Minamisawa K, Uchiumi T, et al. (2002) Complete genomic sequence of nitrogen-fixing symbiotic bacterium Bradyrhizobium japonicum USDA110. DNA research 9:

189–197. pmid:12597275

5. Kaneko T, Maita H, Hirakawa H, Uchiike N, Minamisawa K, et al. (2011) Complete genome sequence of the soybean symbiont Bradyrhizobium japonicum strain USDA6T. Genes 2: 763–787. pmid:24710291
6. Okubo T, Tsukui T, Maita H, Okamoto S, Oshima K, et al. (2011) Complete genome sequence of Bradyrhizobium sp. S23321: insights into symbiosis evolution in soil oligotrophs. Microbes and environments/JSME 27: 306–315.
7. Noisangiam R, Teamtisong K, Tittabutr P, Boonkerd N, Toshiki U, et al. (2012) Genetic diversity, symbiotic evolution, and proposed infection process of Bradyrhizobium strains isolated from root nodules of Aeschynomene americana L. in Thailand. Applied and environmental microbiology 78: 6236–6250. pmid:22752179
8. Teamtisong K, Songwattana P, Noisangiam R, Piromyou P, Boonkerd N, et al. (2014) Divergent Nod-Containing Bradyrhizobium sp. DOA9 with a Megaplasmid and its Host Range. Microbes and Environments.
9. Sadowsky MJ, Tully RE, Cregan PB, Keyser HH (1987) Genetic diversity in Bradyrhizobium japonicum serogroup 123 and its relation to genotype-specific nodulation of soybean. Applied and environmental microbiology 53: 2624–2630. pmid:16347481
10. Wilson K (1987) Preparation of genomic DNA from bacteria. Current protocols in molecular biology: 2.4. 1–2.4. 5.
11. Chevreux B, Wetter T, Suhai S (1999) Genome sequence assembly using trace signals and additional sequence information. pp. 45–56.
12. Grant JR, Stothard P (2008) The CGView Server: a comparative genomics tool for circular genomes. Nucleic acids research 36: W181–W184. pmid:18411202
13. Ohtsubo Y, Ikeda-Ohtsubo W, Nagata Y, Tsuda M (2008) GenomeMatcher: a graphical user interface for DNA sequence comparison. BMC bioinformatics 9: 376. pmid:18793444
14. Thompson JD, Higgins DG, Gibson TJ (1994) CLUSTAL W: improving the sensitivity of progressive multiple sequence alignment through sequence weighting, position-specific gap penalties and weight matrix choice. Nucleic acids research 22: 4673–4680.

pmid:7984417

15. Tamura K, Peterson D, Peterson N, Stecher G, Nei M, et al. (2011) MEGA5: molecular evolutionary genetics analysis using maximum likelihood, evolutionary distance, and maximum parsimony methods. Molecular biology and evolution 28: 2731–2739. pmid:21546353
16. Regensburger B, Hennecke H (1983) RNA polymerase from Rhizobium japonicum. Archives of microbiology 135: 103–109. pmid:6639271
17. Spaink HP, Okker RJ, Wijffelman CA, Pees E, Lugtenberg BJ (1987) Promoters in the nodulation region of the Rhizobium leguminosarum Sym plasmid pRL1JI. Plant Molecular Biology 9: 27–39. pmid:24276795
18. Fischer H-M (1994) Genetic regulation of nitrogen fixation in rhizobia. Microbiological reviews 58: 352. pmid:7968919
19. Dixon R, Kahn D (2004) Genetic regulation of biological nitrogen fixation. Nature Reviews Microbiology 2: 621–631. pmid:15263897
20. Zheng L, White RH, Dean DR (1997) Purification of the Azotobacter vinelandii nifV-encoded homocitrate synthase. Journal of bacteriology 179: 5963–5966. pmid:9294461
21. Krause A, Doerfel A, Göttfert M (2002) Mutational and transcriptional analysis of the type III secretion system of Bradyrhizobium japonicum. Molecular plant-microbe interactions 15: 1228–1235. pmid:12481995
22. Perret X, Viprey V, Freiberg C, Broughton W (1997) Structure and evolution of NGRRS-1, a complex, repeated element in the genome of Rhizobium sp. strain NGR234. Journal of bacteriology 179: 7488–7496. pmid:9393715
23. González V, Santamaría RI, Bustos P, Hernández-González I, Medrano-Soto A, et al. (2006) The partitioned Rhizobium etli genome: genetic and metabolic redundancy in seven interacting replicons. Proceedings of the National Academy of Sciences of the United states of America 103: 3834–3839. pmid:16505379
24. Young JPW, Crossman LC, Johnston AW, Thomson NR, Ghazoui ZF, et al. (2006) The genome of Rhizobium leguminosarum has recognizable core and accessory components. Genome biology 7: R34. pmid:16640791

25. Cytryn EJ, Jitacksorn S, Giraud E, Sadowsky MJ (2008) Insights learned from pBTAi1, a 229-kb accessory plasmid from Bradyrhizobium sp. strain BTAi1 and prevalence of accessory plasmids in other Bradyrhizobium sp. strains. The ISME journal 2: 158–170. pmid:18219284
26. Sullivan JT, Ronson CW (1998) Evolution of rhizobia by acquisition of a 500-kb symbiosis island that integrates into a phe-tRNA gene. Proceedings of the National Academy of Sciences 95: 5145–5149. pmid:9560243
27. Hacker J, Kaper JB (2000) Pathogenicity islands and the evolution of microbes. Annual Reviews in Microbiology 54: 641–679.
28. Geelen D, Mergaert P, Geremia RA, Goormachtig S, Montagu M, et al. (1993) Identification of nodSUIJ genes in Nod locus 1 of Azorhizobium caulinodans: evidence that nodS encodes a methyltransferase involved in Nod factor modification. Molecular microbiology 9: 145–154. pmid:8412659
29. Göttfert M, Hitz S, Hennecke H (1990) Identification of nodS and nodU, two inducible genes inserted between the Bradyrhizobium japonicum nodYABC and nodIJ genes. Mol Plant-Microbe Interact 3: 308–316. pmid:2134855
30. Lewin A, Cervantes E, Wong C, Broughton WJ (1990) nodSU, two new nod genes of the broad host range Rhizobium strain NGR234 encode host-specific nodulation of the tropical tree Leucaena leucocephala. Mol Plant-Microbe Interact 3: 317–326. pmid:2134856
31. Banfalvi Z, Nieuwkoop A, Schell M, Besl L, Stacey G (1988) Regulation of nod gene expression in Bradyrhizobium japonicum. Molecular and General Genetics MGG 214: 420–424. pmid:3146016
32. Ormeño-Orrillo E, Menna P, Almeida LGP, Ollero FJ, Nicolás MF, et al. (2012) Genomic basis of broad host range and environmental adaptability of Rhizobium tropici CIAT 899 and Rhizobium sp. PRF 81 which are used in inoculants for common bean (Phaseolus vulgaris L.). BMC genomics 13: 735. pmid:23270491
33. Debellé F, Plazanet C, Roche P, Pujol C, Savagnac A, et al. (1996) The NodA proteins of Rhizobium meliloti and Rhizobium tropici

specify the N-acylation of Nod factors by different fatty acids. Molecular microbiology 22: 303–314. pmid:8930915

34. Sanjuan J, Grob P, Göttfert M, Hennecke H, Stacey G (1994) NodW is essential for full expression of the common nodulation genes in Bradyrhizobium japonicum. MPMI-Molecular Plant Microbe Interactions 7: 364–369. pmid:8075421
35. Loh J, Stacey G (2003) Nodulation gene regulation in Bradyrhizobium japonicum: a unique integration of global regulatory circuits. Applied and environmental microbiology 69: 10–17. pmid:12513971
36. Terpolilli JJ, Hood GA, Poole PS (2012) What determines the efficiency of N2-fixing Rhizobium-legume symbioses? Advances in microbial physiology 60: 326.
37. Hubber A, Vergunst AC, Sullivan JT, Hooykaas PJ, Ronson CW (2004) Symbiotic phenotypes and translocated effector proteins of the Mesorhizobium loti strain R7A VirB/D4 type IV secretion system. Molecular microbiology 54: 561–574. pmid:15469524
38. González V, Bustos P, Ramírez-Romero MA, Medrano-Soto A, Salgado H, et al. (2003) The mosaic structure of the symbiotic plasmid of Rhizobium etli CFN42 and its relation to other symbiotic genome compartments. Genome biology 4: R36. pmid:12801410
39. Hubber AM, Sullivan JT, Ronson CW (2007) Symbiosis-induced cascade regulation of the Mesorhizobium loti R7A VirB/D4 type IV secretion system. Molecular plant-microbe interactions 20: 255–261. pmid:17378428

SECTION V
VIRULENCE AND ANTIBIOTIC RESISTANCE

CHAPTER

9

GENETIC CHARACTERIZATION OF EXPEC-LIKE VIRULENCE PLASMIDS AMONG A SUBSET OF NMEC

Bryon A. Nicholson[1], Aaron C. West[2], Paul Mangiamele[1], Nicolle Barbieri[1], Yvonne Wannemuehler[1], Lisa K. Nolan[1], Catherine M. Logue[1], Ganwu Li[3,4]

[1]Department of Veterinary Microbiology and Preventive Medicine, College of Veterinary Medicine, Iowa State University, Ames, Iowa, United States of America,

[2]Department of Chemistry, College of Liberal Arts and Sciences, Iowa State University, Ames, Iowa, United States of America,

[3]Department of Veterinary Diagnostic and Production Animal Medicine, College of Veterinary Medicine, 1802 University Blvd, Iowa State University, Ames, Iowa, 50011, United States of America,

[4]Department of Veterinary Preventive Medicine, College of Veterinary Medicine, Nanjing Agricultural University, Nanjing, Jiangsu, 210095, P. R. China

ABSTRACT

Neonatal Meningitis *Escherichia coli* (NMEC) is one of the most common causes of neonatal bacterial meningitis in the US and elsewhere resulting in mortality or neurologic deficits in survivors. Large plasmids have been shown experimentally to increase the virulence of NMEC in the rat model of neonatal meningitis. Here, 9 ExPEC-like plasmids were isolated from NMEC and sequenced to identify the core and accessory plasmid genes of ExPEC-like virulence plasmids in NMEC and create an expanded plasmid phylogeny. Results showed sequenced virulence plasmids carry a strongly conserved core of genes with predicted functions in five distinct categories including: virulence, metabolism, plasmid stability, mobile elements, and unknown genes. The major functions of virulence-associated and plasmid core genes serve to increase *in vivo* fitness by adding multiple iron uptake systems to the genetic repertoire to facilitate NMEC's survival in the host's low iron environment, and systems to enhance bacterial resistance to host innate immunity. Phylogenetic analysis based on these core plasmid genes showed that at least two lineages of ExPEC-like plasmids could be discerned. Further, virulence plasmids from Avian Pathogenic *E. coli* and NMEC plasmids could not be differentiated based solely on the genes of the core plasmid genome.

IMPORTANCE

Neonatal meningitis associated *Escherichia coli* (NMEC) as well as some other types of extra-intestinal pathogenic E. coli (ExPEC) are largely defined by the large plasmid; however relatively few ExPEC-like plasmids in NMEC have been sequenced and relatively few studies have examined their effects on pathogenicity. In this study we sequenced 9 large ExPEC-like plasmids from cases of neonatal bacterial meningitis to determine the core set of genes, and to determine if an evolutionary relationship between ExPEC plasmids of NMEC and other ExPEC exists. Our results show the core plasmid genome of NMEC strains to include genes related to virulence and metabolism as well as a high degree of relatedness to other large ExPEC plasmids.

INTRODUCTION

Escherichia coli is one of the world's most well studied organisms. Ubiquitous in nature and prevalent in biotechnology, it is also one of the most commonly isolated pathogens from human disease. Though some strains are specialized to cause disease inside the intestinal tract, others are equipped to cause extraintestinal disease [1–3]. These extraintestinal pathogenic *E. coli* (ExPEC) include neonatal meningitis *E. coli* (NMEC), one of the predominant agents of neonatal bacterial meningitis [4]. Neonatal bacterial meningitis has a mortality rate between 15 to 40% [5] with 30% of survivors showing serious neurological sequelae including profound intellectual disability, epilepsy, or deafness [4]. Reviews suggest that NMEC are acquired by neonates from their mothers perinatally [5, 6]. From the initial site of colonization, NMEC transcytose the gastrointestinal tract into the bloodstream, and from there, they traverse the blood brain barrier (BBB) (the tight barrier formed by human brain microvascular endothelial cells (HBMEC)) into the central nervous system (CNS). Since successful CNS invasion requires a high bacteremia, NMEC must survive in the bloodstream, a trait that is mediated by two bacterial components, an anti-phagocytic capsule and outer membrane protein A (OmpA), which has an anti-complement role. In addition, NMEC's replication in immune cells may play a role in achieving the requisite bacteremia. Also, NMEC's traversal of HBMEC is preceded by its attachment to the BBB via Type 1 pili and OmpA, with invasion itself being mediated by Ibe proteins, FimH (tip adhesin of Type 1 pili), OmpA, and Cytotoxic Necrotizing Factor 1 (CNF1). K1 capsule may also contribute to invasion by preventing lysosomal fusion, allowing delivery of live bacteria into the CNS [5, 6].

Plasmids also seem to contribute to NMEC virulence. A recent study by Kariyawasam and colleagues indicates that a large plasmid increases the virulence of NMEC RS218 *in vivo* [7], while previous studies have shown that the acquisition of plasmids is one of the primary sources of genetic variance and virulence genes in different *E. coli* pathotypes [8]. The majority of these plasmids has been identified as IncF type plasmids [9], encoding pathogenicity islands, metabolic genes, antimicrobial resistance genes, and hypothetical genes. Acquisition of these plasmids may provide an advantage for survival during infection and in suboptimal environments, as evidenced by multiple studies showing that these plasmids confer a fitness advantage to NMEC during in host tissues [7, 10, 11]. Further, plasmids are associated with an increase in NMEC virulence in animal models [7, 10, 11], and cause cross-species and cross pathogenic group gain of virulence function [12].

Here, we seek to better understand the evolution of large ExPEC-like virulence plasmids in NMEC and ascertain their similarity to plasmids of other ExPEC pathotypes. While plasmids from NMEC strains S88 (O45:K1), S286 (O78:K1), RS218 (O18:K1) and CE10 (O7:K1) have been sequenced and are available in the public domain, it is our contention that these plasmids likely do not account for the full diversity of ExPEC-like virulence plasmid sequences in NMEC. We base this hypothesis on the variability found among NMEC in their possession of plasmid-associated genes [13]. We further contend that this knowledge gap could inhibit future study to identify rational strategies designed to control NMEC-caused disease. We will address some of these shortcomings by sequencing a collection of ExPEC-like virulence plasmids in NMEC strains, identify orthologous genes shared between these large ExPEC-like plasmids in an effort to identify their core and accessory plasmid genomes, ascertain the phylogenetic relationships between NMEC plasmids and determine their similarities to plasmids occurring in other ExPEC.

MATERIALS AND METHODS

Bacterial Strains

A total of 11 *E. coli* isolates from cases of human neonatal meningitis were used as a source of plasmids in this study (Table 1). These isolates are part of a larger collection, containing 91 neonatal meningitis *E. coli* (NMEC) isolated from the Netherlands and United States. The bacterial strains were serogrouped at the *E. coli* Reference Center (Pennsylvania State University), genotyped for chromosomal and plasmid virulence genes and plasmid replicon types, and assigned to phylogenetic groups as previously described [13–15].

Table 1: Strains used in this study

Plasmid	Phylogroup	O:K	Cluster [1]	Size (bp)	GC%	ORFs	Source
pNM14	C	7:80	2	126,885	49.4	136	This Study
pNM15	B2	18:1	8	148,173	51.3	157	This Study
pNM16	B2	83:1	8	130,692	49.3	136	This Study
pNM19	B2	-:1	8	133,585	48.8	142	This Study
pNM26	B2	21:-	6	152,078	49.4	168	This Study
pNM36	B2	14:1	7	129,262	49.2	131	This Study
pNM38	B2	18:1	8	128,645	49.7	136	This Study
pNM49	B2	auto:1	8	121,081	49.6	119	This Study
pNM58	B2	18ac:1	8	157,260	49.7	157	This Study
pS88	B2	45:1	6	133,853	49.3	157	NC_011747.1
pS286	C	78:-	-	97,818	48.6	104	HF922624.1
pAPEC-O1-ColBM	B2	1:1	-	174,241	49.6	199	DQ381420.1
pAPEC-O2-ColV	B2	2:1	-	184,501	49.2	203	AY545598.5
pAPEC-O103-ColBM	B1	103:	-	124,705	50.8	126	NC_011964
p1ColV5155	B2	2:1	-	194,170	49.4	199	CP005930
pVM01	-	-:28	-	151,002	49.5	150	NC_010409
pAPEC-1	B1	78:-	-	103,275	45.1	164	CP000836
pAPEC-O78-ColV	-	78:-	-	144,859	50.4	158	NZ_CP010316

doi:10.1371/journal.pone.0147757.t001

Serogroups, phylogroups and genome statistics of selected sequenced NMEC strains used in the study.

Virulence Gene Amplification

As described in prior work, PCR was performed to determine the presence of chromosomal and plasmid derived virulence genes [13, 16]. Briefly, all strains were removed from freezer stock and struck to MacConkey agar plates. Isolates were then transferred to LB broth and grown at 37°C overnight. Bacterial DNA was harvested using the boil preparation method previously described [17]. For PCR analysis, isolates were tested for 205 genes associated with virulence and their allelic variants, antimicrobial resistance, plasmid replicons, and pathogenicity islands [13]. Primers were obtained from Integrated DNA Technologies (Coralville, Iowa). Replicon typing was carried out using the methodology described by Johnson [8] using multiplex PCR. Phylogrouping was carried out using the revised method of Clermont, assigning strains to 8 phylogenetic groups [18].

Selection of strains for Sequencing

Strains for sequencing were selected based initially on the presence or absence of nine known plasmid virulence genes (*ompTp*, *hlyF*, *cvaC*, *etsA*, *iutA*, *iroN*, *tsh*, *iss*, *and sitC*) reported in previous studies as indicative of virulence plasmids and which may play a role in pathogenesis [9, 19].

To be considered virulence plasmids in this analysis, plasmids must contain at least two known plasmid virulence genes. Candidate plasmid containing strains were further refined by the number of virulence genes detected by PCR. Plasmids containing variable patterns of the previous nine aforementioned virulence genes were selected for sequencing. When presented with multiple strains possessing the same pattern of virulence genes, strains originating from different serogroups were selected followed by, phylogenetic groups, and capsule types in order to increase likelihood of sequencing genetically distinct plasmids.

Plasmid Isolation

The selected strains were grown overnight in 5ml of Luria Bertani (LB) broth, the bacterial cells pelleted by centrifugation, and their plasmid DNA isolated using Qiagen midi plasmid prep kits according to manufacturer's specifications (Venlo, Netherlands). The plasmid DNA pellet was re-suspended in Qiagen buffer EB. Plasmid DNA concentration was measured using a NanoDrop 1000 (Thermo Scientific, Wilmington, DE).

Plasmid separation was achieved using pulse field gel electrophoresis (PFGE) as described by Tivendale et al (6). PFGE was carried out using a CHEF Mapper XA system (BioRad Hercules, California). Plasmids were separated in 1.0% w/v Seakem Gold (Lonza, Bessel, Switzerland) agarose in a 0.5X TBE buffer (Invitrogen Carlsbad, CA). Size controls used were pAPEC-O1-ColBM, pAPEC-O2-ColV, and APECχ7112. Electrophoresis was carried out at 6v/cm with an angle of 120° for 20h with an initial switch time of 1s and a final switch time of 20s [20]. Following electrophoresis, the gels were stained in distilled water containing 0.5mg/ml of ethidium bromide and de-stained in water as necessary. All plasmids were visualized using UV transillumination. Individual plasmid bands were identified and excised using a clean straight razor and placed in 1.5 ml tubes. The plasmids were dissolved in Qiagen buffer QG in an incubator-shaker at 37°C for 30 minutes to remove them from the agarose. The dissolved gel fragments were subjected to sonication to fragment DNA below the maximum threshold for Qiagen column purification using a Diagenode Bioruptor Standard (Diagenode Seraing, Belgium) at high power for 30 seconds on and off time over a 5 minute period. The samples were kept on ice before, during, and post sonication. Post sonication, DNA samples were purified as per the Qiagen protocol for PCR purification using manufacturer's specifications and eluted from purification columns using 11ul of Qiagen buffer EB. After purification, plasmids were quantified using the Q-bit 2.0 Fluorometer using the q-bit HS assay kit (Life Technologies Carlsbad, CA) according to manufacturer's specifications.

Plasmid Sequencing

Paired-end plasmid DNA samples were prepared for sequencing by the Dana Farber/Harvard Cancer Center DNA Resource Core using the Nextera DNA sample preparation kit (Illumina San Diego, CA) according to manufacturer's specifications for indexed multiplexing using 50ng of input DNA. Input DNA was fragmented and tagged with the transposon, adding adapter sequences and allowing for amplification. Tagmented DNA was cleaned to remove any leftover transposon fragments. Purified DNA was then PCR amplified for 8 cycles and purified of enzymes and DNTPs. Library adapter ligation and amplification were tested by qPCR, and libraries were pooled. Library fragments in the size range of 500 to 600bp were selected using a Pippin Prep from Sage Science (Beverly, MA) according to manufacturer's specifications. Sequencing of the tagged plasmids was carried out using an Illumina Mi-Seq machine (Illumina San Diego, CA) using a single lane. Output files ranged in size from 1 to 5 GB and were quality trimmed based on the FASTq PHRED quality scores, cutting off at Q25 and imported into the assembly software. Velvet Optimizer 2.1.4 was used to determine optimal k-mer length and coverage cutoff. Plasmid assembly was achieved using Velvet short read assembler (version 1.2.03; European Bioinformatics Institute) for final assembly [21]. In general, most plasmids assembled into one to three large contigs with few small contigs or contaminating sequences. Plasmid sequences requiring closure were closed using the previously sequenced pS88 [22] as a backbone against which to assemble small contigs left over from the Velvet assembly. Plasmid annotation was carried out for NMEC strains sequenced in this study and elsewhere using the Prokka automated annotation analysis pipeline in order to ensure consistency in our data. The plasmids were annotated using the "-usegenus and–strain" options with a custom database generated from all APEC and NMEC plasmids in the NCBI database.

Plasmid Core and Accessory Genome Analysis

The plasmid genomes of the fully sequenced NMEC plasmids from this study and those available online (11) were entered into an NCBI database, where an all-versus-all blast was performed to ascertain orthologous gene clusters (COGs). Genes showing >90% sequence identity, e-values of lower that 1e-10, and overlap percentages of >90% were determined to be orthologous. Orthologous genes were then sorted into strongly connected components using the algorithm of Tarjan in Perl [23]. Strongly connected components where an orthologous gene was present in each plasmid were grouped as members of a core gene set, while genes present in at least one genome were said to be members of the accessory genome.

Plasmid Phylogenetic Analysis

Phylogenetic analysis was performed for the core genome of the NMEC and two APEC plasmids in order to compare different plasmids from different ExPEC sub-pathotypes. Core genome nucleotide cDNA sequences were extracted from files, and concatenated to form a mega alignment using custom Perl scripts. ClustalOmega [24] was used to align the nucleotide sequences and using default nucleotide settings and interleaved PHYLIP output. The PHYLIP alignment was then used in seqboot from the PHYLIP software package to create 100 bootstrapped replacements in the tree, using the–m weights option. This set of 100 weights was then used in the dnaml program from the PHYLIP software package to create a bootstrapped maximum likelihood tree using the parameters T = 2.5, S, m,1. The bootstrapped trees were then condensed using the consense program from the PHYLIP software package to give a consensus tree with bootstrap values using default parameters. This tree was then reanalyzed in dnaml using the options–u,–t 2.5, to redraw branch lengths assuming the phylogeny was represented by the bootstrap inferred tree.

Plasmid Sequencing Data Accession Numbers

Plasmid sequences for and were deposited in GenBank under accession numbers SAMN03580091 (pNMEC-14-ColV), SAMN03580092 (pNMEC-15-ColV), SAMN03580093 (pNMEC-16-ColV),

SAMN03580094 (pNMEC-19-ColV), SAMN03580095 (pNMEC-26-ColV), SAMN03580096 (pNMEC 36-ColV), SAMN03580097 (pNMEC 38-ColV), SAMN03580098 (pNMEC-49), and SAMN03580099 (pNMEC-58-ColV).

RESULTS

Strain Selection, Purification and Sequence Assembly

A total of 11 plasmids was sequenced from nine NMEC strains, each isolated from different patients. Ten of eleven plasmid-containing strains were of the phylogroup B2, and the dominant serogroup was O18:K1 (Table 1). The source strains of these plasmids were previously assigned to clusters as described elsewhere [13] and were chosen for this study based on the diversity of the host strain serogroup, phylogroup, different virulence gene combinations, and cluster analysis.

Multiple plasmids were recovered from some of the NMEC strains: NMEC 15 and NMEC 36 both carried smaller cryptic plasmids. NMEC 15 contained a 114,753bp cryptic plasmid containing hypothetical and phage genes, while NMEC 36 carried a smaller 41,394bp plasmid. These plasmids contained a large number of hypothetical genes, as well as phage-related genes. Since we wished to focus our study on the plasmids in NMEC that serve a role in virulence and that are considered to be one of the defining traits of the NMEC subpathotype, all plasmids to be sequenced had to contain at least two plasmid-borne virulence genes from a list of nine commonly associated with large ExPEC virulence plasmids (*ompTp*, *hlyF*, *cvaC*, *etsA*, *iutA*, *iroN*, *tsh*, *iss*, *sitC*). Imposition of this selection criterion ensured that only plasmids containing some virulence associated genes would be included, and that cryptic or other plasmids were excluded.

Nine of the 11 plasmids sequenced in this study passed quality control and selection criteria following annotation. Two plasmids from strains NMEC 49 and NMEC 84 had a lower depth of coverage, and contained segments of contaminating DNA, resulting in poor assembly and their subsequent exclusion from analysis [Data not shown]. The plasmids pNM15c and pNM36c both contained fewer than three of the virulence associated genes. Further inspection of these plasmids showed that the majority of the plasmid genome consisted of hypothetical genes, phage genes, and genes of unknown function, and included no known virulence genes, and thus were also excluded from the analysis [Data not shown]. The resulting nine sequenced plasmids were combined with publically available NMEC plasmid sequences meeting the defined criteria described above and included the plasmids pS88 and pS286.

ExPEC-like NMEC Virulence Plasmid Accessory Genome

The virulence plasmid "accessory genome" of these 11 NMEC plasmids is defined as any COGs or single genes not found in the core genome, but present in at least one plasmid. COGs from the accessory genome identified 123 putative genes present in at least two genomes, and 197 putative genes unique to only one plasmid. Each gene's description was derived from the consensus of the genes' annotations. The accessory genome consisted of 142 genes appearing in at least two plasmids and 92 non-orthologous genes, appearing in only one plasmid (Fig 1). A heat map of the 142 genes present in two or more genomes is found.

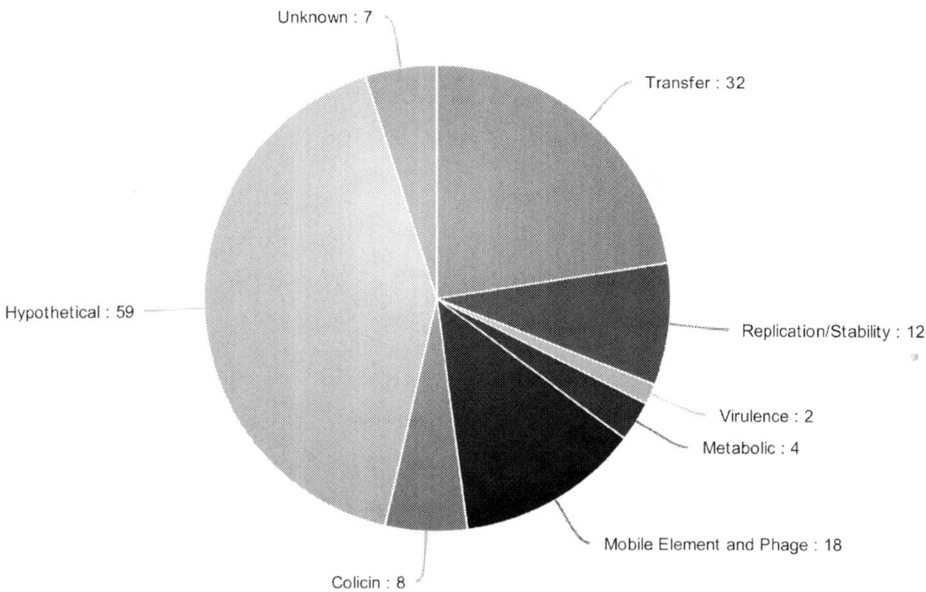

Figure 1: ExPEC-like Plasmid Accessory Genome Functions.

Genes of the ExPEC-like NMEC plasmid accessory genome organized into functional categories based on protein prediction categories.

ExPEC-like NMEC virulence Plasmid Core Genome

To ascertain the range of extant genes in the plasmids of NMEC strains, all sequenced and annotated NMEC virulence plasmids passing selection criteria and two previously sequenced plasmids pS88 [22] and pS286 [25] were concatenated into a cDNA BLAST [26] database. An "all versus all" blast analysis was performed to assign genes to clusters of orthologous groups (COGs). COGs in which an ortholog from each plasmid was found were considered to be "core plasmid" genes and were further subjected to protein function analysis using the CombFunc pipeline [18]. Genes in the core genome were then categorized into eight functional categories based on known function or results from the combFunc pipeline (Table 2). A preliminary core genome of 48 genes was predicted but was reduced to 38 genes by removal of insertion sequences, transposases, and phage genes (Fig 2). The core genome genes can be divided into five categories by putative functions: virulence associated; metabolic and signaling; unknown; plasmid stability and transfer; and mobile elements (Table 2).

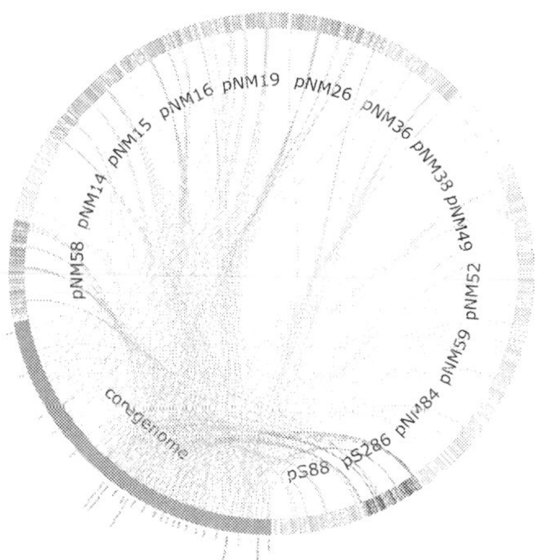

Figure 2: Core Plasmid Phylogeny.

Visualization of ExPEC-like NMEC plasmid core genome sites. Genes within the plasmid core genome are listed in the orange core genome section, while linear representations of the chromosomes of the NMEC plasmids are arranged in the circle. Lines are drawn to the position of the core plasmid genes in their location within the plasmid to show core gene clustering

Table 2: Core genes of ExPEC-like plasmids in NMEC

Gene Name	Description	Function
hlyF	Hemolysin	**Virulence**
ompT	Outer membrane protease	
Bor	bacteriophage lambda bor protein	
iroB	putative glucosyltransferase	
iroC	ATP binding cassette ABC transport homolog	
iroD	putative ferric enterochelin esterase	
iroE	putative hydrolase	
iron	outer membrane receptor fepA	
iucA	aerobactin siderophore biosynthesis protein	
iucB	N(6)-hydroxylysine acetylase	
iucC	aerobactin siderophore biosynthesis protein	
iucD	L-lysine 6 monooxigenase	
iutA	ferric aerobactin receptor precursor	
sitA	iron/manganese transport protein, periplasmic-binding protein	
sitB	iron/manganese transport protein ATP-binding component	
sitC	Iron/manganese transport inner membrane component	
sitD	iron/manganese transport protein, inner membrane component	
putative cobalamin synthase	Putative Cobalamin synthase	**Metabolic and Signaling**
ssbL	DHAP synthase for shikimate pathway	
truncated enolase	putative enolase fragment	
crcB	supercoiling and camphor resistance protein	**Plasmid Stability and Transfer**
repA	RepFIB replication protein	
repA1	Plasmid replication protein	
repA2	Plasmid replication Protein	
traA	conjugal transfer pilin subunit	
traE	conjugal transfer pilus assembly protein	
traL	conjugal transfer pilus assembly protein	
yacB	putative plasmid stabalization system protein	
finO	conjugal transfer fertility inhibition protein	
hypothetical mobile element	putative mobile element protein/integrase	**Mobile Elements**
hypothetical phage protein	hypothetical phage protein	
hypothetical transposase	hypothetical transposase	
orfA	transposase from plasmid origin	
hypothetical protein	DNA Binding Function	
hypothetical protein	DNA polymerase	
hypothetical protein	Helicase activity	
Int	Hypothetical integrase protein	
etsA	putative type I secretion membrane-fusion protein	**Unknown**
etsB	putative type I secretion ATP binding protein	
etsC	putative type I secretion outer membrane protein	
shiF	putative MFS family membrane transport protein	
hypothetical protein	Unknown protein	
hypothetical protein	Unknown protein	
yubQ	putative transglycolylase	
hypothetical siderophore homolog	Putative Siderophore	
mig-14	putative Mig-14 protein	

doi:10.1371/journal.pone.0147757.t002

Core plasmid genome of sequenced plasmids and their putative functions.

The virulence-associated genes of the NMEC plasmid core included all genes of the aerobactin (*iutA/iucABCD*), *sit* (*sitABCD*), and salmochelin (*iroBCDEN)* operons. All three of these operons encode high-affinity iron-transport systems that are used by bacteria to obtain iron in low iron conditions such as those they encounter in host fluids and tissues. These operons have previously been reported in virulence plasmids of uropathogenic *Escherichia coli* (UPEC), APEC, and NMEC with high frequency and have been associated with ExPEC virulence. Another gene found in the core genome of NMEC large virulence plasmids is *iss*. This gene encodes a protein linked with increased serum survival in human *E. coli* isolates. Numerous studies have documented its strong alignment with virulent but not avirulent *E. coli* strains [27]. In addition, the genes *ompT* and *hlyF* are also found in the core genome of APEC's large virulence plasmids. *ompT* is predicted to encode a 42-kDa pro-protein, which is processed in the membrane to a 40k-Da mature form. Mature Iss functions as a narrowly specific outer membrane endoprotease [26] that 1) cleaves paired basic residues and is involved in membrane protein turnover, 2) can degrade interferon-gamma *in-vitro* [28], and 3) cleaves the human defensin LL-37 [29]. The *hlyF* gene is predicted to encode a putative hemolysin gene; however, the exact function of this gene is unknown.

Metabolic genes of the NMEC plasmid core genome consists of three genes, *ssbL*, *eno* and a cobalamin synthesis gene. *ssbL* is a truncated enolase gene and a putative cobalamin (coenzyme B12) synthase gene. *ssbL* encodes a phospho-2-dehydro-3-deoxyheptonate aldolase or DHAP synthase. It is the first enzyme in a series of metabolic reactions known as the shikimate pathway, leading to the production of chorismate, which is responsible for the biosynthesis of the amino acids phenylalanine, tyrosine, and tryptophan and catecholate/phenolate siderophores. This gene co-locates with the *iro* operon in ExPEC plasmids [30]. The putative enolase gene found in the plasmid core genome is a very small, comprising 33% of the *eno* gene from S88 at the C-terminus at 51% identity. The *eno* gene catalyzes conversion of 2-phosphoglycerate to phosphoenolpyruvate (PEP) [31]. PEP is another reactant in the shikimate pathway for aromatic amino acid synthesis [32]. Finally, a putative cobalamin synthesis gene was also found. This putative protein contains two domains homologous to cobalamin synthesis protein p47 metal binding motifs which may act as chapcrones in metal binding. Previous studies of APEC virulence have shown several vitamin synthesis pathways are crucial for APEC's *in vivo* survival [33].

Nine genes comprise the third major gene group of the NMEC plasmid core genome. These genes are responsible for plasmid transfer and stability and include replication genes (*repA*, *repA1 and repA2*), as well as *finO*, encoding an anti-sense RNA molecule inhibiting *traJ* [34] and the supercoiling and camphor-resistance gene *crcB*. The replication genes are responsible for control of plasmid copy number in cells and prevent replication by binding to plasmid sites, thereby inhibiting DNA replication. This inhibition can be reversed by high *dnaG* expression [35–37]. Notably absent from this list are many of the *tra* genes from the type IV secretion system that responsible for plasmid conjugal transfer. They do not show up in the core because they are not present in the plasmid pS286. However, a case can be made for their inclusion since they occur in all others suggesting that pS286 is atypical, incomplete or aberrant. Blast sequence comparison between *crcB* found in this study and all other *E. coli* found that the *crcB*genes of NMEC plasmids and UPEC strains UCI57, UCI58, KTE76, KTE102, KTE240, and *E. coli* 908585 are distinct from the *crcB* gene found in other *E. coli*, showing only 53% amino acid identity. Such differences suggest the possibilities of different binding ligands or function. Upon initial discovery, CrcB was indicated in plasmid supercoiling and resistance to camphor when overexpressed [38]. More recent studies show CrcB to be a putative membrane protein containing a metabolite-binding RNA structure, indicated to be fluoride in these studies, which can alter downstream gene expression [39].

The NMEC plasmid core also includes mobile genetic elements and phage insertions. Mobile element genes consists of phage genes, as well as transposable elements and integration proteins. These elements recognize specific DNA sequences in the plasmid, and can cause DNA transposition for incorporating virulence genes into pathogenicity islands.

The NMEC plasmid core also includes genes of unknown function. The *etsABC* operon encodes a putative ABC transport system. It is predicted that, *etsA* encodes a putative membrane fusion protein, *etsB* encodes ATP-binding/permease protein, and *etsC* encodes an efflux protein. These genes are strongly associated with NMEC pathogenicity islands [13]. The *shiF* gene is a putative member of the COG0477 permease family. The *shiF* gene homolog lies upstream of aerobactin synthesis genes; no evidence that its function is related to this aerobactin system has been found in *E. coli*. The gene *yubQ* encodes an X-polypeptide. It is a protein homologous to lytic transglycosylases (LT) of family four, thought to originate from phage [40]. This class of proteins in general is thought to act on the cell wall, degrading peptidoglycan into metabolites which can be recycled in the cytoplasm. Most *E. coli* contain approximately seven distinct LT encoding proteins with unknown functional differences. LT proteins are essential genes in *E. coli*, deletion of all LT genes results in a lethal phenotype. There are two hypothetical genes of unknown function that encode no known protein domains and the combFunc pipeline was unable to predict any functions of these proteins

Phylogenetic Analysis

The NMEC core plasmid genome was used to develop an ExPEC plasmid phylogeny. To relate the evolution of these plasmids to other *E. coli* plasmids, three APEC plasmids, pAPEC-O1-ColBM, pAPEC-O2-ColV and pAPEC-O103 were included in the analysis. These plasmids meet the same three criteria for inclusion as described above and were included in the analysis to better understand differences between plasmids from different ExPEC subpathotypes. However, UPEC plasmids were not included in this analysis as no publically available sequenced UPEC plasmids met inclusion criteria. The APEC/NMEC core plasmid genome consisted of 15 genes (21,496bp), containing the *iro* and *iut* operons as well as replication proteins *repA* and *repB*, which was used to generate a plasmid phylogeny (Fig 3). Additionally, one outlier was observed in this analysis: pAPEC-O103, which is a hybrid plasmid containing a pathogenicity island and multi-drug resistance island and lacking the complete *tra* operon. As a result of this analysis, it appears that there are at least two lineages of NMEC virulence plasmids. In addition, as a result of the comparative analysis, it appears that APEC and NMEC plasmids cannot be differentiated solely based on the presence or absence of genes in a core genome of plasmid.

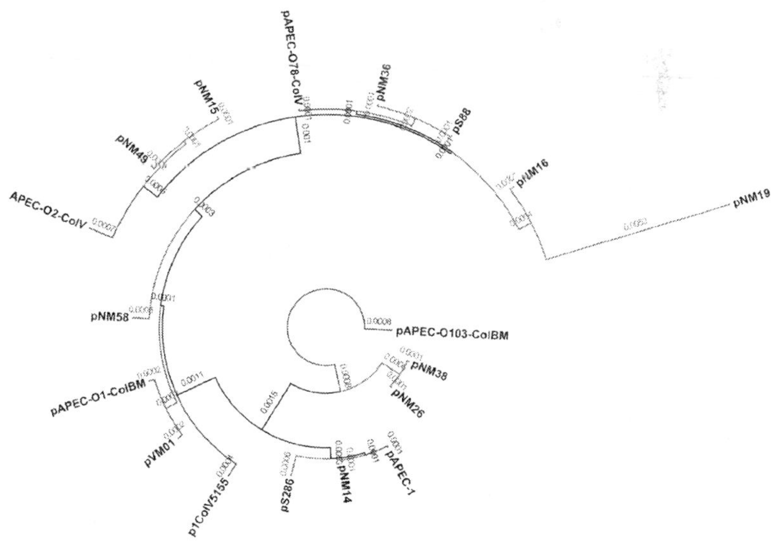

Figure 3: ExPEC-like plasmid phylogeny.

Phylogeny of core plasmid genes of ExPEC-like NMEC virulence plasmids and APEC plasmid genes. Alignment was based on concatenation of core genes shared between APEC and NMEC plasmids. Genes were identified using an all-versus-all nucleotide blast using the blastn algorithm using default parameters. Blast matches sorted into strongly connected components using Tarjan's algorithm. Nucleotide sequences from strongly connected groups were extracted in a randomized order and concatenated. Bootstrap analysis was performed using Phylip seqboot algorithm creating 10 sets of 10 weights using default parameters, which were then used in 10 separate runs of dnaml using the parameters T 2.5, S, M, 1. Phylip consense was used to condense all output trees into a consensus bootstrap analysis using default parameters

DISCUSSION

The contributions made to ExPEC pathogenesis by large virulence plasmids has received increased scrutiny in recent years. In part, this scrutiny is due to creation of the reagents needed for their study, such as plasmid genomic sequences [7, 11, 22, 30], and dissemination of new plasmid isolation protocols that enabled harvesting of large, naturally occurring plasmids from wild-type bacterial strains [41]. Previous to that the presence of large ExPEC plasmids was likely overlooked [8, 16]. Here, in an effort to define the core and accessory genome of NMEC plasmids, nine new plasmids isolated from NMEC were sequenced. Clusters 5 and 9 contained only a single isolate, while clusters 2 and 3 primarily consisted of human fecal *E. coli*biasing the strain selection towards groups 6–8, which were determined to be the most pathogenic and most common clusters for NMEC in the analysis [13]. Efforts were made to sequence strains from different serogroups where multiple options existed, attempting to capture the dominant O-types of O18, O83, and O7, while still sequencing representatives from less common serogroups, such as O14, O21, as well as untypeable NMEC. Selecting for strains outside of the predominant B2 phylogroup is difficult, as our collection of 94 NMEC isolates contains are predominantly B2 strains (78%), with the next most common phylogenetic group being F, representing 6.8% of the isolates in the full 91 sample collection. Finally, K type was used if all other parameters from the analysis were the same, which included one strain of the K80 type in the analysis, as well as multiple untypeable capsules. The previously published NMEC plasmids pRS218 and pCE10 were excluded from the analysis due to the absence of all virulence genes used to define a virulence plasmid in this study. Further analysis of the chromosomes of strain CE10 contained the *iutABCD* and *sitABCDC* genes present in the chromosome, while RS218 did not possess any of these plasmid virulence genes on the plasmid or the chromosome.

Assembled plasmids varied largely in terms of size, but surprisingly had similar GC content and numbers of open reading frames, indicating that a large proportion of these plasmids may not encode functional proteins. This seems to be borne out by the significant number of hypothetical genes and pseudo genes in the accessory genome of the NMEC plasmids. It is unclear whether this result suggests that some plasmids are accumulating genetic information that is not used, or whether they have not yet shed excess genetic information that may not be functional or whether these hypothetical or pseudo genes have an unknown function. Genes related to antibiotic resistance were not found in any of these plasmids with only one plasmid harboring a single putative antibiotic resistance gene, distinct from the APEC hybrid plasmid pAPECO103 encoding both resistance and virulence genes [42]. The lack of resistance genes is corroborated by the lack of multiple drug resistant strains in our NMEC collection [43], as well as those available publically for other NMEC strains. Furthermore a complete and intact transfer region was not found on all of the NMEC plasmids, suggesting some loss of the transmissible capability during the evolution course.

The core genome of the NMEC plasmids contains 47 genes, most of which encode components of iron uptake systems. Iron uptake systems are necessary for the full virulence of septicemia associated *E. coli* [7, 13, 22], uropathogenic *E. coli [44]*, and avian pathogenic *E. coli* [14, 45–47]. Although it has not yet been reported, this study suggests that iron acquisition is important for NMEC virulence since human serum and cerebrospinal fluid (CSF) are also iron-deficient environments. Indeed, the presence of iron acquisition genes in NMEC isolates has been well established [7, 13] and NMEC, UPEC, and APEC share a similar content of iron uptake systems that are not found in non-virulent fecal *E. coli* strains from both human and animal sources [13]. Moreover, unpublished data from our group has found that the expression of several iron uptake systems was significantly upregulated when NMEC were cultured in human serum and CSF compared to strains cultured in LB (unpublished data). The data presented here indicate that three iron uptake systems the ferric aerobactin system (*iutA/iucABCD*), the salmochelin siderophore system (*iroBCDEN)*, and SitABCD system, have been associated with ExPEC virulence, and are encoded by core the genome of NMEC large virulence plasmids. Interestingly, the metabolism gene (*ssbL*) found in the core genome of NMEC virulence plasmids appears to be co-located with these iron genes. This *ssbL* gene encodes a DHAP synthase which may boost production via the shikimate pathway of aromatic amino acids and are necessary for large amount production of enterobactin, salmochelin SX, and yersiniabactin [30]. The finding of *ssbL* gene in the core genome of the NMEC virulence plasmid appears to reinforce the importance of the iron uptake system for NMEC virulence.

Based on the findings published thus far, NMEC's ability to cause meningitis requires at least two major steps and one of them is intravascular multiplication leading to high-level bacteremia. Thus, NMEC must be able to resist or avoid the detrimental impact of the host complement, antimicrobial peptides, polymorphonuclear leukocytes, and other specific and innate immunity factors. In the core genome of NMEC virulence plasmids, the *iss* gene was found, which is homologous to the *bor* gene of bacteriophage λ. The Iss protein increases *E. coli*'s serum survival likely by restricting C3 deposition on the bacterial surface [27, 48] and blockage of the membrane attack complex [49]. Another core genome gene of the NMEC virulence plasmid involved in resistance to innate immunity is *ompT*. The *ompT* gene encodes an outer membrane protease [26] that can degrade interferon-gamma. More recent studies on *ompT* in enteropathogenic *E. coli* (EPEC) and enterohaemorrhagic *E. coli* (EHEC) showed *ompT*cleaves the human defensin LL-37 [29], a pore forming antimicrobial peptide expressed by neutrophils, epithelial cells, and bone marrow that serves both antimicrobial and immunomodulatory functions. The *ompT* gene in the core genome of NMEC virulence plasmids shares approximately 74% of amino acid identity to those encoded by EPEC and EHEC. Thus, it could be inferred that OmpT can degrade interferon and human antimicrobial peptide increasing NMEC's resistance to human innate immunity, thus contributing to NMEC's pathogenesis. In addition, several other accessory genome genes of NMEC virulence plasmids were found to contribute to NMEC's resistance to host immunity. The gene *tsh*, which has been previously reported to be an important virulence factor for closely related APEC strains and is present in 11 to 50 percent of NMEC samples, encodes a protein able to cleave leukocyte glycoproteins, impair chemotaxis and transmigration, and activate leukocytes' programmed cell death [50].

The homolog of CrcB was shown to be involved in the chromosome condensation and plasmid supercoiling, causing a secondary effect of resistance to camphor, which decreases chromosome condensation [38]. Overexpression of the gene increases chromosomal condensation and can correct nucleoid morphology defects in some mutants, and may help these large plasmids partition into daughter cells. As we know the *E. coli* chromosome, like its eukaryotic counterparts, must be condensed approximately 2000-fold to fit inside an *E. coli* cell. Given the large size of the virulence plasmids for NMEC, it is likely that the *crcB* gene is carried on the plasmid is necessary to fit the plasmid into the cells, and may increase the likelihood of the plasmid successfully passing to daughter cells post replication.

In conclusion, this is the first study to define the core and accessory genome of ExPEC-like NMEC plasmid, and perform a phylogenetic analysis. The role of the core genome of NMEC plasmids is primarily to increase the bacteria's *in-vivo* fitness by enhanced uptake of iron in a deficient environment, and resistance to host innate immunity as well as maintain the plasmid in the bacterial population with systems for conjugation and plasmid stability. Phylogenetic analysis of the NMEC plasmids shows that at least two lineages can be discerned and they are nearly indistinguishable from those of the APEC plasmid. This finding supports previous research inferring close genetic relationships between APEC and NMEC plasmids and overlap in a large number of pathogenicity islands [9, 51, 52]. We hope this and future studies lead to a better understanding of the molecular mechanisms by which NMEC large plasmids contribute to virulence and their evolution.

ACKNOWLEDGMENTS

We thankfully acknowledge Kwang S. Kim (Johns Hopkins University), Chitrita DebRoy (E. coli reference center Pennsylvania State University) and Lodewijk Spanjaard (Netherlands Reference Laboratory for Bacterial

Meningitis) for their gift of NMEC strains. Additionally we would like to thank Brett J. Kail his assistance with programming.

AUTHOR CONTRIBUTIONS

Conceived and designed the experiments: LKN CML GL YW NB BAN. Performed the experiments: BAN YW. Analyzed the data: BAN ACW. Contributed reagents/materials/analysis tools: ACW PM BAN. Wrote the paper: BAN CML LKN GL.

REFERENCES

1. Salipante SJ, Roach DJ, Kitzman JO, Snyder MW, Stackhouse B, Butler-Wu SM, et al. Large-scale genomic sequencing of extraintestinal pathogenic Escherichia coli strains. Genome research. 2015;25(1):119–28. pmid:25373147; PubMed Central PMCID: PMCPMC4317167.
2. Jaureguy F, Landraud L, Passet V, Diancourt L, Frapy E, Guigon G, et al. Phylogenetic and genomic diversity of human bacteremic Escherichia coli strains. BMC genomics. 2008;9:560. pmid:19036134; PubMed Central PMCID: PMCPMC2639426.
3. Martin GS, Mannino DM, Eaton S, Moss M. The epidemiology of sepsis in the United States from 1979 through 2000. N Engl J Med. 2003;348(16):1546–54. Epub 2003/04/18. pmid:12700374.
4. Grimwood K, Anderson P, Anderson V, Tan L, Nolan T. Twelve year outcomes following bacterial meningitis: further evidence for persisting effects. Archives of disease in childhood. 2000;83(2):111–6. Epub 2000/07/25. pmid:10906014; PubMed Central PMCID: PMC1718445.
5. Kaper JB, Nataro JP, Mobley HLT. Pathogenic Escherichia coli. Nature Reviews Microbiology. 2004;2(2):123–40. pmid:15040260
6. Croxen MA, Finlay BB. Molecular mechanisms of Escherichia coli pathogenicity. Nat Rev Micro. 2010;8(1):26–38. http://www.nature.com/nrmicro/journal/v8/n1/suppinfo/nrmicro2265_S1.html.
7. Wijetunge DS, Karunathilake KH, Chaudhari A, Katani R, Dudley EG, Kapur V, et al. Complete nucleotide sequence of pRS218, a large virulence plasmid, that augments pathogenic potential of meningitis-associated Escherichia coli strain RS218. BMC

microbiology. 2014;14:203. Epub 2014/08/29. pmid:25164788; PubMed Central PMCID: PMC4155114.

8. Johnson TJ, Wannemuehler YM, Johnson SJ, Logue CM, White DG, Doetkott C, et al. Plasmid replicon typing of commensal and pathogenic Escherichia coli isolates. Applied and environmental microbiology. 2007;73(6):1976–83. Epub 2007/02/06. pmid:17277222; PubMed Central PMCID: PMC1828809.
9. Tivendale KA, Logue CM, Kariyawasam S, Jordan D, Hussein A, Li G, et al. Avian-pathogenic Escherichia coli strains are similar to neonatal meningitis E. coli strains and are able to cause meningitis in the rat model of human disease. Infection and immunity. 2010;78(8):3412–9. pmid:20515929; PubMed Central PMCID: PMCPMC2916289.
10. Skyberg JA, Johnson TJ, Johnson JR, Clabots C, Logue CM, Nolan LK. Acquisition of avian pathogenic Escherichia coli plasmids by a commensal E. coli isolate enhances its abilities to kill chicken embryos, grow in human urine, and colonize the murine kidney. Infection and immunity. 2006;74(11):6287–92. Epub 2006/09/07. pmid:16954398; PubMed Central PMCID: PMC1695531.
11. Cusumano CK, Hung CS, Chen SL, Hultgren SJ. Virulence plasmid harbored by uropathogenic Escherichia coli functions in acute stages of pathogenesis. Infection and immunity. 2010;78(4):1457–67. pmid:20123719; PubMed Central PMCID: PMCPMC2849428.
12. Johnson TJ, Nolan LK. Pathogenomics of the virulence plasmids of Escherichia coli. Microbiology and molecular biology reviews: MMBR. 2009;73(4):750–74. Epub 2009/12/01. pmid:19946140; PubMed Central PMCID: PMC2786578.
13. Logue CM, Doetkott C, Mangiamele P, Wannemuehler YM, Johnson TJ, Tivendale KA, et al. Genotypic and phenotypic traits that distinguish neonatal meningitis-associated Escherichia coli from fecal E. coli isolates of healthy human hosts. Applied and environmental microbiology. 2012;78(16):5824–30. pmid:22706051; PubMed Central PMCID: PMCPMC3406136.
14. Johnson TJ, Wannemuehler Y, Doetkott C, Johnson SJ, Rosenberger SC, Nolan LK. Identification of minimal predictors of avian pathogenic Escherichia coli virulence for use as a rapid diagnostic

tool. Journal of clinical microbiology. 2008;46(12):3987–96. Epub 2008/10/10. pmid:18842938; PubMed Central PMCID: PMC2593276.

15. Clermont O, Bonacorsi S, Bingen E. Rapid and simple determination of the Escherichia coli phylogenetic group. Applied and environmental microbiology. 2000;66(10):4555–8. Epub 2000/09/30. pmid:11010916; PubMed Central PMCID: PMC92342.
16. Rodriguez-Siek KE, Giddings CW, Doetkott C, Johnson TJ, Fakhr MK, Nolan LK. Comparison of Escherichia coli isolates implicated in human urinary tract infection and avian colibacillosis. Microbiology. 2005;151(Pt 6):2097–110. Epub 2005/06/09. pmid:15942016.
17. Johnson JR, Brown JJ. A novel multiply primed polymerase chain reaction assay for identification of variant papG genes encoding the Gal(alpha 1–4)Gal-binding PapG adhesins of Escherichia coli. The Journal of infectious diseases. 1996;173(4):920–6. Epub 1996/04/01. pmid:8603972.
18. Clermont O, Christenson JK, Denamur E, Gordon DM. The Clermont Escherichia coli phylo-typing method revisited: improvement of specificity and detection of new phylo-groups. Environmental microbiology reports. 2013;5(1):58–65. Epub 2013/06/13. pmid:23757131.
19. Wijetunge DS, Gongati S, DebRoy C, Kim KS, Couraud PO, Romero IA, et al. Characterizing the pathotype of neonatal meningitis causing Escherichia coli (NMEC). BMC microbiology. 2015;15(1):211. pmid:26467858; PubMed Central PMCID: PMCPMC4606507.
20. Tivendale KA, Allen JL, Browning GF. Plasmid-borne virulence-associated genes have a conserved organization in virulent strains of avian pathogenic Escherichia coli. Journal of clinical microbiology. 2009;47(8):2513–9. Epub 2009/06/26. pmid:19553581; PubMed Central PMCID: PMC2725677.
21. Zerbino DR, Birney E. Velvet: algorithms for de novo short read assembly using de Bruijn graphs. Genome research. 2008;18(5):821–9. Epub 2008/03/20. pmid:18349386; PubMed Central PMCID: PMC2336801.
22. Peigne C, Bidet P, Mahjoub-Messai F, Plainvert C, Barbe V, Médigue C, et al. The Plasmid of Escherichia coli Strain S88

(O45:K1:H7) That Causes Neonatal Meningitis Is Closely Related to Avian Pathogenic E. coli Plasmids and Is Associated with High-Level Bacteremia in a Neonatal Rat Meningitis Model. Infection and immunity. 2009;77(6):2272–84. pmid:19307211

23. Tarjan R. Depth-First Search and Linear Graph Algorithms. SIAM Journal on Computing. 1972;1(2):15.
24. Sievers F, Wilm A, Dineen D, Gibson TJ, Karplus K, Li W, et al. Fast, scalable generation of high-quality protein multiple sequence alignments using Clustal Omega. Molecular systems biology. 2011;7:539. pmid:21988835; PubMed Central PMCID: PMCPMC3261699.
25. Lemaitre C, Mahjoub-Messai F, Dupont D, Caro V, Diancourt L, Bingen E, et al. A conserved virulence plasmidic region contributes to the virulence of the multiresistant Escherichia coli meningitis strain S286 belonging to phylogenetic group C. PloS one. 2013;8(9):e74423. pmid:24086343; PubMed Central PMCID: PMCPMC3784414.
26. Guindon S, Delsuc F, Dufayard JF, Gascuel O. Estimating maximum likelihood phylogenies with PhyML. Methods in molecular biology. 2009;537:113–37. Epub 2009/04/21. pmid:19378142.
27. Johnson TJ, Wannemuehler YM, Nolan LK. Evolution of the iss gene in Escherichia coli. Applied and environmental microbiology. 2008;74(8):2360–9. Epub 2008/02/19. pmid:18281426; PubMed Central PMCID: PMC2293169.
28. Grodberg J, Dunn JJ. Ompt Encodes the Escherichia-Coli Outer-Membrane Protease That Cleaves T7-Rna Polymerase during Purification. Journal of bacteriology. 1988;170(3):1245–53. pmid:ISI:A1988M260100031.
29. Thomassin JL, Brannon JR, Gibbs BF, Gruenheid S, Le Moual H. OmpT outer membrane proteases of enterohemorrhagic and enteropathogenic Escherichia coli contribute differently to the degradation of human LL-37. Infection and immunity. 2012;80(2):483–92. Epub 2011/12/07. pmid:22144482; PubMed Central PMCID: PMC3264287.
30. Lemaitre C, Bidet P, Benoist JF, Schlemmer D, Sobral E, d'Humieres C, et al. The ssbL gene harbored by the ColV plasmid

of an Escherichia coli neonatal meningitis strain is an auxiliary virulence factor boosting the production of siderophores through the shikimate pathway. Journal of bacteriology. 2014;196(7):1343–9. pmid:24443535; PubMed Central PMCID: PMCPMC3993347.

31. Klein M, Sprenger GA, Freudl R. Cloning, nucleotide sequence, and functional expression of the Escherichia coli enolase (eno) gene in a temperature-sensitive eno mutant strain. DNA Seq. 1996;6(6):351–5. pmid:8988374.
32. Tribe DE, Camakaris H, Pittard J. Constitutive and repressivle enzymes of the common pathway of aromatic biosynthesis in Escherichia coli K-12: regulation of enzyme synthesis at different growth rates. Journal of bacteriology. 1976;127(3):1085–97. pmid:8426; PubMed Central PMCID: PMCPMC232899.
33. Li G, Laturnus C, Ewers C, Wieler LH. Identification of genes required for avian Escherichia coli septicemia by signature-tagged mutagenesis. Infection and immunity. 2005;73(5):2818–27. Epub 2005/04/23. pmid:15845486; PubMed Central PMCID: PMC1087346.
34. Arthur DC, Ghetu AF, Gubbins MJ, Edwards RA, Frost LS, Glover JN. FinO is an RNA chaperone that facilitates sense-antisense RNA interactions. The EMBO journal. 2003;22(23):6346–55. pmid:14633993; PubMed Central PMCID: PMCPMC291848.
35. Ingmer H, Miller C, Cohen SN. The RepA protein of plasmid pSC101 controls Escherichia coli cell division through the SOS response. Molecular microbiology. 2001;42(2):519–26. Epub 2001/11/13. pmid:11703672.
36. Manen D, Upegui-Gonzalez LC, Caro L. Monomers and dimers of the RepA protein in plasmid pSC101 replication: domains in RepA. Proceedings of the National Academy of Sciences of the United States of America. 1992;89(19):8923–7. pmid:1409587; PubMed Central PMCID: PMCPMC50036.
37. Xia G, Manen D, Yu Y, Caro L. In vivo and in vitro studies of a copy number mutation of the RepA replication protein of plasmid pSC101. Journal of bacteriology. 1993;175(13):4165–75. pmid:8320230; PubMed Central PMCID: PMCPMC204846.

38. Sand O, Gingras M, Beck N, Hall C, Trun N. Phenotypic characterization of overexpression or deletion of the Escherichia coli crcA, cspE and crcB genes. Microbiology. 2003;149(Pt 8):2107–17. pmid:12904550.

39. Baker JL, Sudarsan N, Weinberg Z, Roth A, Stockbridge RB, Breaker RR. Widespread genetic switches and toxicity resistance proteins for fluoride. Science. 2012;335(6065):233–5. pmid:22194412; PubMed Central PMCID: PMCPMC4140402.

40. Scheurwater E, Reid CW, Clarke AJ. Lytic transglycosylases: bacterial space-making autolysins. The international journal of biochemistry & cell biology. 2008;40(4):586–91. pmid:17468031.

41. Williams LE, Detter C, Barry K, Lapidus A, Summers AO. Facile recovery of individual high-molecular-weight, low-copy-number natural plasmids for genomic sequencing. Applied and environmental microbiology. 2006;72(7):4899–906. pmid:16820486; PubMed Central PMCID: PMCPMC1489313.

42. Johnson TJ, Jordan D, Kariyawasam S, Stell AL, Bell NP, Wannemuehler YM, et al. Sequence analysis and characterization of a transferable hybrid plasmid encoding multidrug resistance and enabling zoonotic potential for extraintestinal Escherichia coli. Infection and immunity. 2010;78(5):1931–42. pmid:20160015; PubMed Central PMCID: PMCPMC2863545.

43. Johnson TJ, Wannemuehler Y, Kariyawasam S, Johnson JR, Logue CM, Nolan LK. Prevalence of avian-pathogenic Escherichia coli strain O1 genomic islands among extraintestinal and commensal E. coli isolates. Journal of bacteriology. 2012;194(11):2846–53. Epub 2012/04/03. pmid:22467781; PubMed Central PMCID: PMC3370642.

44. Garcia EC, Brumbaugh AR, Mobley HL. Redundancy and specificity of Escherichia coli iron acquisition systems during urinary tract infection. Infection and immunity. 2011;79(3):1225–35. pmid:21220482; PubMed Central PMCID: PMCPMC3067483.

45. Ewers C, Janßen T, Kießling S, Philipp H-C, Wieler LH. Molecular epidemiology of avian pathogenic Escherichia coli (APEC) isolated from colisepticemia in poultry. Veterinary Microbiology. 2004;104(1–2):91–101. pmid:15530743

46. La Ragione RM, Woodward MJ. Virulence factors of Escherichia coli serotypes associated with avian colisepticaemia. Res Vet Sci. 2002;73(1):27–35. pmid:12208104.
47. Rodriguez-Siek KE, Giddings CW, Doetkott C, Johnson TJ, Nolan LK. Characterizing the APEC pathotype. Veterinary Research. 2005;36(2):241–56. pmid:15720976
48. Lynne AM, Skyberg JA, Logue CM, Doetkott C, Foley SL, Nolan LK. Characterization of a series of transconjugant mutants of an avian pathogenic Escherichia coli isolate for resistance to serum complement. Avian Dis. 2007;51(3):771–6. pmid:17992940.
49. Binns MM, Mayden J, Levine RP. Further characterization of complement resistance conferred on Escherichia coli by the plasmid genes traT of R100 and iss of ColV,I-K94. Infection and immunity. 1982;35(2):654–9. pmid:7035371; PubMed Central PMCID: PMCPMC351091.
50. Ruiz-Perez F, Wahid R, Faherty CS, Kolappaswamy K, Rodriguez L, Santiago A, et al. Serine protease autotransporters from Shigella flexneri and pathogenic Escherichia coli target a broad range of leukocyte glycoproteins. Proceedings of the National Academy of Sciences. 2011;108(31):12881–6.
51. Ewers C, Li G, Wilking H, Kiessling S, Alt K, Antao EM, et al. Avian pathogenic, uropathogenic, and newborn meningitis-causing Escherichia coli: how closely related are they? Int J Med Microbiol. 2007;297(3):163–76. pmid:17374506.
52. Johnson TJ, Wannemuehler Y, Johnson SJ, Stell AL, Doetkott C, Johnson JR, et al. Comparison of extraintestinal pathogenic Escherichia coli strains from human and avian sources reveals a mixed subset representing potential zoonotic pathogens. Applied and environmental microbiology. 2008;74(22):7043–50. pmid:18820066; PubMed Central PMCID: PMCPMC2583479.

SECTION VI
PLASMID ECOLOGY AND EVOLUTION

CHAPTER

10

GENOMICS OF MICROBIAL PLASMIDS: CLASSIFICATION AND IDENTIFICATION BASED ON REPLICATION AND TRANSFER SYSTEMS AND HOST TAXONOMY

Masaki Shintani[1,2] Zoe K. Sanchez[1] and Kazuhide Kimbara[1]

[1]Department of Applied Chemistry and Biochemical Engineering, Graduate School of Engineering, Shizuoka University, Shizuoka, Japan

[2]Department of Bioscience, Graduate School of Science and Technology, Shizuoka University, Shizuoka, Japan

Plasmids are important "vehicles" for the communication of genetic information between bacteria. The exchange of plasmids transmits pathogenically and environmentally relevant traits to the host bacteria, promoting their rapid evolution and adaptation to various environments. Over the past six decades, a large number of plasmids have been identified and isolated from different microbes. With the revolution of sequencing technology, more than 4600 complete sequences of plasmids found in bacteria, archaea, and eukaryotes have been determined. The classification of a wide variety of plasmids is not only important to understand their features, host ranges, and microbial evolution but is also necessary to effectively use them as genetic tools for microbial engineering. This review summarizes the current situation of the classification of fully sequenced plasmids based on their host taxonomy and their features of replication and conjugative transfer. The majority of the fully sequenced plasmids are found in bacteria in the *Proteobacteria*, *Firmicutes*, *Spirochaetes*, *Actinobacteria*, *Cyanobacteria* and *Euryarcheota* phyla, and key features of each phylum are included. Recent advances in the identification of novel types of plasmids and plasmid transfer by culture-independent methods using samples from natural environments are also discussed.

INTRODUCTION

Plasmids are circular or linear extrachromosomal replicons that are found in many microorganisms in the domains *Bacteria*, *Archaea*, and *Eukaryota* (Funnell and Phillips, 2004). Plasmids are transmissible by conjugation (Frost et al., 2005; Sota and Top, 2008; Frost and Koraimann, 2010). Smillie et al. (2010)reported that about 14% of the full-sequenced plasmids were predicted to be conjugative. Conjugation is one of the most effective mechanisms to spread genetic elements among bacteria (Guglielmini et al., 2011). It is therefore one of the most important "vehicles" for bacterial communication of genetic information, facilitating the rapid evolution and adaptation abilities seen in bacteria (Aminov, 2011).

Plasmids are also important genetic tools used to manipulate and analyze microorganisms through the introduction, modification or removal of target genes (Frost et al., 2005; Sota and Top, 2008). New plasmids have been reported with the recent revolution in nucleotide sequencing technology. Currently, there are 4602 complete sequences of plasmids in the NCBI Plasmid Genome database: 4418 are from bacteria, 137 are from archaea, and the remaining 47 are from eukaryota (Figure 1A; based on the NCBI database, http://ftp.ncbi.nih.gov/genomes/Plasmids, Aug. 2014).

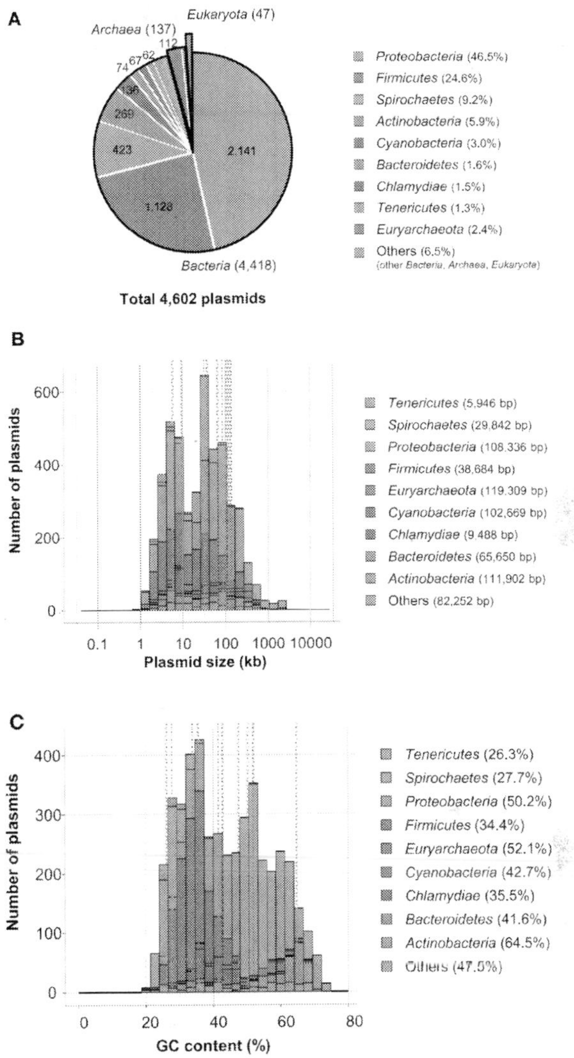

Figure 1: The phylogenetic distribution of fully sequenced plasmids (A), histograms of plasmid size (B), and GC content (%) (C). The ratios of each phylum (A), the average size of plasmids (B), and the average GC contents (C) are shown in parentheses and as dotted lines.

Knowledge of the relationships between plasmid features and host taxonomy is important in order to understand how the plasmids have been spread among microbes. The classification of known plasmids is necessary to identify newly isolated plasmids in the future. Information about the host range of these plasmids is critical to effectively use them as genetic tools for microbial engineering. This information is also important to detect, isolate and identify novel types of plasmids in environmental samples.

In this review, known plasmids for which the complete sequence was available were classified by their host and their (putative) replication or transfer systems. The classification may help to predict which types of plasmids could be replicated in or be transferred to the host microbes. Several insights into the mechanisms of replication and conjugative transfer used by plasmids are summarized first, followed by a description of the features of representative plasmids in several phyla. Recent advances in the identification of novel types of plasmids and plasmid transfers by culture-independent methods using samples from natural environments are also described.

Replication and Transfer Mechanisms

In plasmids, replication occurs at a specific site known as the origin of vegetative replication (*oriV*). Well-known replication systems of circular plasmids include theta-type replication, rolling-circle replication, and strand displacement-type replication (del Solar et al., 1998). Many theta-type replicating plasmids contain repeated DNA sequences, or iterons, which bind the replication initiation protein, Rep (reviewed in Espinosa et al., 2000; Krüger et al., 2004). The mechanisms of replication control have been extensively studied in two iteron-containing plasmids, pPS10 and R6K (Giraldo and Fernandez-Tresguerres, 2004; Rakowski and Filutowicz, 2013). ColE1-family plasmids are another group of theta-type replicating plasmids whose replication is strictly controlled by an antisense RNA (Polisky, 1988; Cesareni et al., 1991; Espinosa et al., 2000; Brantl, 2004). The rolling-circle replication mechanism is found in many small multi-copy plasmids (Espinosa et al., 2000; Khan, 2005). The representative plasmids using rolling-circle replication are pT181 (Inc14; see below for the discussion of Incompatibility groups), pC194 (Inc8), pMV158 (Inc11), and pUB110 (Inc13), all identified in staphylococci (Khan, 2005). Similar plasmids are also found in *Bacillus* (Guglielmetti et al., 2007) and several genera in *Actinobacteria* (Ventura et al., 2007).

Ravin recently reviewed the replication mechanisms of a prophage of *Escherichia coli*, N15, which was the first linear plasmid identified with covalently closed ends (Ravin, 2011). Most linear plasmids have conserved "telomeres" containing inverted repeat sequences (Ravin, 2011). The 5′ telomeric ends are blocked by covalently attached telomere terminal proteins (Ravin, 2011). Linear plasmids have sets of conserved telomere replication genes known as *tpg* and *tap* (Bao and Cohen, 2001, 2003), or *tpc* and *tac*(Huang et al., 2007). Many linear type plasmids have been found in *Actinobacteria*, especially in the genera *Mycobacteria*, *Rhodococcus*, and *Streptomyces* (Ventura et al., 2007).

Conjugative transfer is another important mechanism by which plasmids spread DNA among different bacteria. Self-transmissible plasmids in Gram-negative bacteria generally carry complete sets of genes required for transfer, the origin of transfer (*oriT*), the relaxase protein, the type IV coupling protein (T4CP), and the type IV secretion system (T4SS). Garcillán-Barcia et al. (2009, 2011) and Smillie et al. (2010) classified the conjugative, or mobilizable, plasmids in the GenBank database into six mobility (MOB) types (MOB_C, MOB_F, MOB_H, MOB_P, MOB_Q, and MOB_V) according to the amino acid sequences of their relaxase proteins. An additional classification was performed based on the plasmids' T4SS involved in mating pair formation (MPF) during conjugation. Smillie et al. proposed four classes of MPF (MPF_F, MPF_G, MPF_I, and MPF_T) according to the T4SS amino acid sequences (Smillie et al., 2010). They also investigated the presence of two key elements of plasmid mobility, type IV pili coupling protein (T4CP) and the ATPase VirB4 (Smillie et al., 2010). During conjugation, double-stranded plasmid DNA is cleaved at the *oriT* site by a relaxase protein, which then covalently binds to the *oriT* DNA. The resultant DNA-protein complex is transported to the recipient cell by T4SS. This single-stranded DNA is transferred into the recipient cell by the T4CP. The mobilizable plasmids only have *oriT*, relaxase, and sometimes T4CP (Garcillán-Barcia et al., 2009, 2011).

Gram-positive bacteria transfer plasmids by two methods, although the detailed mechanisms are not well understood. First, a single strand of plasmid DNA is transported via a T4SS, which seems to be widely used as a means for transferring plasmids in Gram positive bacteria (Goessweiner-Mohr et al., 2013). Several plasmids of the order *Actinomycetales* have conjugative systems that function in a manner similar to the segregation of chromosomal DNA during bacterial cell division and sporulation. The translocation of double-stranded DNA to the recipient cell is mediated by an FtsK-homologous protein (Goessweiner-Mohr et al., 2013). As for archaeal plasmids, only the plasmids in *Sulfolobales* are known to be transferred (Greve et al., 2004); however, the mechanisms are still not well understood.

Distribution and Classification of Plasmids

Wide variations in both size and GC contents were observed among the 4602 plasmids found in the GenBank database. The average size was 80 kb (range: 744 bp-2.58 Mb), and the average GC content was 44.1% (range: 19.3–75.6%;). More than 90% of the plasmids in the database were identified in 22 phyla: *Proteobacteria* (2142 sequences, 47%), *Firmicutes* (1129 sequences, 25%), *Spirochaetes* (423 sequences, 9.2%), *Actinobacteria* (269 sequences, 5.8%), *Cyanobacteria* (136 sequences, 3.0%) *Bacteroidetes* (74 sequences, 1.6%), *Chlamydiae* (67 sequences, 1.5%), and *Tenericutes* (62 sequences, 1.3%; Figure 1A). The relationships between the phyla of plasmid hosts and plasmid size or GC content are shown in Figures 1B,C. The bimodal distribution of plasmid sizes was plotted as previously shown (Smillie et al., 2010; Garcillán-Barcia et al., 2011) with peaks at approximately 4–8 kb and 32 kb, whereas the number of sequenced plasmids increased from 1730 plasmids to 4602 plasmids. This fact indicates that the distribution of plasmid sizes is conserved if the number of sequenced plasmids increases. The average size of the 4602 plasmids (79.8 kb) was larger than those of 1730 plasmids (63.5 kb). The difference was due to the number of plasmids more than 1 Mb in 4602 plasmids (45 plasmids) was much larger than those in 1730 plasmids (9 plasmids). The average sizes of plasmids in the phyla *Actinobacteria*, *Bacteroidetes*, *Cyanobacteria*, *Proteobacteria*, and *Euryarcheota* were similarly distributed (650–1200 kb), while those in *Chlamydiae*, *Firmicutes*, *Spirocaetes*, and *Tenericutes* were rather small (6–40 kb, Figure 1B). However, the average GC content of plasmids was different in each host phylum (26.3–64.5%, Figure 1C). Nishida reported that the GC contents of the majority of plasmids were lower than those of their host chromosomes, although the difference was less than 10% (Nishida, 2012). Thus, it was reasonable that plasmids found in *Actinobacteria* had the highest GC content (64.5%; Figure 1C), because their host genomes ranged from 51% to greater than 70% GC (Ventura et al., 2007). It was suggested that the lower GC contents of plasmids than those of host chromosome might be due to the higher energy cost to maintain G and C than A and T (Rocha and Danchin, 2002). On the other hand, the replication of plasmids depending on the host cell's respiration machinery might be expected to have the same GC content as the host (Rocha and Danchin, 2002). Therefore, the GC content of the plasmid is likely to be important for determining its host range. It was suggested that the nucleotide composition of plasmid might be progressively altered toward the average nucleotide composition of the host genome (Lawrence and Ochman, 1998; Rocha and Danchin, 2002). Based on the hypothesis, the relationships of GC contents between the plasmid and the host genome might be also important for predicting when the plasmid was introduced into the host species. Indeed, even in the identical host species, the ranges of GC contents of plasmids were wider than those of the host genome sequence. For example, the GC contents of the whole nucleotide sequences of 10 plasmids in *Pseudomonas aeruginosa* ranged from 45.8 to 63.8% (the average GC contents were 58.7%). According to the *Pseudomonas* genome database (http://www.pseudomonas.com/), the GC contents of *Pseudomonas aeruginosa* ranged from 66.1 to 66.6% (average GC contents = 66.4%). The plasmid with 63.8% GC content might be introduced into strains *P. aeruginosa* earlier than the one with 45.8%.

Plasmids have been classified based on incompatibility since the 1970s. Incompatibility (Inc) is defined as the inability of plasmids sharing similar replication and partition systems to be propagated stably in the same host cell line. Inc groups have been independently classified in three different genera; there are 27 Inc groups in *Enterobacteriaceae*, 14 Inc groups of *Pseudomonas*, and approximately 18 Inc groups in *Staphylococcus* (Udo and Grubb, 1991; Lawley et al., 2004; Taylor et al., 2004; Thomas and Haines, 2004; Sota and Top, 2008; Carattoli, 2009). Several Inc groups of *Pseudomonas* are identical to those in enterobacteria, such as IncP-1 (equivalent to IncP), IncP-3 (equivalent to IncA/C), IncP-4 (equivalent to IncQ), and IncP-6(equivalent to IncG/U). Classification into an Inc group is always based on the amino acid sequence of the replication initiation (Rep) protein (replicon typing), and it is not necessarily confirmed by conventional methods whether the plasmid shows incompatibility with the same Inc group plasmid in the same host cell line. The classification based on the replicon typing is also useful for grouping the plasmids in unidentified Inc groups. As described above, classification of plasmids based on their MOB types and MPF classes has previously been reported (Garcillán-Barcia et al., 2009, 2011; Smillie et al., 2010). Plasmid classification by replicon typing is based on the molecular characteristics of the replicons and has been quite successful (Carattoli et al., 2005). There are difficulties with this method: (i) plasmids frequently carry multiple replicons, and it is therefore difficult to classify the plasmid into single replicon group. (ii) Detailed information about Inc groups or Rep types is limited among several microbial taxonomies especially enterobacteria, and it is difficult to identify replication regions for the other types of plasmids. The classification of plasmids based on their mobility (Garcillán-Barcia et al., 2009; Smillie et al., 2010; Garcillán-Barcia et al., 2011) can overcome these problems because (i) classification by MOB types can cover the whole microbial plasmids, and (ii) plasmids rarely carry more than one relaxase gene (Garcillán-Barcia et al., 2009). However, this classification is not appropriate for non-transmissible plasmids.

In this review, 4602 plasmids with complete sequences listed in the GenBank database were classified based on the genes encoding (putative) Rep proteins, MOB classes, and MPF types. Amino acid sequences of previously identified Rep proteins were used; their accession numbers in the DDBJ/EMBL/GenBank database are listed. The local TBLASTN program was used for the classification of Rep proteins with the following parameters: *e*-value $<10^{-5}$, >50% identity, and >0.5 query coverage. The parameters for identity and coverage were chosen based on variations in the amino acid sequences of the replication initiation protein TrfA from the IncP-1 plasmid (data not shown). The IncP-1 plasmid is one of the best-studied plasmids distributed across many bacterial classes (Adamczyk and Jagura-Burdzy, 2003), and thus, its TrfA protein sequence was used to set the criteria for analysis. For the ColE1 family plasmids, nucleotide sequences of RNA II were used as queries for the local BLASTN program using an *e*-value $<10^{-5}$. Notably, the Rep proteins of the IncB, IncFII (RepA1), IncI, and IncK plasmids showed greater than 87% identity with one another. Similarly, the Rep proteins of Inc4, Inc9, Inc10, and Inc14 plasmids have 311–314 amino acid sequences and share a high degree of identity (>75%) with each other. These findings suggest that plasmids from these Inc groups are closely related to one another. In the case that one open reading frame simultaneously showed identity with the Rep proteins of these Inc groups, the one with the smaller *e*-value was used for the classification. Among the 4602 plasmids analyzed, 1845 plasmids (40.0%) in *Proteobacteria*, *Firmicutes*, *Actinobacteria*, *Cyanobacteria*, *Bacteroidetes*, *Tenericutes*, *Euryarchaeota*and other phyla were classified into previously known Inc groups or unidentified Inc groups with other known Rep types (Figures 2A,B), although several plasmids were classified into multiple Inc groups.

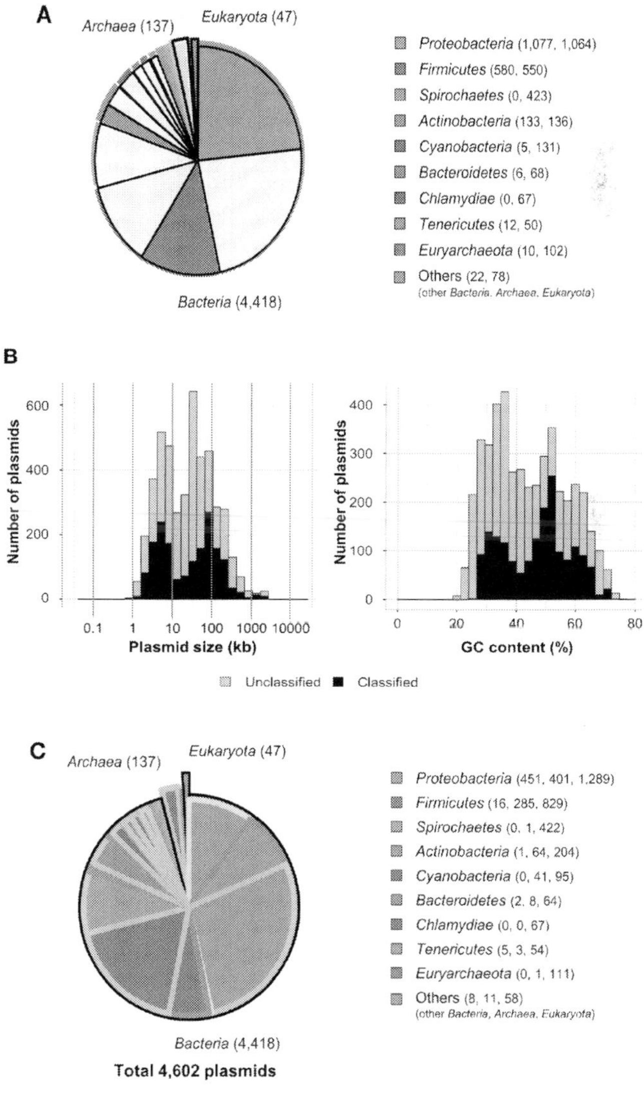

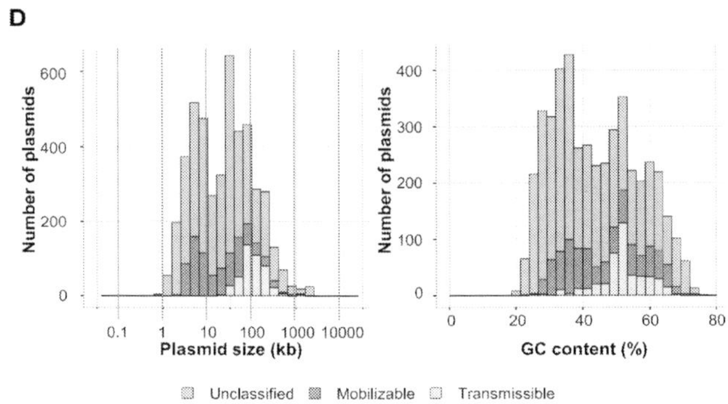

Figure 2: Ratios of classified plasmids in each phylum (A) and histograms of their size (B, left) and GC content (B, right) are shown. The unclassified plasmids are shown as shaded, and the numbers of classified plasmids and unclassified plasmids are shown in parentheses (A). Ratios of putative transmissible plasmids (yellow), putative mobilizable plasmids (green), and others (gray) in each phylum (C) are shown. Histograms of their size (D, left) and GC content (D, right) are shown.

The classification of plasmids into MOB classes and MPF types using T4CPs and VirB4 homologs was performed as proposed by de la Cruz's group (Garcillán-Barcia et al., 2009; Smillie et al., 2010). Instead of using the local PSI-BLAST program (ver. 2.2.24, http://ftp.ncbi.nlm.nih.gov/blast/executables/blast+/LATEST/) as described previously (Garcillán-Barcia et al., 2009), we used the local TBLASTN program with >50% identity and >0.7 query coverage for the alignment of MOB and MPF gene sequences. The MPF type of a plasmid was determined by the presence of more than two MPF genes on the plasmid. There are five MPF types of plasmids, as defined by de la Cruz's group: (i) non-transmissible plasmids that do not code for a relaxase ("non-mob"); (ii) plasmids that do not contain a relaxase but contain T4CP, VirB4, and MPF, or any two of these three elements ("non-mob, protein export"); (iii) mobilizable plasmids that contain a relaxase gene but lack VirB4 and MPF ("mob"); (iv) conjugative plasmids that contain a known type of T4SS (MPF_F, MPF_T, MPF_I or MPF_G), plus relaxase and T4CP ("determined conjugative"); and (v) conjugative plasmids that contain genes for relaxase, T4CP, and VirB4, but whose T4SS does not belong to a specific MPF ("undetermined conjugative"). Because the relationship between the plasmid and each MOB class or MPF type was previously reviewed in detail (Garcillán-Barcia et al., 2009, 2011; Smillie et al., 2010), the distribution of "mob" or putative mobilizable, "determined conjugative" or putative transmissible, and the other plasmids were discussed in this review. As shown in Figure 2C, putative transmissible plasmids were primarily found in the phylum *Proteobacteria*, with smaller numbers found in *Firmicutes*, *Bacteroidetes*, *Tenericutes*, and others. Many putative mobilizable plasmids were found in the phyla *Proteobacteria* (401/2141 plasmids, 18.7%), *Firmicutes* (285/1130 plasmids, 25.2%), *Actinobacteria* (64/269 plasmids, 23.8%), and *Cyanobacteria* (41/136 plasmids, 30.1%) (Figure 2C). The mobilizable plasmids had a mean peak at a smaller size (around 5 kb) than that of the transmissible plasmids (around 100 kb; Figure 2D, left). The two major peaks of mobilizable and transmissible plasmids were similarly found as those in previous report (Smillie et al., 2010). This fact indicates that the size distribution of mobilizable and transmissible plasmids is conserved if the number of full-sequenced plasmids increases. As for the distribution of the GC content of the plasmids, both types showed a wide range of values (23–74%; Figure 2D, right), probably because various hosts harbor both mobilizable and conjugative plasmids.

Plasmids in *Proteobacteria*

Plasmids within *Proteobacteria* are one of the most extensively studied groups because they include those harboring the pathogenic genes of bacteria known to infect both animals and plants. Indeed, about half of the plasmids in the database were found in *Proteobacteria* (Figure 1A). Although many catabolic plasmids have also been identified in the phylum *Proteobacteria*, they are not described in detail here, as they have been reviewed elsewhere (Shintani et al., 2010; Shintani and Nojiri, 2013). The majority of the plasmids within this phylum were found in the classes *Gammaproteobacteria* (1389; 63%), *Alphaproteobacteria*(451; 22%), and *Betaproteobacteria* (187; 8.7%). The size distribution was bimodal, with one group at around 5 kb and the other at 65–131 kb (Figure 3A, left), although the average size of plasmids in each major class was different: 58.7 kb in *Gammaproteobacteria*, 218 kb in *Alphaproteobacteria*, and 259 kb in *Betaproteobacteria*. The average GC contents of plasmids in these groups were 48.0% in *Gammaproteobacteria*,

56.5% in *Alphaproteobacteria*, and 58.5% in *Betaproteobacteria* (Figure 3A, right). Within this phylum, 1077 plasmids were classified using previously known Rep proteins.

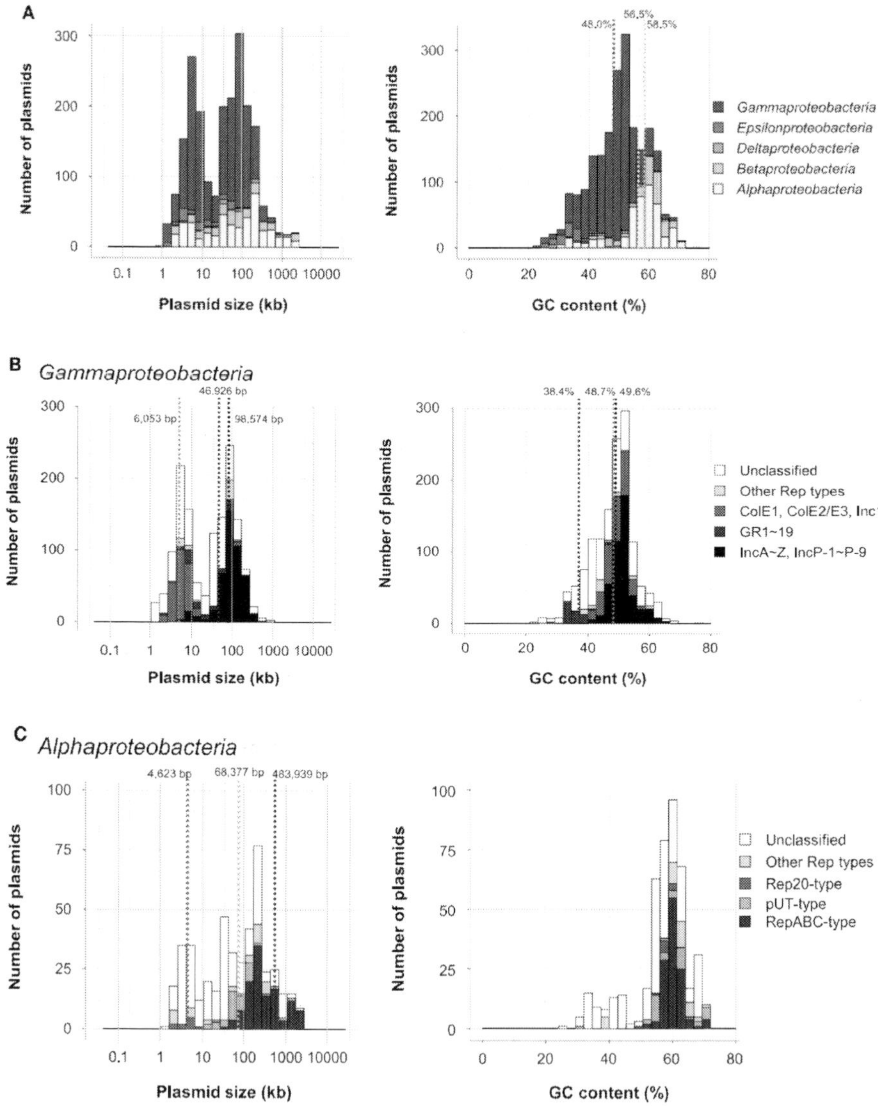

Figure 3: (A) Histograms of plasmid size (left) and GC content (right) of *Proteobacteria* in each class. **(B)** Histograms of plasmid size (left) and GC content (right) in *Gammaproteobacteria*. The distribution of classified plasmids are shown in black (IncA-Z and IncP-1 to IncP-9), blue (GR1 to GR19), dark gray (ColE1, ColE2/E3, and Inc13), light gray (other Rep types), and white

(unclassified). **(C)** Histograms of plasmid size (left) and GC content (right) in *Alphaproteobacteria*. The distribution of classified plasmids are shown in blue (RepABC-type), light blue (pUT-type), dark gray (Rep20-type), light gray (other Rep types), and white (unclassified). The average sizes of Rep20-type, pUT-type, and RepABC-type plasmids are represented by dotted lines.

Plasmids in *Gammaproteobacteria*

Among the 1396 plasmids in *Gammaproteobacteria*, 830 were classified into Inc groups (IncA to IncZ, IncP-1 to IncP-9, and Inc13; 485 plasmids), the ColE1-type plasmid group (210 plasmids), the ColE2/E3-type plasmid group (38 plasmids), and other groups with known Rep (119 plasmids), while several plasmids were classified into multiple Inc or Rep type groups. The average sizes of plasmids classified into Inc13, ColE1-type, and ColE2/E3-type plasmids (6053 bp) were smaller than those of other classified plasmids (98,574 bp) (Figure 3B, left), although the average GC contents were similar between the smaller and larger groups (48.7% and 49.6%, respectively; Figure 3B, right). This was probably because the sizes of the plasmids are closely related to their replication systems.

The majority of plasmids in *Gammaproteobacteria* were from *Enterobacteriales* (927) and *Pseudomonadales* (183). The plasmids from *Enterobacteriales* contain antibiotic resistance plasmids, which cause the rapid development of resistance to antibiotics in bacterial populations. Thus, more than 75% of the plasmids in *Enterobacteriales* (697/927) were classified into Inc groups or Rep types, including IncF, IncA/C (equivalent to IncP-3), IncL/M, IncI, IncHI2/S, and IncN, containing plasmid families already known to carry resistance genes (Carattoli, 2009; Wang et al., 2013). Carattoli and Partridge have separately reviewed the emergence and spread of antibiotic resistance via plasmids among *Enterobacteriaceae* (Carattoli, 2009; Partridge, 2011). The spread of plasmids carrying genes for extended-spectrum β-lactamases (ESBLs) has been a particular problem in the treatment of *Enterobacteriaceae* infections (Paterson, 2006). Plasmid groups found in *Enterobacteriales*, such as *Enterobacter*, *Escherichia*, *Salmonella*, *Shigella*, and *Yersinia*, include a large number of virulence plasmids. It should be noted that virulence-associated traits of *E. coli* are almost exclusively found on IncF-family plasmids (Johnson and Nolan, 2009). In virulent *Yersinia* species, extensive research has been carried out on "low calcium ion (Ca^{2+})

response plasmids," which regulate the growth and expression of several virulence-associated properties by Ca^{2+} and temperature. This group includes the pCD1, pYVe8081, and pYVe227 series, which encode a set of secreted anti-host proteins and a type III secretion system (Straley et al., 1993). Notably, proteins responsible for replication initiation encoded on pCD-type plasmids showed high similarity with those encoded on IncZ plasmids, and 186 of the 976 plasmids of *Enterobacteriales* carried genes encoding these types of Rep proteins.

Although the IncP-1 to IncP-14 groups have been found in *Pseudomonas*, only 21 plasmids in the order *Pseudomonadales* were classified into these groups; the other 93 plasmids classified into these Inc groups were in other orders (including *Enterobacteriales* and *Thiotrichales*), other classes (including *Alphaproteobacteria* and *Betaproteobacteria*), or even other phyla (including *Actinobacteria* and *Thermotogae*;). Recently, Xiong et al. determined the complete sequence of pOZ176, an IncP-2 plasmid which is resistant to heavy metals (Xiong et al., 2013). Ten additional plasmids have been identified which carry a gene encoding a putative Rep protein with >50% identity and >0.5 query coverage, suggesting that these plasmids may have IncP-2-like replication systems. The IncP-2 group contains degradative plasmids whose sizes are typically around 500 kb (Shintani et al., 2010; Shintani and Nojiri, 2013); the average size of the ten plasmids was 425 kb. Two plasmids were found to share >83% identity with the sequence of the Rep protein from pOZ176: pPNAP01, a plasmid of *Polaromonas naphthalenivorans* CJ2, encodes proteins involved in the degradation of naphthalene (Jeon et al., 2003, 2006; Yagi et al., 2009), and pBB1 (GenBank/EMBL/DDBJ accession no. CP002879), a plasmid of *Cupriavidus necator* N-1, encodes proteins putatively involved in the degradation of catechols (Poehlein et al., 2011). Taken together, these findings suggest that the nucleotide sequences of the replication systems used by previously-known Inc group plasmids could be important tools for classifying other plasmids. It will be therefore helpful for the classification to determine the nucleotide sequences of the other unsequenced Inc group plasmids including IncP-5, IncP-8, IncP-10, IncP-11, IncP-12, IncP-13, and IncP-14 plasmids.

Acinetobacter baumannii is known to be an important pathogen, and the plasmids in the genus *Acinetobacter* are considered to be key genetic factors in the spread of multi-drug resistance (Dijkshoorn et al., 2007; Evans and Amyes, 2014). Bertini et al. proposed PCR-based replicon typing of 19 groups (GRs) of resistance plasmids in *A. baumannii* on the basis of nucleotide sequence similarities between their replication initiation proteins (Bertini et al., 2010). Using the amino acid sequences of Rep from plasmids in the GR1 to GR19 groups, we were able to classify 78 of the 4602 plasmids found in the database, including 60 of the 90 plasmids in genus *Acinetobacter*. The remaining 18 plasmids were also found in *Gammaproteobacteria*, and, notably, these plasmids were all classified into GR3. The average GC contents of plasmids classified into GR1-19 (38.4%) were lower than those of other plasmids in *Gammaproteobacteria* (Figure 3B). These findings indicate that the plasmids with GR1- to GR19-type Rep proteins were primarily distributed in *Acinetobacter* and that GR3 group plasmids may have a broader host range than plasmids in other GR groups.

Plasmids in *Alphaproteobacteria*

The majority of the 451 plasmids in *Alphaproteobacteria* were found in *Rhizobiales* (143 plasmids), *Rhodospirillales* (122 plasmids), *Rhodobacteriales* (94 plasmids), and *Sphingomonadales* (65 plasmids). Many plasmids in *Alphaproteobacteria* carry genes encoding RepABC proteins (Cevallos et al., 2008). The RepAB proteins are involved in the partitioning of the plasmid, and RepC is associated with replication initiation (Cevallos et al., 2008; Pinto et al., 2012). Indeed, 123 plasmids in *Alphaproteobacteria* had RepABC type genes. Although plasmids containing these types of Rep genes have an average size of 484 kb, their sizes vary widely from 30 to 2430 kb (Figure 3C).

Hosts in the *Rhizobiales* are known to be symbiotic bacteria, and they can usually fix nitrogen only upon establishing a mutualistic interaction with plants, particularly those of the *Leguminosae* family (Cevallos et al., 2008; Pinto et al., 2012). Their 143 plasmids are characterized by a highly variable plasmid number (from 0 to 11) and size (from 150 to 1683 kb). Of these, 101 plasmids were classified into the RepABC type group. Notably, most of the genes for nodulation and nitrogen fixation are located on a single plasmid, called the symbiotic plasmid (pSym). pSym contains the structural genes for nitrogenase (*nifHDK*) and/or genes essential to the production of the nodulation factor (*notABC*).

Introduction of pSym into a plasmid-free *A. tumefaciens* strain leads to nodulation of a specific host and, sometimes, to nitrogen fixation, albeit at modest levels. Conversely, elimination of pSym impairs both the nodulation and nitrogen fixation capacities of the original bacterial strain (Torres et al., 2011). The presence of pSym is common to members of the fast-growing *Rhizobium* and *Sinorhizobium* species, including symbionts of many plants of agronomic interest. RepABC plasmids were also found in classes *Rhodospirillales* and *Rhodobacteriales* as previously reported (Petersen et al., 2011).

Approximately half of the plasmids in class *Rhodosprillales* were found in *Acetobacter* (60 of 122 plasmids). The acetic acid bacteria are important for the industrial oxidation of various compounds, and thus, the construction of shuttle vector(s) to facilitate functional gene studies of these bacteria was reported 30 years ago (Fukaya et al., 1985; Okumura et al., 1985). Nevertheless, a well-defined method for the classification of *Acetobacter* plasmids has not been developed. Two groups have recently reported the characterization of Rep proteins, Rep7 and Rep20, from small plasmids pGR7 (Grones and Grones, 2012) and pAG20 (Babic et al., 2014), respectively, found in *Acetobacter* (class *Rhodospirillales*). The Rep20 gene was also found in plasmids in *Rhodospirillales* and *Rhodobacteriales*(Figure 3C). The average size of these plasmids was 4.6 kb, while that of RepABC-type plasmids was 483 kb (Figure 3C), suggesting that the newly characterized Rep proteins may be appropriate for the replication of small plasmids.Several plasmids in *Sphingomonadales* have been found in bacteria which degrade biphenyl (pNL1 and pNL2 in *Novosphingobium aromaticivorans* DSM 12444; (Stillwell et al., 1995; Romine et al., 1999), gamma-hexachlorocyclohexane (pCHQ1, pUT1, and pUT2 in *Sphingobium japonicum* UT26S; (Nagata et al., 2006, 2010, 2011), and dibenzo-*p*-dioxin (pSWIT01 and pSWIT02 in *Sphingomonas wittichii* RW1; (Miller et al., 2010). The (putative) Rep proteins of these plasmids (pCHQ1, pUT1, pUT2, pSWIT01, pNL1, and pNL2) were found in 59 other plasmids in *Alphaproteobacteria*, and 24 plasmids were found in *Sphingomonadales*. Notably, 42 plasmids in *Alphaproteobacteria* were classified as pUT1- or pUT2 type plasmids, and 36 of these were the pUT1 type (Figure 3C). The average size of the pUT-type plasmids was 68.4 kb, which was larger than Rep20 type plasmids and smaller than RepABC-type plasmids (Figure 3C). Most of the plasmids classified into these three Rep types were in *Alphaproteobacteria*, whereas two pUT1-type plasmids and six RepABC-type plasmids were found in other classes or phyla. Thus, they are important vehicles among *Alphaproteobacteria*, although the plasmids with smaller sizes and lower GC contents were still unable to be classified (Figure 3C).

Plasmids in *Firmicutes*

Plasmids found in *Firmicutes* were smaller (1.3–627 kb, 39 kb average) and had lower GC contents (23–63%, 34.4% average) than those in *Proteobacteria* (0.74–2580 kb, 103 kb average; 22–76%, 50.2% average; Figures 1B,C). Among the 1129 plasmids in *Firmicutes*, 1021 plasmids, or 90%, were found in the class *Bacilli* (Figure 4A). Based on comparisons with the amino acids sequences of (putative) Rep genes of the Inc groups of staphylococci, 475 plasmids in *Firmicutes* were shown to have putative Rep genes of Inc1 to Inc18 plasmids (Figure 4B), and 311 plasmids were from *Bacilliales*, including *Bacillus* (57 plasmids) and *Staphylococcus* (220 plasmids;). A subset of 158 plasmids classified into Inc1 to Inc18 groups was found in *Lactobacillales*, including *Enterococcus* (36 plasmids), *Lactobacilllus* (56 plasmids), *Lactococcus* (15 plasmids), and *Streptococcus* (34 plasmids). The Rep genes of the Inc4, Inc8, Inc9, Inc10, Inc11, Inc13 and Inc14 group plasmids are known as rolling-circle replication plasmids (Smith and Thomas, 2004; Khan, 2005; Guglielmetti et al., 2007), while those of Inc1, Inc7, and Inc18 are theta-type replication plasmids (Bruand et al., 1991; Jensen et al., 2010b; Liu et al., 2013). Another classification system based on the nucleotide sequences of plasmids was proposed for enterococci and staphylococci, which included 26 Rep families and 10 unique families (Jensen et al., 2010a; Lozano et al., 2012). Of the 4602 plasmids included in our analysis, 388 plasmids could be classified using the 26 Rep families, and 383 of them were found in *Firmicutes*. Several plasmids were classified both a Rep family and an Inc group, but 91 plasmids were only classified into the Rep family (Figure 4B). The distribution of plasmid size was bimodal, and the 269 putative rolling-circle replication plasmids had a smaller average size (5713 bp), while the average size of the 200 theta-type replication plasmids was larger (45,498 bp; Figure 4B, left). The GC contents of both groups were similar, 32.3% and 34.1%, respectively (Figure 4B, right).

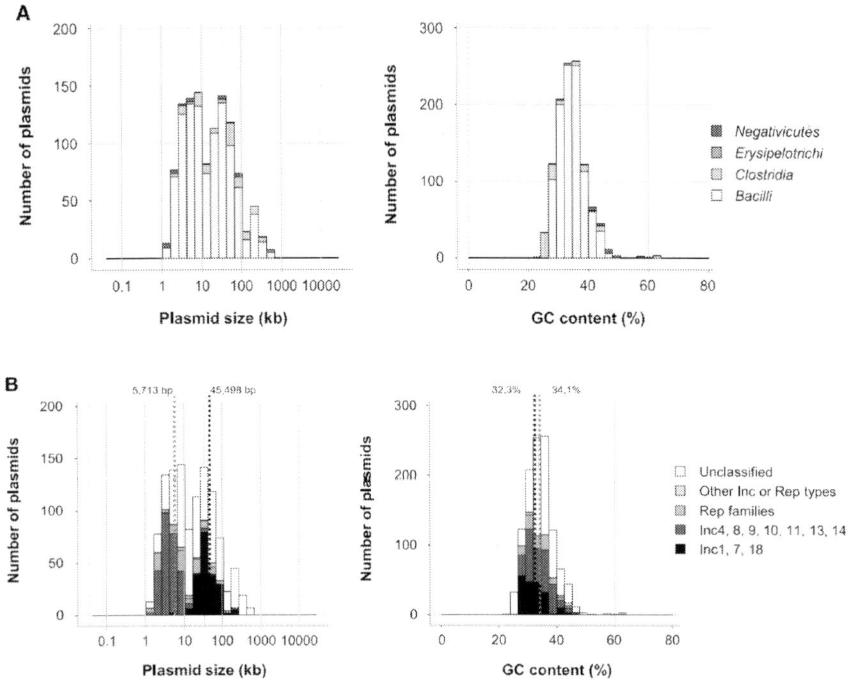

Figure 4: (A) Histograms of plasmid size (left) and GC content (right) of *Firmicutes* in each class. **(B)** Histograms of plasmid size (left) and GC content (right) in *Bacilli*. The distribution of classified plasmids is shown in black (Inc1, Inc7, and Inc18), dark gray (Inc4, Inc8, Inc9, Inc10, Inc11, Inc13, and Inc14), light blue (Rep families number 1–24 containing 7b, 10b), light gray (other Inc or Rep types) and white (unclassified). The average sizes and GC contents of groups Inc4, 8, 9, 10, 11, 13, 14 and those of Inc1, 7, 18 are represented by dotted lines.

Plasmids conferring resistance and virulence to *Bacilli*, including *Enterococcus* (*Bacilliales*) and *Staphylococcus* (*Lactobacilliales*), are causative agents of hospital infection outbreaks. The transmission of antibiotic resistance to *Enterococcus* is mediated by the pheromone-responsiveness and broad host range of Inc18 group plasmids (Palmer et al., 2010). The pheromone-responsive plasmids have mainly been described in *Enterococcus faecalis*. The representative examples of these plasmids are pCF10 and pAD1 (Clewell, 2007; Dunny, 2007). Their transfers are induced by pheromones generated from lipoprotein signal peptides encoded on the host chromosome (Clewell, 2007; Dunny,

2007). Pheromone receptors are encoded on the plasmid, and binding of the pheromone to its receptor causes an effective pair formation between donor and recipient cells. Rosvoll *et al.* have reported an in-depth investigation of Inc18 plasmids, which are found mostly in *E. faecium* (Rosvoll et al., 2010). Of these, only pAMβ1 and pIP501 have been well-characterized and were originally identified in *E. faecalis* and *Streptococcus agalactiae*, respectively. Within *Staphylococcus*, the pSK1 family (Inc1, Rep 20 family), which includes a multidrug-resistance plasmid like pSK4 and pSK7, and the pSK41 family (Inc7, Rep 15 family), including a conjugative multidrug-resistance plasmid like pGO1, have been investigated along with the previously discussed rolling-circle replicating plasmid (Jensen et al., 2010b; Liu et al., 2013).

The majority of the lactic acid bacteria (LAB) that belong to *Bacilli*, including *Lactococcus*, *Lactobacillus*, *Leuconostoc*, *Pediococcus*, and *Streptococcus*, as well as *Bifidobacterium* (phylum *Actinobacteria*), are important in food fermentation. Recently, the plasmids from *Lactococcus lactis* were reviewed in detail regarding plasmid-encoded traits of biotechnological significance (Ainsworth et al., 2014). Genetic engineering technology, including the development of cloning vectors for the LAB, is critical for the effective application of these bacteria in the food industry (Shareck et al., 2004). Versatile vectors have been developed for high-level, inducible gene expression in *Lactobacillus sakei* and *Lactobacillus plantarum* (Sorvig et al., 2005). In the GenBank database, 166 plasmids were found in *Lactobacillus*, and notably, multiple (five to ten) plasmids were found in individual strains, such as *Lactobacillus fermentum*MTCC 8711 (Jayashree et al., 2013), *Lactobacillus plantarum* 16 (Crowley et al., 2013), and *Lactobacillus reuteri* I5007 (Hou et al., 2014). Fukao et al. reported that one of the LAB, *L. brevis* KB290, has nine plasmids with RepABC systems (Fukao et al., 2013). Understanding how these plasmids could be maintained simultaneously in a single host cell will be important to the development of genetic tools for LAB.

Plasmids in *Spirochaetes*

Although 423 plasmids were found in *Spirochaetes* (Figure 1A), the majority of them were identified in genus *Borrelia* (416 out of 423), which is known to be an important pathogenic bacterium. The average GC contents of these plasmids were comparatively low (27.7%; Figure

1C). A unique feature of *Borrelia* is its segmented genome consisting of multiple circular and linear plasmids in addition to its linear chromosome within a single cell (Casjens et al., 2000). The features of *Borrelia* have been studied in detail (Chaconas and Kobryn, 2010; Chaconas and Norris, 2013). *B. burgdorferi* type strain B31, the causative agent of Lyme disease, possesses twelve linear plasmids and nine circular plasmids (Fraser et al., 1997; Casjens et al., 2000). Tilly et al. recently characterized the indispensable elements involved in the maintenance of cp26, a 26-kb circular plasmid found in *B. burgdorferi* (Tilly et al., 2012). Most plasmids can be lost without affecting the host growth; however, cp26 remains in all isolates of *B. burgdorferi* (Tilly et al., 2012). This plasmid carries genes required for survival of its host, including the *resT* gene, which encodes a telomere resolvase involved in the resolution of the replicated telomeres of the linear chromosome and plasmids in *B. burgdorferi* (Tilly et al., 2012).

Notably, several circular plasmids in B31 are homologous throughout almost their entire lengths (cp32s) and carry many genes encoding lipoproteins located on cell surfaces, although very few of these are metabolic or housekeeping genes (Casjens et al., 2000). Bunikis et al. proposed an efficient method to identify each plasmid based on multiplex PCR (Bunikis et al., 2011). The mechanisms used by the host strain to replicate and maintain similar plasmids have been studied in one of the cp32 plasmids (Eggers et al., 2002). There are still many interesting characteristics that require further investigation of plasmids in *Spirochetes*, and comparative genomic analyses of *Borrelia* in particular will reveal important information. Casjens et al. compared the genomic structures of plasmids identified in four different *Borrelia* strains (Casjens et al., 2012). They suggested that there are at least 28 plasmid compatibility types among the four strains and that several inter-plasmid genomic rearrangements may have occurred. More genomic information will be necessary for understanding the genetic evolution of these plasmids and their host chromosome.

Plasmids in *Actinobacteria*

All of the plasmids in the phylum *Actinobacteria* were found in class *Actinobacteria*, and, as shown in Figures 5A,B, they were found in nine suborders and one order (*Bifidobacteriales*). The plasmids in

Actinobacteria were large and had high GC contents (Figures 1B,C). The phylum *Actinobacteria* includes a wide variety of bacteria with different morphologies, physiologies, and metabolic properties, and they have various kinds of plasmids (Ventura et al., 2007). In a review of the genomics of *Actinobacteria*, Ventura et al. showed the phylogenetic relationships between Rep proteins from actinobacterial plasmids (Ventura et al., 2007). In this review, sequences from 42 different types of Rep proteins from actinobacterial plasmids were used as queries, and of these proteins, 27 were found in more than two plasmids. Of the 269 plasmids in *Actinobacteria*, 110 were found in *Corynebacterineae*, 69 in *Streptomycineae*, 32 in *Micrococcineae*, and 29 in *Bifidobacteriales*. The average sizes and GC contents of these plasmids varied widely (4631–200,330 bp; 56.1–70.3%; Figures 5A,B).

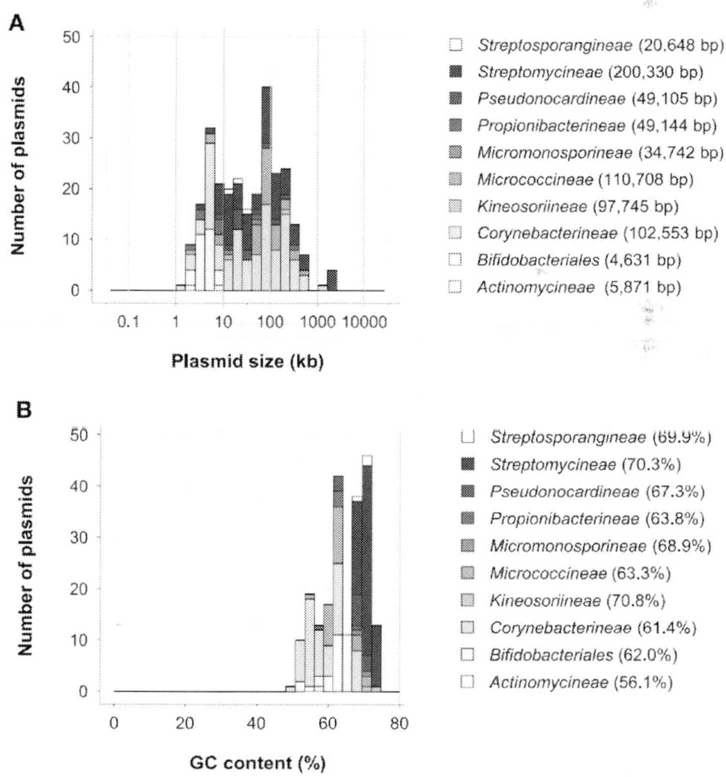

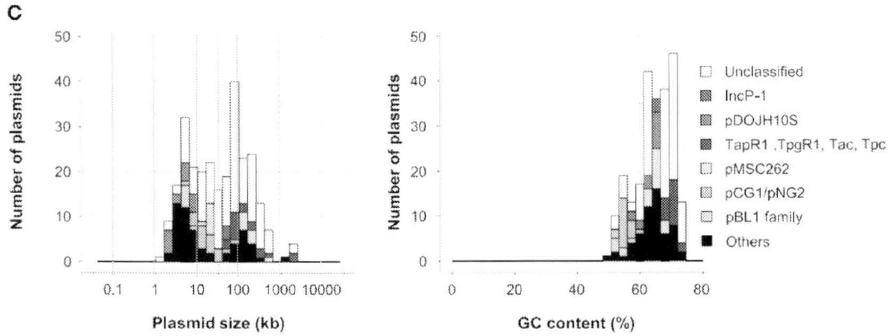

Figure 5: Histograms of plasmid size (A) and GC content (B) of *Actinobacteria* in each suborder and order (*Bifidobacteriales*). The average size of plasmids **(A)** and the average GC content **(B)** are shown in parentheses. **(C)** Histograms of plasmid size (left) and GC content (right) in phylum *Actinobacteria*. The distribution of classified plasmids is shown in beige (pBL1 family), spring green (pCG1/pNG2 family), yellow (pMSC262 family), brown (plasmids with TapR1, TpgR1 Tac, and/or Tpc), light brown (pDOJH10S-type), and gray (IncP-1 plasmids), black (other Rep types), and white (unclassified plasmids).

The majority of the plasmids from *Corynebacterineae* were found in *Corynebacterium* (38 plasmids), *Mycobacterium* (29 plasmids), and *Rhodococcus* (31 plasmids). Because *C. glutamicum* is widely used as an industrial producer of amino acids, plasmids of the amino acid-producing isolates were classified to develop efficient recombinant DNA techniques (Tauch et al., 2003a). Several rolling-circle replication plasmids were found in *C. glutamicum*, and the pBL1 (including pAG3 and pCG2) and pCG1 (including pCG4 and pGA2) families were proposed (Tauch et al., 2003a). A plasmid found in the human pathogen *C. diphtheria* S601, pNG2, has a Rep protein similar to those of the pCG1 family (Tauch et al., 2003a,b).

There were six pBL1-family plasmids and 22 pCG1/pNG2 plasmids in the GenBank database. While plasmids in the pBL1 family had small sizes (4.4–6.8 kb), those in the pCG1/pNG2 family had a wide range of sizes (4.1–85 kb, Figure 5C). Tauch et al. also reported a theta-type replication plasmid family in *Corynebacterium*, the pCRY4 family, but among the 4602 plasmids analyzed, only one plasmid was classified into this family. Of the plasmids found in *Mycobacterium*, only the low copy number plasmid pAL5000 has been widely used (Labidi et al., 1985). Plasmids in the pMSC262 family are compatible with pAL5000, including pVT2, pMUM001, and linear plasmid pCLP (Le Dantec et al., 2001), which had a wide range of sizes (24–615 kb, Figure 5C). Among the 29 plasmids of *Mycobacterium*, 16 plasmids were classified into the pMSC262 family. Plasmids in *Rhodococcus* were found in the biphenyl degrader, *R. jostii* RHA1 (Seto et al., 1995). This strain has three plasmids, pRHL1, pRHL2, and pRHL3 (Masai et al., 1997; Shimizu et al., 2001; McLeod et al., 2006). The *R. erythropolis* PR4 strain is known to utilize n-alkanes and alkylbenzene, and it has three plasmids, pREL1, pREC1, and pREC2 (Sekine et al., 2006). Only a small number of plasmids in *Rhodococcus* were predicted to have Rep proteins similar to those of pRHL1, pRHL3, pREC1, or pREC2.

Plasmids in *Streptomycieae* are found in *Streptomyces* species, and several of them are large linear plasmids. These linear plasmids have conserved "telomeres" containing inverted repeat sequences (Chen et al., 1993; Pandza et al., 1998). The 5′ telomeric ends are blocked by covalently attached telomere terminal proteins (Bao and Cohen, 2001). They have sets of conserved telomere replication genes known as *tpgR1* and *tapR1* or *tpc* and *tac* (Bao and Cohen, 2001, 2003). In *Streptomyces*, 13 plasmids, all linear, carry these telomere replication gene sets. Their sizes were relatively large (54–1797 kb, Figure 5C) and their GC contents were high (68.3–71.9%, Figure 5C). A rolling circle replication plasmid, pIJ101, from *S. lividans* has been used as a cloning vector in *Streptomyces* (Kendall and Cohen, 1988; Ventura et al., 2007), but only two other plasmids have been found to carry a similar Rep gene.

Plasmids in *Bifidobacteriales* were small, with an average size of 4.6 kb (Figure 5A). Among the 29 plasmids in *Bifidobacteriales*, 28 of them were classified into the known Rep types pKJ50 (Park et al., 1999), pNAC2, pNAC3 (Corneau et al., 2004), and pDOJH10S (Lee and O'Sullivan, 2006). Rep proteins of pKJ50, pNAC2, and pNAC3 were found only in *Bifidobacterium*; however, those of pDOJH10S were found in plasmids of other suborders of *Actinobacteria*, including *Corynebacterineae*, *Micrococcinaea*, and *Propionibacterineae*. It is likely that plasmids with Rep types of pDOJH10S have a broad host range in the phylum *Actinobacteria*.

Plasmids in *Cyanobacteria*

Cyanobacteria are known for the ability to undergo oxygenic photosynthesis and are a promising platform for the production of renewable chemicals and fuels. Thus, a plasmid vector to introduce exogenous genes into cyanobacterial cells has been developed. Taton et al. have developed a vector system for plasmids with a broad host range using the previously known plasmid origin of *Cyanobacteria*(http://golden.ucsd.edu/CyanoVECTOR/) (Taton et al., 2014). The origins of plasmids were from pDU1 in *Nostoc* sp. PCC 7524 (*Anabaena* sp. PCC 7120; Walton et al., 1992), pANS (pUH24) in *Synechococus elongatus* PCC 7942 (Golden and Sherman, 1983), pDC1 in *Nostoc* sp. PCC 8009 (*Nostoc* sp. ATCC 29133; Lambert and Carr, 1983), and pFDA in *Fremyella diplosiphon* PCC 7601 (Cobley et al., 1993). Although vectors containing these origins could be a useful module to construct new vectors for *Cyanobacteria*, the homologous rep genes were not found in other plasmids in our database. This finding indicates that *Cyanobacteria* may contain a wider variety of plasmids than other phyla. Smillie et al. (2010) proposed that cyanobacteria may use an as-yet-uncharacterized system to conjugate because, within the phylum, the plasmids encoding relaxase and/or T4CP did not possess known T4SS or VirB4. Indeed, plasmid transfer from *Proteobacteria* to *Cyanobacteria* has been previously reported (Encinas et al., 2014). Further in-depth analyses and classification of *Cyanobacteria* plasmids are necessary to understand the spread of plasmids in this phylum.

Plasmids in *Archaea*

Among the 4602 plasmids in the database, 137 plasmids were found in *Archaea*: 112 plasmids were in *Euryarcheoata* (Figure 1A) and 23 plasmids were in *Crenarchaeota*. There were large differences in the average sizes and GC contents between the two phyla. The plasmids in *Euryarcheoata* had an average size of 119 kb and GC content of 52.1%, while those in *Crenarchaeota* were 20 kb and 39.2%, respectively. *Euryarcheoata* includes all methanogens, haloarchaea, and some hyperthermophilic genera such as *Pyrococcus* and *Thermococcus*. Most known archaeal plasmids have cryptic phenotypes. The best characterized plasmids among methanogens are pME2001, pURB500, and pC2A (Greve et al., 2004). The putative Rep genes have been reported for pME2001 (Luo et al., 2012) and pURB500 (Tumbula et al., 1997), although multiple regions (or genes) were required for replication of pURB500. The putative gene involved in plasmid replication identified in pC2A showed limited homology with Rep proteins encoded by rolling-circle plasmids (del Solar et al., 1993; Metcalf et al., 1997). In *Thermococcales*, there are two well-characterized plasmids, pGT5 from *Pyrococcus abyssi* and pTN1 from *Thermococcus nautilus*, and their respective replication initiation proteins have been identified. Both pGT5 and pTN1 have been shown to use the rolling circle replication system (Arnold et al., 1999; Soler et al., 2007). The majority of plasmids in *Crenarchaeota* were found in *Sulfolobales* plasmids (21/23 plasmids), and they were classified into pRN-type and pNOB-type groups (Greve et al., 2004). Joshua *et al.* have reported the functional characterization of pRN1 (Joshua et al., 2013). A pRN-type plasmid, pSSVx, is known to be a virus-plasmid hybrid that coexists intracellularly with the fusellovirus SSV1 and can be packaged into viral particles (Arnold et al., 1999). The pNOB8-type plasmids are larger in size (around 30 kb) than pRN-type plasmids (less than 10 kb) and are known to be conjugative (She et al., 1998). The putative Rep gene for the pNOB8-type group has not been identified (Greve et al., 2004). Based on the TBLASTN analysis, these Rep genes of pRUB500, pME2001, pC2A, pGT5, pTN1, and pRN1 were found in only 15 plasmids of *Archaea*, suggesting that a wider variety of plasmids may exist in this kingdom than in others.

Genetic tools used in the analysis of archaea are currently being developed. Many archaeal species are resistant to conventional antibiotics, a factor that limits the use of genetic manipulation in archaea (Atomi et al., 2012). Several shuttle vectors are available for certain genera within *Euryarcheoata*, such as halophiles (*Haloarcula*), methanogens (*Methaonococcus* and *Methonosarcina*), and *Thermococcales*(*Pyrococcus*). However, those specific to *Crenarcheota* are only useful for the genus *Sulfolobales*. Methanogens are known to generate methane and were the first microorganisms identified as *Archaea*. Halophiles are the *Archaea* living in the most saline environments on the earth, while *Thermococcales* are found in high temperature environments, mostly above 80°C (Heuer and Smalla, 2012). *Sulfolobales* are the only members of *Crenarcheota* whose genetic manipulation methods have been established thus far (Leigh et al., 2011; Atomi et al., 2012). The plasmids of *Archaea* are typically introduced by transformation; conjugative transfer has not been reported with the exception of a small number of plasmids in *Sulfolobus* (Greve et al., 2004; Smillie et al., 2010).

Recent Work to Identify other Environmental Plasmids

A large number of plasmids have been identified by chance during the analysis of the host bacteria based on specific phenotypes. Indeed, in our recent report of the whole genome sequence of a biphenyl-degrading bacterium, *Geobacillus* sp. JF8, we showed that it carried a plasmid, pBt40, with biphenyl-degradative genes (Shintani et al., 2014b). The putative Rep gene of pBt40 was identified, and its homologs were recently found in plasmids isolated from other strains within the genus *Geobacillus*, although detailed characterizations of these plasmids have not been performed. Considering that most environmental bacteria cannot be cultivated, culture-dependent methods to detect transconjugants of plasmids are known to be highly biased. Smalla and Sobecky (2002) proposed multiphasic approaches for the characterization of plasmids and other mobile genetic elements (Smalla and Sobecky, 2002). Culture-independent methods have recently been developed to isolate and identify novel types of plasmids and to detect plasmid transfers (Heuer and Smalla, 2012).

Methods for Plasmid Isolation and Identification

Notably, 28 plasmids in the database have been identified by culture-independent methods from unidentified host strains or uncultivated microbes. PCR using plasmid-specific sequences is one of the most common methods used to detect plasmids in environmental samples. Various primers and PCR methods to identify plasmids have been designed, such as those for *Enterobacteriaceae* (Carattoli et al., 2005), enterococci (Rosvoll et al., 2010), and other Gram-positive bacteria, mostly staphylococci (Clewell, 2007; Jensen et al., 2010a; Lozano et al., 2012). Methods of direct detection by PCR using degenerate primers have been reported for IncP-1, IncP-7, and IncP-9 plasmids from environmental samples (Dealtry et al., 2014) and for plasmids of different MOB families (Alvarado et al., 2012). Many plasmids have been identified from environmental samples, including the human gut, by metagenomic analyses (Elsaied et al., 2011; Kristiansson et al., 2011; Zhang et al., 2011; Brolund et al., 2013; Song et al., 2013). Brown Kav et al. characterized the overall plasmid population in the bovine rumen (termed rumen plasmidomes) by sequencing the extracted circular plasmids (Brown Kav et al., 2012, 2013). They developed a bioinformatics pipeline that could successfully detect low abundance plasmids and remove contamination of chromosomal DNA (Brown Kav et al., 2012). Sentchilo et al. compared the plasmid metagenomes from two separate activated sludge systems and found that the plasmids from the two environments were strongly different (Sentchilo et al., 2013). Jorgensen et al. reported an *in silico*procedure for identifying small plasmids in metagenomics datasets of the rat cecum (Jorgensen et al., 2014). They successfully identified 160 circular sequences carrying a gene with a plasmid replication domain, and the majority of these sequences were novel (Jorgensen et al., 2014).

Exogenous plasmid isolation is a culture-independent method used to capture plasmids directly from microbial communities. This method is based on the ability of the plasmid mobility and replication in a recipient. Transconjugants are selected by their phenotypes, such as antibiotic resistances, conferred by plasmid carriage. This method enables researchers to search for plasmids in a wider microcosm containing the uncultivated microbial fraction (Heuer et al., 2012). Therefore, the potential to obtain novel types of plasmids is higher. Indeed, this method has been used to isolate a number of novel plasmids (Miyazaki et al., 2006; Sobecky and Hazen, 2009; Sen et al., 2011; Eikmeyer et al., 2012; Brown et al., 2013; Oliveira et al., 2013; Norberg et al., 2014). A transposon-aided capture (TRACA) method has been developed to study and isolate plasmids independent of plasmid-encoded traits in various bacterial habitats (Jones and Marchesi, 2007). This method has also facilitated the successful identification of several novel types of plasmids (Jones and Marchesi, 2007; Zhang et al., 2011; Burmolle et al., 2012).

Methods to Detect Plasmid Transfer

There are multiple reviews of the conventional methods to detect plasmid transfer in various environments based on culture-dependent approaches (Smalla and Sobecky, 2002; Heuer and Smalla, 2007, 2012; Shintani et al., 2010). Systems able to directly detect plasmid transfers by culture-independent methods have been developed using fluorescent protein markers, such as GFP, DsRed, mCherry, GusA,LuxAB, with antibiotic resistance genes (Amann and Fuchs, 2008). Fluorescence *in situ*hybridization (FISH) using specific probes targeting 16S rRNA sequences and *in situ* PCR in transconjugant cells are able to detect transconjugants at the single-cell level (Amann and Fuchs, 2008; Cenciarini-Borde et al., 2009; Wagner and Haider, 2012). Fluorescent cells are also detectable and separable with the use of flow cytometry or a micromanipulator at the single cell level; both methods have a strong potential to identify actual host ranges of plasmids (Musovic et al., 2010; Shintani et al., 2014a).

In silico analyses have been developed to identify horizontal gene transfer phenomena including plasmid transfer. Yamashita et al. carried out a plasmidome network analysis of all available complete bacterial plasmids to identify and characterize the most recent horizontal gene transfer or plasmid transfer (Yamashita et al., 2014). de Been *et al.* reported a novel approach for the reconstruction of mobile genetic elements from whole-genome sequence data, with which they discovered specific plasmid lineages shared between farm animals and humans (de Been et al., 2014).

Over the next decade, the number of fully-sequenced plasmids will greatly increase through the use of deep sequencing methods and *in silico* analyses of datasets. The classification methods using Rep genes and mobility types based on their nucleotide sequences are useful to identify newly isolated plasmids. In this review, it is discussed genes for replication and conjugative transfer, but not the other gene(s) on plasmids including genes for resistance to antibiotics, metabolism of natural and synthetic compounds, pathogenicity, and host symbiosis. The classification lists in this review will be helpful to understand how these "other" genes on plasmids could be spread among microbes. The classification will be also of help to predict host candidates of plasmids found by a metagenomics approach, rather than mostly a collection of results obtained.

Considering that less than half of the plasmids in the database were able to be classified, there may still be unknown or novel types of plasmid replication or transfer systems in microbes. Therefore, characterizations of plasmid features involved in replication, maintenance, transfer, and host range based on molecular biological and biochemical methods are still necessary. Culture-independent methods to detect plasmids and their transfers are currently available for analyzing non-culturable or uncultivated microbes. These studies will not only enable us to understand the evolution of microbes mediated by plasmid transfers but will also provide us with useful tools for genetic analyses of both culturable and non-culturable microbes.

CONFLICT OF INTEREST STATEMENT

The authors declare that the research was conducted in the absence of any commercial or financial relationships that could be construed as a potential conflict of interest.

ACKNOWLEDGMENTS

This work was partially supported by JSPS KAKENHI grant number 24780087 to MS.

REFERENCES

1. Adamczyk, M., and Jagura-Burdzy, G. (2003). Spread and survival of promiscuous IncP-1 plasmids. *Acta Biochim. Pol.* 50, 425–453.
2. Ainsworth, S., Stockdale, S., Bottacini, F., Mahony, J., and van Sinderen, D. (2014). The *Lactococcus lactis* plasmidome: much learnt, yet still lots to discover. *FEMS Microbiol. Rev.* 38, 1066–1088. doi: 10.1111/1574-6976.12074
3. Alvarado, A., Garcillan-Barcia, M. P., and de la Cruz, F. (2012). A degenerate primer MOB typing (DPMT) method to classify gamma-proteobacterial plasmids in clinical and environmental settings. *PLoS ONE* 7:e40438. doi: 10.1371/journal.pone.0040438
4. Amann, R., and Fuchs, B. M. (2008). Single-cell identification in microbial communities by improved fluorescence *in situ* hybridization techniques. *Nat. Rev. Microbiol.* 6, 339–348. doi: 10.1038/nrmicro1888
5. Aminov, R. I. (2011). Horizontal gene exchange in environmental microbiota. *Front. Microbiol.* 2:158. doi: 10.3389/fmicb.2011.00158
6. Arnold, H. P., She, Q., Phan, H., Stedman, K., Prangishvili, D., Holz, I., et al. (1999). The genetic element pSSVx of the extremely thermophilic crenarchaeon *Sulfolobus* is a hybrid between a plasmid and a virus. *Mol. Microbiol.* 34, 217–226. doi: 10.1046/j.1365-

2958.1999.01573.x

7. Atomi, H., Imanaka, T., and Fukui, T. (2012). Overview of the genetic tools in the Archaea. *Front. Microbiol.* 3:337. doi: 10.3389/fmicb.2012.00337

8. Babic, M., Reskova, Z., Bugala, J., Cimova, V., Grones, P., and Grones, J. (2014). The Rep20 replication initiator from the pAG20 plasmid of *Acetobacter aceti. Mol. Biotechnol.* 56, 1–11. doi: 10.1007/s12033-013-9680-6

9. Bao, K., and Cohen, S. N. (2001). Terminal proteins essential for the replication of linear plasmids and chromosomes in *Streptomyces. Genes Dev.* 15, 1518–1527. doi: 10.1101/gad.896201

10. Bao, K., and Cohen, S. N. (2003). Recruitment of terminal protein to the ends of *Streptomyces* linear plasmids and chromosomes by a novel telomere-binding protein essential for linear DNA replication. *Genes Dev.* 17, 774–785. doi: 10.1101/gad.1060303

11. Bertini, A., Poirel, L., Mugnier, P. D., Villa, L., Nordmann, P., and Carattoli, A. (2010). Characterization and PCR-based replicon typing of resistance plasmids in *Acinetobacter baumannii. Antimicrob. Agents Chemother.* 54, 4168–4177. doi: 10.1128/AAC.00542-10

12. Brantl, S. (2004). "Plasmid replication control by antisense RNA," in *Plasmid Biology*, eds G. Phillips and B. Funnell (Washington, DC: ASM Press), 47–62.

13. Brolund, A., Franzen, O., Melefors, O., Tegmark-Wisell, K., and Sandegren, L. (2013). Plasmidome-analysis of ESBL-producing *Escherichia coli* using conventional typing and high-throughput sequencing. *PLoS ONE* 8:e65793. doi: 10.1371/journal.pone.0065793

14. Brown, C. J., Sen, D., Yano, H., Bauer, M. L., Rogers, L. M., Van der Auwera, G. A., et al. (2013). Diverse broad-host-range plasmids from freshwater carry few accessory genes. *Appl. Environ. Microbiol.* 79, 7684–7695. doi: 10.1128/AEM.02252-13

15. Brown Kav, A., Benhar, I., and Mizrahi, I. (2013). A method for purifying high quality and high yield plasmid DNA for metagenomic and deep sequencing approaches. *J. Microbiol. Methods* 95, 272–279. doi: 10.1016/j.mimet.2013.09.008

16. Brown Kav, A., Sasson, G., Jami, E., Doron-Faigenboim, A.,

Benhar, I., and Mizrahi, I. (2012). Insights into the bovine rumen plasmidome. *Proc. Natl. Acad. Sci. U.S.A*. 109, 5452–5457. doi: 10.1073/pnas.1116410109

17. Bruand, C., Ehrlich, S. D., and Janniere, L. (1991). Unidirectional theta replication of the structurally stable *Enterococcus faecalis*plasmid pAM beta 1. *EMBO J.* 10, 2171–2177.
18. Bunikis, I., Kutschan-Bunikis, S., Bonde, M., and Bergström, S. (2011). Multiplex PCR as a tool for validating plasmid content of *Borrelia burgdorferi. J. Microbiol. Methods* 86, 243–247. doi: 10.1016/j.mimet.2011.05.004
19. Burmolle, M., Norman, A., Sorensen, S. J., and Hansen, L. H. (2012). Sequencing of IncX-plasmids suggests ubiquity of mobile forms of a biofilm-promoting gene cassette recruited from *Klebsiella pneumoniae. PLoS ONE* 7:e41259. doi: 10.1371/journal.pone.0041259
20. Carattoli, A. (2009). Resistance plasmid families in *Enterobacteriaceae. Antimicrob. Agents Chemother*. 53, 2227–2238. doi: 10.1128/AAC.01707-08
21. Carattoli, A., Bertini, A., Villa, L., Falbo, V., Hopkins, K. L., and Threlfall, E. J. (2005). Identification of plasmids by PCR-based replicon typing. *J. Microbiol. Methods* 63, 219–228. doi: 10.1016/j.mimet.2005.03.018
22. Casjens, S., Palmer, N., Van Vugt, R., Huang, W. M., Stevenson, B., Rosa, P., et al. (2000). A bacterial genome in flux: the twelve linear and nine circular extrachromosomal DNAs in an infectious isolate of the Lyme disease spirochete *Borrelia burgdorferi. Mol. Microbiol*. 35, 490–516. doi: 10.1046/j.1365-2958.2000.01698.x
23. Casjens, S. R., Mongodin, E. F., Qiu, W. G., Luft, B. J., Schutzer, S. E., Gilcrease, E. B., et al. (2012). Genome stability of Lyme disease spirochetes: comparative genomics of *Borrelia burgdorferi* plasmids. *PLoS ONE* 7:e33280. doi: 10.1371/journal.pone.0033280
24. Cenciarini-Borde, C., Courtois, S., and La Scola, B. (2009). Nucleic acids as viability markers for bacteria detection using molecular tools. *Future Microbiol*. 4, 45–64. doi: 10.2217/17460913.4.1.45
25. Cesareni, G., Helmer-Citterich, M., and Castagnoli, L. (1991). Control of ColE1 plasmid replication by antisense RNA. *Trends*

Genet. 7, 230–235. doi: 10.1016/0168-9525(91)90143-E

26. Cevallos, M. A., Cervantes-Rivera, R., and Gutierrez-Rios, R. M. (2008). The *repABC* plasmid family. *Plasmid* 60, 19–37. doi: 10.1016/j.plasmid.2008.03.001
27. Chaconas, G., and Kobryn, K. (2010). Structure, function, and evolution of linear replicons in *Borrelia*. *Annu. Rev. Microbiol*. 64, 185–202. doi: 10.1146/annurev.micro.112408.134037
28. Chaconas, G., and Norris, S. J. (2013). Peaceful coexistence amongst *Borrelia* plasmids: getting by with a little help from their friends? *Plasmid* 70, 161–167. doi: 10.1016/j.plasmid.2013.05.002
29. Chen, C. W., Yu, T. W., Lin, Y. S., Kieser, H. M., and Hopwood, D. A. (1993). The conjugative plasmid SLP2 of *Streptomyces lividans* is a 50 kb linear molecule. *Mol. Microbiol*. 7, 925–932. doi: 10.1111/j.1365-2958.1993.tb01183.x
30. Clewell, D. B. (2007). Properties of *Enterococcus faecalis* plasmid pAD1, a member of a widely disseminated family of pheromone-responding, conjugative, virulence elements encoding cytolysin. *Plasmid* 58, 205–227. doi: 10.1016/j.plasmid.2007.05.001
31. Cobley, J. G., Zerweck, E., Reyes, R., Mody, A., Seludo-Unson, J. R., Jaeger, H., et al. (1993). Construction of shuttle plasmids which can be efficiently mobilized from *Escherichia coli* into the chromatically adapting cyanobacterium, *Fremyella diplosiphon*. *Plasmid* 30, 90–105. doi: 10.1006/plas.1993.1037
32. Corneau, N., Emond, E., and Lapointe, G. (2004). Molecular characterization of three plasmids from *Bifidobacterium longum*. *Plasmid*51, 87–100. doi: 10.1016/j.plasmid.2003.12.003
33. Crowley, S., Bottacini, F., Mahony, J., and van Sinderen, D. (2013). Complete genome sequence of *Lactobacillus plantarum* strain 16, a broad-spectrum antifungal-producing lactic acid bacterium. *Genome Announc*. 1, e00533–e00513. doi: 10.1128/genomeA.00533-13
34. Dealtry, S., Holmsgaard, P. N., Dunon, V., Jechalke, S., Ding, G. C., Krogerrecklenfort, E., et al. (2014). Shifts in abundance and diversity of mobile genetic elements after the introduction of diverse pesticides into an on-farm biopurification system over the course of a year. *Appl. Environ. Microbiol*. 80, 4012–4020. doi: 10.1128/AEM.04016-13

35. de Been, M., Lanza, V. F., de Toro, M., Scharringa, J., Dohmen, W., Du, Y., et al. (2014). Dissemination of cephalosporin resistance genes between *Escherichia coli* strains from farm animals and humans by specific plasmid lineages. *PLoS Genet.* 10:e1004776. doi: 10.1371/journal.pgen.1004776

36. del Solar, G., Giraldo, R., Ruiz-Echevarria, M. J., Espinosa, M., and Diaz-Orejas, R. (1998). Replication and control of circular bacterial plasmids. *Microbiol. Mol. Biol. Rev.* 62, 434–464.

37. del Solar, G., Moscoso, M., and Espinosa, M. (1993). Rolling circle-replicating plasmids from gram-positive and gram-negative bacteria: a wall falls. *Mol. Microbiol.* 8, 789–796. doi: 10.1111/j.1365-2958.1993.tb01625.x

38. Dijkshoorn, L., Nemec, A., and Seifert, H. (2007). An increasing threat in hospitals: multidrug-resistant *Acinetobacter baumannii*. *Nat. Rev. Microbiol.* 5, 939–951. doi: 10.1038/nrmicro1789

39. Dunny, G. M. (2007). The peptide pheromone-inducible conjugation system of *Enterococcus faecalis* plasmid pCF10: cell-cell signalling, gene transfer, complexity and evolution. *Philos. Trans. R. Soc. Lond. B Biol. Sci.* 362, 1185–1193. doi: 10.1098/rstb.2007.2043

40. Eggers, C. H., Caimano, M. J., Clawson, M. L., Miller, W. G., Samuels, D. S., and Radolf, J. D. (2002). Identification of loci critical for replication and compatibility of a *Borrelia burgdorferi* cp32 plasmid and use of a cp32-based shuttle vector for the expression of fluorescent reporters in the lyme disease spirochaete. *Mol. Microbiol.* 43, 281–295. doi: 10.1046/j.1365-2958.2002.02758.x

41. Eikmeyer, F., Hadiati, A., Szczepanowski, R., Wibberg, D., Schneiker-Bekel, S., Rogers, L. M., et al. (2012). The complete genome sequences of four new IncN plasmids from wastewater treatment plant effluent provide new insights into IncN plasmid diversity and evolution. *Plasmid* 68, 13–24. doi: 10.1016/j.plasmid.2012.01.011

42. Elsaied, H., Stokes, H. W., Kitamura, K., Kurusu, Y., Kamagata, Y., and Maruyama, A. (2011). Marine integrons containing novel integrase genes, attachment sites, attI, and associated gene cassettes in polluted sediments from Suez and Tokyo Bays. *ISME J.* 5, 1162–1177. doi: 10.1038/ismej.2010.208

43. Encinas, D., Garcillan-Barcia, M. P., Santos-Merino, M., Delaye, L., Moya, A., and de la Cruz, F. (2014). Plasmid conjugation from proteobacteria as evidence for the origin of xenologous genes in cyanobacteria. *J. Bacteriol.* 196, 1551–1559. doi: 10.1128/JB.01464-13

44. Espinosa, M., Cohen, S., Couturier, M., Del Solar, G., Diaz-Orejas, R., Giraldo, R., et al. (2000). "Plasmid replication and copy number control," in *The Horizontal Gene Pool: Bacterial Plasmids and Gene Spread*, ed C. M. Thomas (Amsterdam: Harwood Academic Publishers), 1–47.

45. Evans, B. A., and Amyes, S. G. (2014). OXA beta-lactamases. *Clin. Microbiol. Rev.* 27, 241–263. doi: 10.1128/CMR.00117-13

46. Fraser, C. M., Casjens, S., Huang, W. M., Sutton, G. G., Clayton, R., Lathigra, R., et al. (1997). Genomic sequence of a Lyme disease spirochaete, *Borrelia burgdorferi*. *Nature* 390, 580–586. doi: 10.1038/37551

47. Frost, L., Leplae, R., Summers, A., and Toussaint, A. (2005). Mobile genetic elements: the agents of open source evolution. *Nat. Rev. Microbiol.* 3, 722–732. doi: 10.1038/nrmicro1235

48. Frost, L. S., and Koraimann, G. (2010). Regulation of bacterial conjugation: balancing opportunity with adversity. *Future Microbiol.* 5, 1057–1071. doi: 10.2217/fmb.10.70

49. Fukao, M., Oshima, K., Morita, H., Toh, H., Suda, W., Kim, S. W., et al. (2013). Genomic analysis by deep sequencing of the probiotic *Lactobacillus brevis* KB290 harboring nine plasmids reveals genomic stability. *PLoS ONE* 8:e60521. doi: 10.1371/journal.pone.0060521

50. Fukaya, M., Okumura, H., Masai, H., Uozumi, T., and Beppu, T. (1985). Construction of new shuttle vectors for *Acetobacter*. *Agric. Biol. Chem.* 49, 2083–2090. doi: 10.1271/bbb1961.49.2083

51. Funnell, B., and Phillips, G. (2004). "Preface," in *Plasmid Biology*, eds B. Funnell and G. Phillips (Washington, DC: ASM Press), xi.

52. Garcillán-Barcia, M. P., Alvarado, A., and de la Cruz, F. (2011). Identification of bacterial plasmids based on mobility and plasmid population biology. *FEMS Microbiol. Rev.* 35, 936–956. doi: 10.1111/j.1574-6976.2011.00291.x

53. Garcillán-Barcia, M. P., Francia, M. V., and de la Cruz, F. (2009). The diversity of conjugative relaxases and its application in plasmid classification. *FEMS Microbiol. Rev.* 33, 657–687. doi: 10.1111/j.1574-6976.2009.00168.x

54. Giraldo, R., and Fernandez-Tresguerres, M. E. (2004). Twenty years of the pPS10 replicon: insights on the molecular mechanism for the activation of DNA replication in iteron-containing bacterial plasmids. *Plasmid* 52, 69–83. doi: 10.1016/j.plasmid.2004.06.002

55. Goessweiner-Mohr, N., Arends, K., Keller, W., and Grohmann, E. (2013). Conjugative type IV secretion systems in Gram-positive bacteria. *Plasmid* 70, 289–302. doi: 10.1016/j.plasmid.2013.09.005

56. Golden, S. S., and Sherman, L. A. (1983). A hybrid plasmid is a stable cloning vector for the cyanobacterium *Anacystis nidulans* R2. *J. Bacteriol.* 155, 966–972.

57. Greve, B., Jensen, S., Brügger, K., Zillig, W., and Garrett, R. A. (2004). Genomic comparison of archaeal conjugative plasmids from *Sulfolobus*. *Archaea* 1, 231–239. doi: 10.1155/2004/151926

58. Grones, P., and Grones, J. (2012). Characterization of the theta replication plasmid pGR7 from *Acetobacter aceti* CCM 3610. *Res. Microbiol.* 163, 419–426. doi: 10.1016/j.resmic.2012.07.002

59. Guglielmetti, S., Mora, D., and Parini, C. (2007). Small rolling circle plasmids in *Bacillus subtilis* and related species: organization, distribution, and their possible role in host physiology. *Plasmid* 57, 245 264. doi: 10.1016/j.plasmid.2006.09.002

60. Guglielmini, J., Quintais, L., Garcillán-Barcia, M. P., de la Cruz, F., and Rocha, E. P. (2011). The repertoire of ICE in prokaryotes underscores the unity, diversity, and ubiquity of conjugation. *PLoS Genet.* 7:e1002222. doi: 10.1371/journal.pgen.1002222

61. Heuer, H., Binh, C. T., Jechalke, S., Kopmann, C., Zimmerling, U., Krogerrecklenfort, E., et al. (2012). IncP-1epsilon plasmids are important vectors of antibiotic resistance genes in agricultural systems: diversification driven by class 1 integron gene cassettes. *Front. Microbiol.* 3:2. doi: 10.3389/fmicb.2012.00002

62. Heuer, H., and Smalla, K. (2007). Horizontal gene transfer between bacteria. *Environ. Biosafety Res.* 6, 3–13. doi: 10.1051/ebr:2007034

63. Heuer, H., and Smalla, K. (2012). Plasmids foster diversification

and adaptation of bacterial populations in soil. *FEMS Microbiol. Rev*. 36, 1083–1104. doi: 10.1111/j.1574-6976.2012.00337.x

64. Hou, C., Wang, Q., Zeng, X., Yang, F., Zhang, J., Liu, H., et al. (2014). Complete genome sequence of *Lactobacillus reuteri* I5007, a probiotic strain isolated from healthy piglet. *J. Biotechnol*. 179, 63–64. doi: 10.1016/j.jbiotec.2014.03.019

65. Huang, C. H., Tsai, H. H., Tsay, Y. G., Chien, Y. N., Wang, S. L., Cheng, M. Y., et al. (2007). The telomere system of the *Streptomyces*linear plasmid SCP1 represents a novel class. *Mol. Microbiol*. 63, 1710–1718. doi: 10.1111/j.1365-2958.2007.05616.x

66. Jayashree, S., Pooja, S., Pushpanathan, M., Vishnu, U., Sankarasubramanian, J., Rajendhran, J., et al. (2013). Genome sequence of *Lactobacillus fermentum* strain MTCC 8711, a probiotic bacterium isolated from yogurt. *Genome Announc*. 1, e00770–e00713. doi: 10.1128/genomeA.00770-13

67. Jensen, L. B., Garcia-Migura, L., Valenzuela, A. J., Løhr, M., Hasman, H., and Aarestrup, F. M. (2010a). A classification system for plasmids from enterococci and other Gram-positive bacteria. *J. Microbiol. Methods* 80, 25–43. doi: 10.1016/j.mimet.2009.10.012

68. Jensen, S. O., Apisiridej, S., Kwong, S. M., Yang, Y. H., Skurray, R. A., and Firth, N. (2010b). Analysis of the prototypical *Staphylococcus aureus* multiresistance plasmid pSK1. *Plasmid* 64, 135–142. doi: 10.1016/j.plasmid.2010.06.001

69. Jeon, C. O., Park, M., Ro, H. S., Park, W., and Madsen, E. L. (2006). The naphthalene catabolic (*nag*) genes of *Polaromonas naphthalenivorans* CJ2: evolutionary implications for two gene clusters and novel regulatory control. *Appl. Environ. Microbiol*. 72, 1086–1095. doi: 10.1128/AEM.72.2.1086-1095.2006

70. Jeon, C. O., Park, W., Padmanabhan, P., Derito, C., Snape, J. R., and Madsen, E. L. (2003). Discovery of a bacterium, with distinctive dioxygenase, that is responsible for *in situ* biodegradation in contaminated sediment. *Proc. Natl. Acad. Sci. U.S.A*. 100, 13591–13596. doi: 10.1073/pnas.1735529100

71. Johnson, T. J., and Nolan, L. K. (2009). Pathogenomics of the virulence plasmids of *Escherichia coli*. *Microbiol. Mol. Biol. Rev*. 73, 750–774. doi: 10.1128/MMBR.00015-09

72. Jones, B. V., and Marchesi, J. R. (2007). Transposon-aided capture (TRACA) of plasmids resident in the human gut mobile metagenome. *Nat. Methods* 4, 55–61. doi: 10.1038/nmeth964

73. Jorgensen, T. S., Xu, Z., Hansen, M. A., Sorensen, S. J., and Hansen, L. H. (2014). Hundreds of circular novel plasmids and DNA elements identified in a rat cecum metamobilome. *PLoS ONE* 9:e87924. doi: 10.1371/journal.pone.0087924

74. Joshua, C. J., Perez, L. D., and Keasling, J. D. (2013). Functional characterization of the origin of replication of pRN1 from *Sulfolobus islandicus* REN1H1. *PLoS ONE* 8:e84664. doi: 10.1371/journal.pone.0084664

75. Kendall, K. J., and Cohen, S. N. (1988). Complete nucleotide sequence of the *Streptomyces lividans* plasmid pIJ101 and correlation of the sequence with genetic properties. *J. Bacteriol.* 170, 4634–4651.

76. Khan, S. A. (2005). Plasmid rolling-circle replication: highlights of two decades of research. *Plasmid* 53, 126–136. doi: 10.1016/j.plasmid.2004.12.008

77. Kristiansson, E., Fick, J., Janzon, A., Grabic, R., Rutgersson, C., Weijdegård, B., et al. (2011). Pyrosequencing of antibiotic-contaminated river sediments reveals high levels of resistance and gene transfer elements. *PLoS ONE* 6:e17038. doi: 10.1371/journal.pone.0017038

78. Krüger, R., Rakowski, S. A., and Filutowicz, M. (2004). "Participating elements in the replication of iteron-containing plasmids," in *Plasmid Biology*, eds G. Phillips and B. Funnell (Washington, DC: ASM Press), 25–45.

79. Labidi, A., David, H. L., and Roulland-Dussoix, D. (1985). Restriction endonuclease mapping and cloning of *Mycobacterium fortuitum*var. *fortuitum* plasmid pAL5000. *Ann. Inst. Pasteur Microbiol.* 136B, 209–215. doi: 10.1016/S0769-2609(85)80045-4

80. Lambert, G., and Carr, N. (1983). A restriction map of plasmid pDC1 from the filamentous cyanobacterium *Nostoc* sp. *MAC PCC 8009. Plasmid* 10, 196–198. doi: 10.1016/0147-619X(83)90072-0

81. Lawley, T., Wilkins, B. M., and Frost, L. S. (2004). "Bacterial conjugation in Gram-negative bacteria," in *Plasmid Biology*, eds G.

Phillips and B. Funnell (Washington, DC: ASM Press), 203–226.

82. Lawrence, J. G., and Ochman, H. (1998). Molecular archaeology of the *Escherichia coli* genome. *Proc. Natl. Acad. Sci. U.S.A.* 95, 9413–9417. doi: 10.1073/pnas.95.16.9413

83. Le Dantec, C., Winter, N., Gicquel, B., Vincent, V., and Picardeau, M. (2001). Genomic sequence and transcriptional analysis of a 23-kilobase mycobacterial linear plasmid: evidence for horizontal transfer and identification of plasmid maintenance systems. *J. Bacteriol.* 183, 2157–2164. doi: 10.1128/JB.183.7.2157-2164.2001

84. Lee, J. H., and O'Sullivan, D. J. (2006). Sequence analysis of two cryptic plasmids from *Bifidobacterium longum* DJO10A and construction of a shuttle cloning vector. *Appl. Environ. Microbiol.* 72, 527–535. doi: 10.1128/AEM.72.1.527-535.2006

85. Leigh, J. A., Albers, S. V., Atomi, H., and Allers, T. (2011). Model organisms for genetics in the domain *Archaea*: methanogens, halophiles, *Thermococcales* and *Sulfolobales*. *FEMS Microbiol. Rev.* 35, 577–608. doi: 10.1111/j.1574-6976.2011.00265.x

86. Liu, M. A., Kwong, S. M., Jensen, S. O., Brzoska, A. J., and Firth, N. (2013). Biology of the staphylococcal conjugative multiresistance plasmid pSK41. *Plasmid* 70, 42–51. doi: 10.1016/j.plasmid.2013.02.001

87. Lozano, C., García-Migura, L., Aspiroz, C., Zarazaga, M., Torres, C., and Aarestrup, F. M. (2012). Expansion of a plasmid classification system for Gram-positive bacteria and determination of the diversity of plasmids in *Staphylococcus aureus* strains of human, animal, and food origins. *Appl. Environ. Microbiol.* 78, 5948–5955. doi: 10.1128/AEM.00870-12

88. Luo, Y. R., Kang, S. G., Kim, S. J., Kim, M. R., Li, N., Lee, J. H., et al. (2012). Genome sequence of benzo(a)pyrene-degrading bacterium *Novosphingobium pentaromativorans* US6-1. *J. Bacteriol.* 194, 907. doi: 10.1128/JB.06476-11

89. Masai, E., Sugiyama, K., Iwashita, N., Shimizu, S., Hauschild, J. E., Hatta, T., et al. (1997). The *bphDEF* meta-cleavage pathway genes involved in biphenyl/polychlorinated biphenyl degradation are located on a linear plasmid and separated from the initial *bphACB* genes in *Rhodococcus* sp. strain RHA1. *Gene* 187, 141–149. doi:

10.1016/S0378-1119(96)00748-2

90. McLeod, M. P., Warren, R. L., Hsiao, W. W., Araki, N., Myhre, M., Fernandes, C., et al. (2006). The complete genome of *Rhodococcus* sp. RHA1 provides insights into a catabolic powerhouse. *Proc. Natl. Acad. Sci. U.S.A.* 103, 15582–15587. doi: 10.1073/pnas.0607048103
91. Metcalf, W. W., Zhang, J. K., Apolinario, E., Sowers, K. R., and Wolfe, R. S. (1997). A genetic system for *Archaea* of the genus *Methanosarcina*: liposome-mediated transformation and construction of shuttle vectors. *Proc. Natl. Acad. Sci. U.S.A.* 94, 2626–2631. doi: 10.1073/pnas.94.6.2626
92. Miller, T. R., Delcher, A. L., Salzberg, S. L., Saunders, E., Detter, J. C., and Halden, R. U. (2010). Genome sequence of the dioxin-mineralizing bacterium *Sphingomonas wittichii* RW1. *J. Bacteriol.* 192, 6101–6102. doi: 10.1128/JB.01030-10
93. Miyazaki, R., Sato, Y., Ito, M., Ohtsubo, Y., Nagata, Y., and Tsuda, M. (2006). Complete nucleotide sequence of an exogenously isolated plasmid, pLB1, involved in gamma-hexachlorocyclohexane degradation. *Appl. Environ. Microbiol.* 72, 6923–6933. doi: 10.1128/AEM.01531-06
94. Musovic, S., Dechesne, A., Sørensen, J., and Smets, B. F. (2010). Novel assay to assess permissiveness of a soil microbial community toward receipt of mobile genetic elements. *Appl. Environ. Microbiol.* 76, 4813–4818. doi: 10.1128/AEM.02713-09
95. Nagata, Y., Kamakura, M., Endo, R., Miyazaki, R., Ohtsubo, Y., and Tsuda, M. (2006). Distribution of gamma-hexachlorocyclohexane-degrading genes on three replicons in *Sphingobium japonicum* UT26. *FEMS Microbiol. Lett.* 256, 112–118. doi: 10.1111/j.1574-6968.2005.00096.x
96. Nagata, Y., Natsui, S., Endo, R., Ohtsubo, Y., Ichikawa, N., Ankai, A., et al. (2011). Genomic organization and genomic structural rearrangements of *Sphingobium japonicum* UT26, an archetypal γ-hexachlorocyclohexane-degrading bacterium. *Enzyme Microb. Technol.* 49, 499–508. doi: 10.1016/j.enzmictec.2011.10.005
97. Nagata, Y., Ohtsubo, Y., Endo, R., Ichikawa, N., Ankai, A., Oguchi, A., et al. (2010). Complete genome sequence of the representative γ-hexachlorocyclohexane-degrading bacterium *Sphingobium*

japonicum UT26. *J. Bacteriol.* 192, 5852–5853. doi: 10.1128/JB.00961-10

98. Nishida, H. (2012). Comparative analyses of base compositions, DNA sizes, and dinucleotide frequency profiles in archaeal and bacterial chromosomes and plasmids. *Int. J. Evol. Biol.* 2012:342482. doi: 10.1155/2012/342482

99. Norberg, P., Bergstrom, M., and Hermansson, M. (2014). Complete nucleotide sequence and analysis of two conjugative broad host range plasmids from a marine microbial biofilm. *PLoS ONE* 9:e92321. doi: 10.1371/journal.pone.0092321

100. Okumura, H., Uozumi, T., and Beppu, T. (1985). Construction of plasmid vectors and a genetic transformation system for *Acetobacter aceti*. *Agric. Biol. Chem.* 49, 1011–1017. doi: 10.1271/bbb1961.49.1011

101. Oliveira, C. S., Moura, A., Henriques, I., Brown, C. J., Rogers, L. M., Top, E. M., et al. (2013). Comparative genomics of IncP-1epsilon plasmids from water environments reveals diverse and unique accessory genetic elements. *Plasmid* 70, 412–419. doi: 10.1016/j.plasmid.2013.06.002

102. Palmer, K. L., Kos, V. N., and Gilmore, M. S. (2010). Horizontal gene transfer and the genomics of enterococcal antibiotic resistance. *Curr. Opin. Microbiol.* 13, 632–639. doi: 10.1016/j.mib.2010.08.004

103. Pandza, S., Biukovic, G., Paravic, A., Dadbin, A., Cullum, J., and Hranueli, D. (1998). Recombination between the linear plasmid pPZG101 and the linear chromosome of *Streptomyces rimosus* can lead to exchange of ends. *Mol. Microbiol.* 28, 1165–1176. doi: 10.1046/j.1365-2958.1998.00877.x

104. Park, M. S., Shin, D. W., Lee, K. H., and Ji, G. E. (1999). Sequence analysis of plasmid pKJ50 from *Bifidobacterium longum*. *Microbiology* 145(Pt 3), 585–592. doi: 10.1099/13500872-145-3-585

105. Partridge, S. R. (2011). Analysis of antibiotic resistance regions in Gram-negative bacteria. *FEMS Microbiol. Rev.* 35, 820–855. doi: 10.1111/j.1574-6976.2011.00277.x

106. Paterson, D. L. (2006). Resistance in gram-negative bacteria: *Enterobacteriaceae*. *Am. J. Infect. Control* 34, S20–S28. discussion:

S64–S73. doi: 10.1016/j.ajic.2006.05.238

107. Petersen, J. M., Zielinski, F. U., Pape, T., Seifert, R., Moraru, C., Amann, R., et al. (2011). Hydrogen is an energy source for hydrothermal vent symbioses. *Nature* 476, 176–180. doi: 10.1038/nature10325

108. Pinto, U. M., Pappas, K. M., and Winans, S. C. (2012). The ABCs of plasmid replication and segregation. *Nat. Rev. Microbiol.* 10, 755–765. doi: 10.1038/nrmicro2882

109. Poehlein, A., Kusian, B., Friedrich, B., Daniel, R., and Bowien, B. (2011). Complete genome sequence of the type strain *Cupriavidus necator* N-1. *J. Bacteriol.* 193, 5017. doi: 10.1128/JB.05660-11

110. Polisky, B. (1988). ColE1 replication control circuitry: sense from antisense. *Cell* 55, 929–932. doi: 10.1016/0092-8674(88)90235-8

111. Rakowski, S. A., and Filutowicz, M. (2013). Plasmid R6K replication control. *Plasmid* 69, 231–242. doi: 10.1016/j.plasmid.2013.02.003

112. Ravin, N. V. (2011). N15: the linear phage-plasmid. *Plasmid* 65, 102–109. doi: 10.1016/j.plasmid.2010.12.004

113. Rocha, E. P., and Danchin, A. (2002). Base composition bias might result from competition for metabolic resources. *Trends Genet.* 18, 291–294. doi: 10.1016/S0168-9525(02)02690-2

114. Romine, M. F., Stillwell, L. C., Wong, K. K., Thurston, S. J., Sisk, E. C., Sensen, C., et al. (1999). Complete sequence of a 184-kilobase catabolic plasmid from *Sphingomonas aromaticivorans* F199. *J. Bacteriol.* 181, 1585–1602.

115. Rosvoll, T. C., Pedersen, T., Sletvold, H., Johnsen, P. J., Sollid, J. E., Simonsen, G. S., et al. (2010). PCR-based plasmid typing in *Enterococcus faecium* strains reveals widely distributed pRE25-, pRUM-, pIP501- and pHTbeta-related replicons associated with glycopeptide resistance and stabilizing toxin-antitoxin systems. *FEMS Immunol. Med. Microbiol.* 58, 254–268. doi: 10.1111/j.1574-695X.2009.00633.x

116. Sekine, M., Tanikawa, S., Omata, S., Saito, M., Fujisawa, T., Tsukatani, N., et al. (2006). Sequence analysis of three plasmids harboured in *Rhodococcus erythropolis* strain PR4. *Environ. Microbiol.* 8, 334–346. doi: 10.1111/j.1462-2920.2005.00899.x

117. Sen, D., Van der Auwera, G. A., Rogers, L. M., Thomas, C. M., Brown, C. J., and Top, E. M. (2011). Broad-host-range plasmids from agricultural soils have IncP-1 backbones with diverse accessory genes. *Appl. Environ. Microbiol.* 77, 7975–7983. doi: 10.1128/AEM.05439-11

118. Sentchilo, V., Mayer, A. P., Guy, L., Miyazaki, R., Green Tringe, S., Barry, K., et al. (2013). Community-wide plasmid gene mobilization and selection. *ISME J.* 7, 1173–1186. doi: 10.1038/ismej.2013.13

119. Seto, M., Kimbara, K., Shimura, M., Hatta, T., Fukuda, M., and Yano, K. (1995). A novel transformation of polychlorinated biphenyls by *Rhodococcus* sp. strain RHA1. *Appl. Environ. Microbiol.* 61, 3353–3358.

120. Shareck, J., Choi, Y., Lee, B., and Miguez, C. B. (2004). Cloning vectors based on cryptic plasmids isolated from lactic acid bacteria: their characteristics and potential applications in biotechnology. *Crit. Rev. Biotechnol.* 24, 155–208. doi: 10.1080/07388550490904288

121. She, Q., Phan, H., Garrett, R. A., Albers, S. V., Stedman, K. M., and Zillig, W. (1998). Genetic profile of pNOB8 from *Sulfolobus*: the first conjugative plasmid from an archaeon. *Extremophiles* 2, 417–425. doi: 10.1007/s007920050087

122. Shimizu, S., Kobayashi, H., Masai, E., and Fukuda, M. (2001). Characterization of the 450-kb linear plasmid in a polychlorinated biphenyl degrader, *Rhodococcus* sp. strain RHA1. *Appl. Environ. Microbiol.* 67, 2021–2028. doi: 10.1128/AEM.67.5.2021-2028.2001

123. Shintani, M., Matsui, K., Inoue, J., Hosoyama, A., Ohji, S., Yamazoe, A., et al. (2014a). Single-cell analyses revealed transfer ranges of IncP-1, IncP-7, and IncP-9 plasmids in a soil bacterial community. *Appl. Environ. Microbiol.* 80, 138–145. doi: 10.1128/AEM.02571-13

124. Shintani, M., and Nojiri, H. (2013). "Mobile genetic elements (MGEs) carrying catabolic genes," in *Management of Microbial Resources in the Environment*, eds A. Malik, E. Grohmann, and M. Alves (Dordrecht: Springer), 167–214. doi: 10.1007/978-94-007-5931-2_8

125. Shintani, M., Ohtsubo, Y., Fukuda, K., Hosoyama, A., Ohji, S., Yamazoe, A., et al. (2014b). Complete genome sequence of the

thermophilic polychlorinated biphenyl degrader *Geobacillus* sp. strain JF8 (NBRC 109937). *Genome Announc* 2, e01213–13. doi: 10.1128/genomeA.01213-13

126. Shintani, M., Takahashi, Y., Yamane, H., and Nojiri, H. (2010). The behavior and significance of degradative plasmids belonging to Inc groups in *Pseudomonas* within natural environments and microcosms. *Microbes Environ.* 25, 253–265. doi: 10.1264/jsme2.ME10155
127. Smalla, K., and Sobecky, P. A. (2002). The prevalence and diversity of mobile genetic elements in bacterial communities of different environmental habitats: insights gained from different methodological approaches. *FEMS Microbiol. Ecol.* 42, 165–175. doi: 10.1111/j.1574-6941.2002.tb01006.x
128. Smillie, C., Garcillán-Barcia, M. P., Francia, M. V., Rocha, E. P., and de la Cruz, F. (2010). Mobility of plasmids. *Microbiol. Mol. Biol. Rev.* 74, 434–452. doi: 10.1128/MMBR.00020-10
129. Smith, M. C., and Thomas, C. D. (2004). An accessory protein is required for relaxosome formation by small staphylococcal plasmids. *J. Bacteriol.* 186, 3363–3373. doi: 10.1128/JB.186.11.3363-3373.2004
130. Sobecky, P. A., and Hazen, T. H. (2009). Horizontal gene transfer and mobile genetic elements in marine systems. *Methods Mol. Biol.* 532, 435–453. doi: 10.1007/978-1-60327-853-9_25
131. Soler, N., Justome, A., Quevillon-Cheruel, S., Lorieux, F., Le Cam, E., Marguet, E., et al. (2007). The rolling-circle plasmid pTN1 from the hyperthermophilic archaeon *Thermococcus nautilus*. *Mol. Microbiol.* 66, 357–370. doi: 10.1111/j.1365-2958.2007.05912.x
132. Song, X., Sun, J., Mikalsen, T., Roberts, A. P., and Sundsfjord, A. (2013). Characterisation of the plasmidome within *Enterococcus faecalis* isolated from marginal periodontitis patients in Norway. *PLoS ONE* 8:e62248. doi: 10.1371/journal.pone.0062248
133. Sorvig, E., Mathiesen, G., Naterstad, K., Eijsink, V. G., and Axelsson, L. (2005). High-level, inducible gene expression in *Lactobacillus sakei* and *Lactobacillus plantarum* using versatile expression vectors. *Microbiology* 151, 2439–2449. doi: 10.1099/mic.0.28084-0

134. Sota, M., and Top, E. (2008). "Horizontal gene transfer mediated by plasmids," in *Plasmids: Current Research and Future Trends*, ed G. Lipps (Norfolk, VA: Caister Academic Press; Horizon Scientific Press), 111–181.

135. Stillwell, L. C., Thurston, S. J., Schneider, R. P., Romine, M. F., Fredrickson, J. K., and Saffer, J. D. (1995). Physical mapping and characterization of a catabolic plasmid from the deep-subsurface bacterium *Sphingomonas* sp. strain F199. *J. Bacteriol.* 177, 4537–4539.

136. Straley, S. C., Plano, G. V., Skrzypek, E., Haddix, P. L., and Fields, K. A. (1993). Regulation by Ca^{2+} in the *Yersinia* low-Ca^{2+} response. *Mol. Microbiol.* 8, 1005–1010. doi: 10.1111/j.1365-2958.1993.tb01644.x

137. Taton, A., Unglaub, F., Wright, N. E., Zeng, W. Y., Paz-Yepes, J., Brahamsha, B., et al. (2014). Broad-host-range vector system for synthetic biology and biotechnology in cyanobacteria. *Nucleic Acids Res.* 42:e136. doi: 10.1093/nar/gku673

138. Tauch, A., Bischoff, N., Brune, I., and Kalinowski, J. (2003a). Insights into the genetic organization of the *Corynebacterium diphtheriae*erythromycin resistance plasmid pNG2 deduced from its complete nucleotide sequence. *Plasmid* 49, 63–74. doi: 10.1016/S0147-619X(02)00115-4

139. Tauch, A., Pühler, A., Kalinowski, J., and Thierbach, G. (2003b). Plasmids in *Corynebacterium glutamicum* and their molecular classification by comparative genomics. *J. Biotechnol.* 104, 27–40. doi: 10.1016/S0168-1656(03)00157-3

140. Taylor, D. E., Gibreel, A., Lawley, T. D., and Tracz, D. M. (2004). "Anitibiotic resistance plasmids," in *Plasmid Biology*, eds G. Phillips and B. Funnell (Washington, DC: ASM Press), 473–491.

141. Thomas, C. M., and Haines, A. S. (2004). "Plasmids of the genus *Pseudomonas*," in *Pseudomonas*, ed J. L. Ramos (New York, NY: Plenum Publishing Corporation), 197–231.

142. Tilly, K., Checroun, C., and Rosa, P. A. (2012). Requirements for *Borrelia burgdorferi* plasmid maintenance. *Plasmid* 68, 1–12. doi: 10.1016/j.plasmid.2012.01.009

143. Torres, M. J., Rubia, M. I., Bedmar, E. J., and Delgado, M. J. (2011).

Denitrification in *Sinorhizobium meliloti*. *Biochem. Soc. Trans*. 39, 1886–1889. doi: 10.1042/BST20110733

144. Tumbula, D. L., Bowen, T. L., and Whitman, W. B. (1997). Characterization of pURB500 from the archaeon *Methanococcus maripaludis*and construction of a shuttle vector. *J. Bacteriol*. 179, 2976–2986.

145. Udo, E. E., and Grubb, W. B. (1991). A new incompatibility group plasmid in *Staphylococcus aureus*. *FEMS Microbiol. Lett*. 62, 33–36. doi: 10.1111/j.1574-6968.1991.tb04412.x

146. Ventura, M., Canchaya, C., Tauch, A., Chandra, G., Fitzgerald, G. F., Chater, K. F., et al. (2007). Genomics of *Actinobacteria*: tracing the evolutionary history of an ancient phylum. *Microbiol. Mol. Biol. Rev*. 71, 495–548. doi: 10.1128/MMBR.00005-07

147. Wagner, M., and Haider, S. (2012). New trends in fluorescence *in situ* hybridization for identification and functional analyses of microbes. *Curr. Opin. Biotechnol*. 23, 96–102. doi: 10.1016/j.copbio.2011.10.010

148. Walton, D. K., Gendel, S. M., and Atherly, A. G. (1992). Nucleotide sequence of the replication region of the *Nostoc* PCC 7524 plasmid pDU1. *Nucleic Acids Res*. 20:4660. doi: 10.1093/nar/20.17.4660

149. Wang, J., Stephan, R., Karczmarczyk, M., Yan, Q., Hachler, H., and Fanning, S. (2013). Molecular characterization of bla ESBL-harboring conjugative plasmids identified in multi-drug resistant *Escherichia coli* isolated from food-producing animals and healthy humans. *Front. Microbiol*. 4:188. doi: 10.3389/fmicb.2013.00188

150. Xiong, J., Alexander, D. C., Ma, J. H., Deraspe, M., Low, D. E., Jamieson, F. B., et al. (2013). Complete sequence of pOZ176, a 500-kilobase IncP-2 plasmid encoding IMP-9-mediated carbapenem resistance, from outbreak isolate *Pseudomonas aeruginosa* 96. *Antimicrob. Agents Chemother*. 57, 3775–3782. doi: 10.1128/AAC.00423-13

151. Yagi, J. M., Sims, D., Brettin, T., Bruce, D., and Madsen, E. L. (2009). The genome of *Polaromonas naphthalenivorans* strain CJ2, isolated from coal tar-contaminated sediment, reveals physiological and metabolic versatility and evolution through extensive horizontal gene transfer. *Environ. Microbiol*. 11, 2253–2270. doi:

10.1111/j.1462-2920.2009.01947.x

152. Yamashita, A., Sekizuka, T., and Kuroda, M. (2014). Characterization of antimicrobial resistance dissemination across plasmid communities classified by network analysis. *Pathogens* 3, 356–376. doi: 10.3390/pathogens3020356

153. Zhang, T., Zhang, X. X., and Ye, L. (2011). Plasmid metagenome reveals high levels of antibiotic resistance genes and mobile genetic elements in activated sludge. *PLoS ONE* 6:e26041. doi: 10.1371/journal.pone.0026041

SECTION VII
PLASMIDS AS GENETIC TOOLS

CHAPTER

11

A PLASMID-BASED SYSTEM FOR EXPRESSING SMALL INTERFERING RNA LIBRARIES IN MAMMALIAN CELLS

Ajamete Kaykas and Randall T Moon

Howard Hughes Medical Institute, Department of Pharmacology, and Center for Developmental Biology, University of Washington School of Medicine, Seattle, WA 98195, USA

ABSTRACT

Background

RNA interference (RNAi) is an evolutionarily conserved process that functions to inhibit gene expression. The use of RNAi in mammals as a tool to study gene function has rapidly developed in the last couple of years since the discovery that the function-inhibiting units of RNAi are short 21–25 nt double-stranded RNAs (siRNAs) derived from their longer template. The use of siRNAs allows for gene-specific knock-down without induction of the non-specific interferon response in mammalian cells. Multiple systems have been developed to introduce siRNAs into mammals. One of the most appealing of these techniques is the use of vectors containing polymerase III promoters to drive expression of hairpin siRNAs. However, there are multiple limitations to using hairpin siRNA vectors including the observation that some are unstable in bacteria and are difficult to sequence.

Results

To circumvent the limitation of hairpin siRNA vectors we have developed a convergent opposing siRNA expression system called pHippy. We have generated pHippy vectors or expression cassettes that knock down the expression of both reporter and endogenous genes. As a proof of principle that pHippy can be used to generate random siRNA libraries, we generated a small siRNA library against PGL3 luciferase and demonstrated that we could recover functional siRNAs that knock down PGL3 luciferase.

Conclusions

siRNA is a powerful tool to study gene function. We have developed a new vector with opposing convergent promoters for the expression of siRNAs, which can be used to knock down endogenous genes in a high throughput manner or to perform functional screening with random or cDNA-derived siRNA libraries.

BACKGROUND

RNA interference is an evolutionarily conserved process that functions to inhibit gene expression [1, 2]. The phenomenon of RNAi was first described by Fire et al. six years ago. In seminal experiments they injected double stranded RNA (dsRNA) into *Caenorhabditis elegans*, which led to efficient sequence-specific gene silencing of the mRNA that was complimentary to the dsRNA [3]. RNAi has also been described in plants as a phenomenon called post-transcriptional gene silencing (PTGS) which is likely used as a viral defense mechanism [4, 5, 6]. Introduction of long dsRNA into a variety of organisms such as *Drosophila*, *Trypanosoma*, and pre-implanted mouse oocytes has been shown to specifically inhibit the complimentary mRNA [5, 6, 7]. However, in somatic mammalian cells long dsRNA also induces the interferon response, which globally inhibits translation by induction of the kinase, PKR, and 2', 5'-oligoadenylate synthetase. This has limited the use of RNAi as a tool to study gene function in mammalian cells [8, 9].

The first indication that the molecules that regulate PTGS were short RNAs processed from longer dsRNA was the identification of short 21 nt to 22 nt dsRNA derived from the longer dsRNA in plants [6]. This observation was recapitulated in *Drosophila* embryo extracts where long dsRNA was found processed into 21–25 nt short RNA by the RNase III type enzyme, Dicer [9, 10, 11]. These observations led Elbashir et al. to test if synthetic 21–25 nt synthetic dsRNAs function to specifically inhibit gene expression in *Drosophila* embryo lysates and mammalian cell culture [9, 10, 11]. They demonstrated that siRNA had the ability to specifically inhibit gene expression in mammalian cell culture without induction of the interferon response. These observations led to the development of many techniques for the specific knockdown of genes in mammalian cell culture.

Of these techniques, plasmid-based systems that generate hairpin siRNAs are very appealing [12, 13, 14, 15]. These vectors are fairly inexpensive and have been shown to inhibit multiple genes both transiently and in long-term experiments. However, hairpin vectors suffer from multiple limitations. Hairpins can be hard to synthesize in bacteria, difficult to sequence, and the oligos needed to generate them can be costly and error-prone [14, 16]. In addition, the hairpin length and sequence can affect the ability of the siRNA to inhibit gene expression [12, 17]. Although cDNA-based siRNA libraries have been generated in hairpin vectors, generation of these libraries is technically difficult and heterologous hairpin loops have to be added to the siRNA [18, 19]. The latter may increase off-target effects of the siRNAs generated from these strategies.

To circumvent the limitations of hairpin siRNA expression vectors, we have generated a siRNA expression vector, which expresses siRNAs from convergent opposing promoters. We have named this vector «HI inverted U6 p romoter p lasmid,» or pHippy. Short double-stranded oligonucleotides can be easily and efficiently introduced into this vector to knock down any mRNA, or siRNA expression cassettes can be generated in a high throughput manner by PCR. We demonstrate that pHippy can generate functional siRNAs in mammalian cells that knock down the expression of both ectopic reporters and endogenous genes. Because both strands of the siRNA are transcribed from the same template DNA, random or cDNA-derived siRNA libraries can easily be generated in pHippy. As a proof of principle we generated a random pool of siRNAs against PGL3 luciferase and screened this library for silencing activity. This library was deconvoluted to identify functional vectors that have the original siRNA sequence against PGL3 luciferase. Although this is only a partial random library to a previous efficacious siRNA, this demonstrates that random siRNA libraries can be used to functionally identify genes.

RESULTS AND DISCUSSION

Design of an opposing convergent siRNA expression vector, pHippy

pHippy contains two opposing polymerase III (pol. III) promoters to drive the expression of both strands of a template DNA cloned in between the promoters. To circumvent generating an inverted repeat, which can cause plasmid instability in *E. coli*, the human H1 and human U6 promoters were used instead of two H1 or U6 promoters. Both the H1 and U6 promoter were modified to contain a five Thymidine pol. III termination sequence at the -5 to -1 position and a BsmBl site at the -12 to -6 position (Fig. 1a). This does not appear to affect the ability of either promoter to drive expression of RNA nor does it appear to affect the transcriptional start site. The BsmBl cuts 5 bases downstream of its recognition site and in the case of pHippy will leave 4 Thymidine 3' overhangs at -1 to -4 of both U6 and H1 promoters. Digestion of pHippy with BsmB1 renders the two ends incompatible for ligation and hence gives a very low level of background when ligated. More than 25 double stranded oligos have been cloned into pHippy with nearly 100% cloning efficiency (data not shown). The target sequence for any mRNA can be cloned into pHippy and the DNA will be transcribed from both strands and form a double stranded RNA with two 5'Uridine overhangs, which closely resembles functional siRNA produced by Dicer [11, 12, 20, 21]. This vector system can be used to generate siRNA to target any transcript or open reading frame (ORF) (Fig. 1b). Pools of siRNAs for specific mRNAs can be generated by restriction enzyme or DNase digestion of a cDNA for any gene and subsequent cloning of the small fragments into pHippy

(Fig. 1b). This circumvents the need to find specific targets, which can be time consuming and costly. In addition, libraries of siRNAs can be generated by restriction enzyme or DNase digestion of cDNA libraries or by cloning of random oligos into pHippy (Fig. 1b). These libraries can be used to screen populations of transfected cells for phenotypic changes and the cells with the desired phenotype can be recovered, allowing the siRNA construct to be rescued and characterized. This type of screen allows an unbiased means to identify genes involved in diverse biological processes.

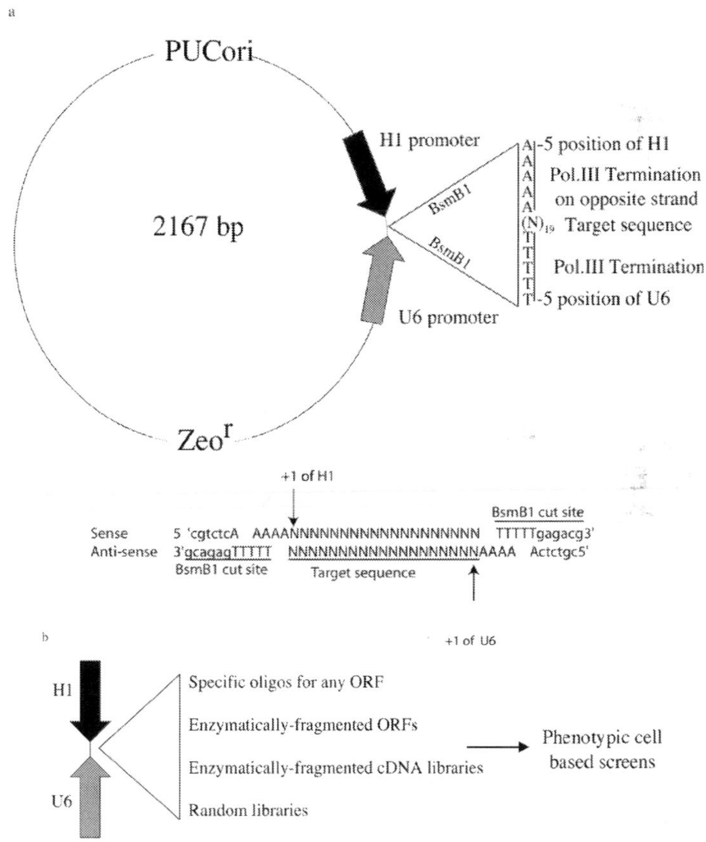

Figure 1: Depiction of the pHippy dual siRNA expression vector and some of its possible uses. **a**, pHippy has convergent opposing Human H1 and U6 polymerase III promoters that drive expression of both strands of any template cloned in between the BsmB 1 cloning sites. pHippy also

contains the *Puc* origin and the Zeocin-resistance gene for propagation and replication in bacteria. As depicted the H1 and U6 promoters were modified to contain a polymerase III termination signal (TTTTT) between the -5 to -1 position of the promoter, and BsmBl sites. BsmBl is a type II restriction enzyme, which cuts outside of its recognition sequence and will in the case of pHippy leaves 3› TTTT overhangs on both strands of the plasmid as depicted. 19 nt target siRNA can be cloned into pHippy as double stranded oligos by addition of AAAA to the 5› ends of the oligos as depicted b. **b**, Proposed uses for the pHippy siRNA vector system.

pHippy efficiently inhibits the ectopic expression of reporter genes

To determine if pHippy produces functional siRNAs, we generated a pHippy vector against PGL3 luciferase (pHippyPGL3luc) from a previously reported synthetic siRNA directed against PGL3 luciferase [9]. Using the same target sequence against PGL3 luciferase, we made two hairpin siRNA vectors, with one driven from the U6 promoter (U6PGL31ucHP) and one driven from the H1 promoter (HIPGL3lucHP). To determine the efficiency of inhibition of the PGL3 luciferase target by the various siRNA vectors, 293T cells were transfected with a cocktail of PGL3 luciferase, Renilla luciferase, and a siRNA vector driving a hairpin or a pHippy siRNA. After normalization for transfection with Renilla luciferase, the empty pHippy vector gave similar luciferase levels as cells transfected with luciferase alone, and this level was set to 100% luciferase activity (Fig 2a). Both of the vectors that express hairpin RNAs against PGL3 luciferase significantly inhibited luciferase expression (Fig. 2a). The pHippy vector with the target sequence against PGL3 luciferase also inhibited expression of PGL3 luciferase. Moreover, at the same concentration of DNA transfected, pHippyPGL31uc is more efficient at inhibiting Luciferase expression than either of the hairpin vectors. Specifically, at 30 ng of DNA transfected PHippyPGL31uc is 2–4 times more efficient than inhibition of PGL3 luciferase by the vectors that generate hairpin RNAs. Although pHippy-generated siRNA appears to be more efficient than hairpin generated siRNA, a more detailed side by side comparison would have to be conducted to establish specific activities.

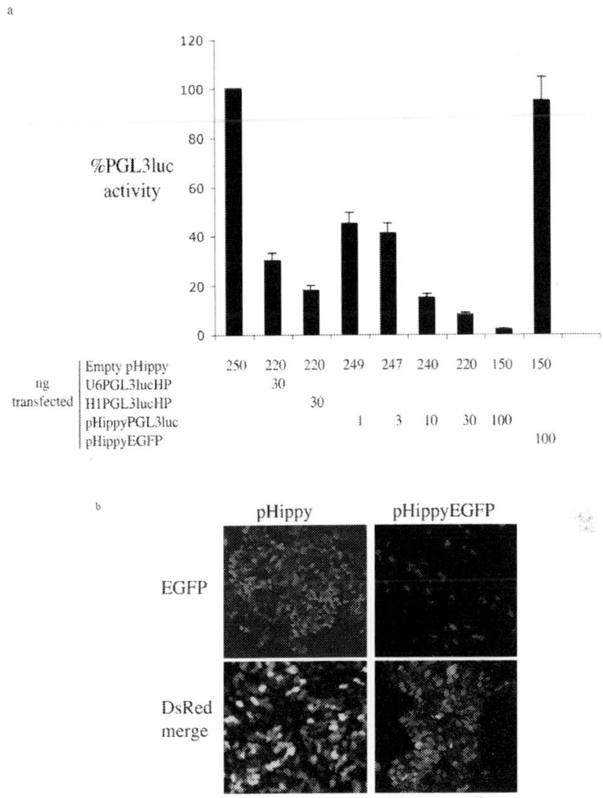

Figure 2: **a**, Luciferase assay to determine the efficiency of target suppression by the pHippy vector. Hairpin siRNA expression vectors were generated against PGL3 luciferase with either the U6 (U6GPL31ucHP) or II1 (IIlPGL31ucHP) promoters. For comparison, pHippy siRNA vectors were generated against PGL3 luciferase (pHippyPGL31uc) or against EGFP (pHippyEGFP). 293T cells were transfected with the constructs shown and assayed for luciferase activity 24 hours later. The absolute levels of PGL3 luciferase were ~100,000 relative light units (RLUs) which was set at 100%. All experiments were normalized for transfection efficiency with an expression vector for Renilla luciferase. The average normalized PGL3 luciferase levels and standard deviation are shown for 3 experiments. **b**, EGFP fluorescent assay to monitor the suppression of protein expression by pHippyEGFP. 293T cells were co-transfected with empty pHippy or pHippyEGFP and expression vectors for DsRED and EGFP. Confocal image of EGFP and EGFP merged with DsRED are shown. One representative experiment is shown.

To determine if pHippy generates sequence-specific inhibitory siRNAs, a vector was designed against EGFP. pHippyEGFP did not inhibit the expression of PGL3 Luciferase (Fig. 2a). However, it inhibited the expression of EGFP (Fig. 2b). This not only establishes that the inhibition of expression of EGFP and Luciferase is gene-specific, but also demonstrates that pHippy can theoretically be used to knock down the expression of any gene. Supportingly, to date we have knocked down expression of more than 10 unique genes with pHippy (data not shown). We have not observed off-target effect of these siRNAs, though we have not systematically screened for such potential effects. We speculate that siRNAs generated by pHippy are less likely to generate off-target effects because there are no heterologous sequences included in the siRNA, such as a hairpin loop. On average we have found that 3–4 constructs with unique target sequences have to be tested to obtain functional siRNA, which is very similar to other siRNA systems.

pHippy efficiently inhibits the expression of endogenous genes

To determine if pHippy can be used to inhibit the expression of endogenous genes, five pHippy constructs were generated against the human Low-density Lipoprotein Receptor-related Protein 6 (LRP6). To first demonstrate that the constructs could inhibit expression of ectopic LRP6, they were screened for their ability to inhibit expression of a fusion protein consisting of LPR6 and Renilla luciferase (LRP6Rluc). This fusion of a target gene to Renilla luciferase allows for rapid quantitative assessment of the efficiency of any given siRNA construct. Three of the five pHippy constructs generated against LRP6 inhibit LRP6Rluc expression by more than 50% (Fig. 3a). To test if expression of endogenous LRP6 was also inhibited, we determined if any of the pHippy constructs altered an indirect readout of the biological function of LRP6. Briefly, LRP6 is part of the Wnt receptor signaling complex, and it is required to receive and transduce Wnt signaling to downstream components in the Wnt signaling cascade [22]. Thus, the biological function of LRP6 culminates in the activation of β-catenin-mediated transcription, which can be efficiently measured by activation of the Super(8X)Topflash reporter [23]. To test if LRP6 is required for Wnt signal transduction in human cells, 293T cells were co-transfected with a cocktail of Super(8X)Topflash, Renilla luciferase, and the five pHippy constructs directed against LRP6 or control pHippy constructs. 24 hours after transfection the cells were treated with Wnt3a-conditioned media to activate Wnt signaling and thus the expression of Super(8X) Topflash, then cultured for an additional 24 hours. After normalization for transfection with Renilla luciferase, the cells transfected with empty pHippy or pHippyEGFP were set to 1 fold activation (Fig. 3b). Treatment with Wnt3a increases the reporter activation about 100-fold (Fig. 3b). Co-transfection of some of the pHippyLPR6 constructs inhibited Wnt3a activation of Super(8X)Topflash more than 50% (Fig 3b). Strikingly, there was a strict correlation between the ability of the pHippyLRP6 constructs to inhibit the expression of LRP6Rluc and the ability to inhibit Wnt3a activation of Super(8X)Topflash (Fig. 3a,3b). These experiments demonstrate that pHippy constructs have the ability to inhibit the expression and thus the biological function of endogenous genes.

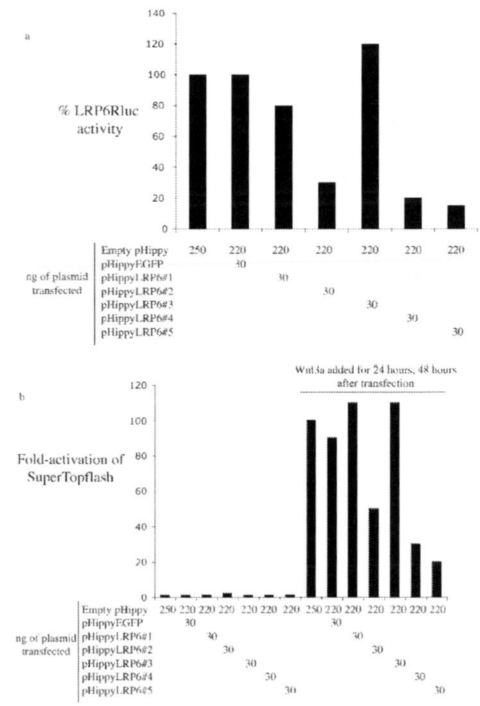

Figure 3: Assays to determine if pHippy can be used to inhibit endogenous genes. **a**, Renilla luciferase assays to determine if any of 5 unique pHippy constructs against Human LRP6 inhibit activity of a fusion of LRP6 to

Renilla luciferase. 293T were co-transfected with the vectors shown and assayed for luciferase activity 48 hours later. Cells co-transfected with LRP6Rluc and empty pHippy were set to 100%. All experiments were normalized for transfection with PGL3 luciferase. The average normalized Renilla luciferase levels and standard deviation are shown for 3 experiments. **b**, Determination of whether inhibition of endogenous LRP6 inhibits Wnt3a activation of the Wnt/β-catenin signaling pathway. 293T cells were co-transfected with the vectors shown, Super(8X)Topflash (to measure Wnt/β-catenin-signaling), and a constitutive expression vector for Renilla luciferase. 24 hours after transfection the cells were treated with Wnt3a conditioned media for 24 hours and then assayed for luciferase activity. Untreated cells transfected with empty vector were set to 1 fold activation of Super(8X)Topflash, which corresponds to ~10,000 RLUs. Cells treated with Wnt3a induce about 100 fold-activation of luciferase as shown. All experiments were normalized for transfection efficiency with the Renilla luciferase expression vector. One representative experiment is shown.

pHippy expression cassettes can easily be generated in a high throughput manner by PCR

To generate pHippy siRNA constructs in a rapid manner, a PCR method was devised that incorporates the U6 and H1 promoters from pHippy on either end of a PCR product. A gene-specific primer for a target gene can be sandwiched between the two convergent promoters. To develop this system three oligonucleotides were synthesized: 1) a 97 nt primer consisting of the entire H1 promoter from pHippy, 2) a gene-specific primer containing 18 nt of complimentary sequence to both the H1 and U6 promoters and 21 nt of gene-specific (PGL3 luciferase) or random control sequences, and 3) an 18 nt anti-sense primer to the human U6 promoter (Fig. 4a). After PCR with H1, U6, and limiting amounts of the gene-specific primer using a U6 template, a single robust band was generated for all of the different gene-specific primers of the appropriate size (Fig. 4b). These PCR products were cleaned with a nucleospin column and co-transfected into 293T cells with the PGL3 luciferase reporter and Renilla luciferase as a normalization control. Neither of the control PCR products significantly inhibited PGL3 luciferase activity (Fig. 4c). However, the PCR product generated against PGL3 luciferase inhibited luciferase in a dose-dependent manner (Fig. 4c). Although the pHippy PCR cassette against PGL3 luciferase inhibited PGL3 luciferase expression, it was not as efficient as the pHippy vector directed against PGL3 luciferase (Fig. 4c). This may be due to faster degradation of the PCR product and might be circumvented by additional sequences on the 5' and 3' ends of the PCR product.

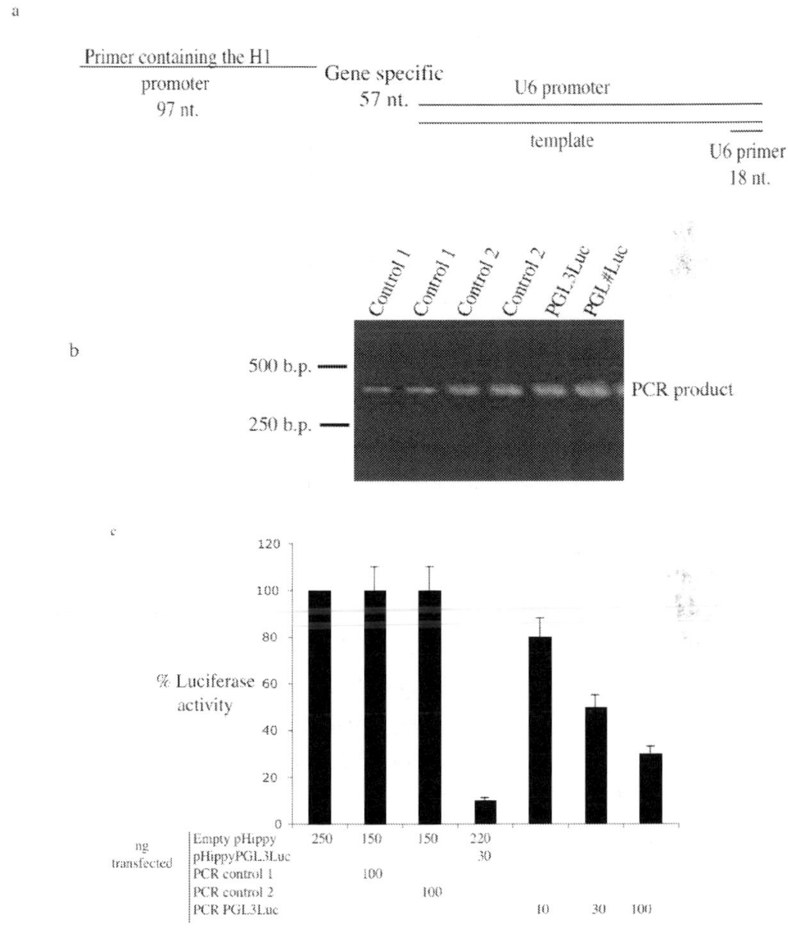

Figure 4: A high throughput system to generate pHippy expression cassettes. **a**, A multiple primer PCR system was used to generate PCR products that express siRNA after transfection into cells. A 5' primer (U6 primer) was designed to human U6. A second primer (gene specific) was designed with sequences that are complementary to the 3' end of the

human 6 promoter, the target siRNA sequence, and a sequence that is complementary to the 3' of the human H1 promoter. A third primer was designed that contains the human H1 promoter and is complementary to the gene specific primer. **b**, An agarose gel stained with ethidium bromide demonstrates the robustness of the PCR method for several different gene specific primers. **c**, A luciferase assay as described in the legend to Figure 3b to determine if the PCR products inhibit PGL3 luciferase activity.

Generation of random siRNA-libraries in pHippy

The pHippy system is well-suited for generation of cDNA or random siRNA libraries because both strands of DNA template are transcribed to generate siRNA. To determine if pHippy could in principle be used for a random screen, a random library of siRNAs based on PGL3 luciferase was generated (Fig. 5a). This library was generated by randomizing 3 nt in the functional PGL3 siRNA described in Figure 2 and corresponds to a library of 64 possible siRNAs. To determine if this library could be screened to recover siRNA activity, 130 randomly chosen clones from *E. coli* containing the library were picked and pooled in groups of 10. These 13 pools were screened for their ability to inhibit PGL3 luciferase activity. The library consists of a maximum of 64 possible siRNAs and only 2 of the pools from the 130 clones would be predicted to inhibit PGL3 luciferase activity. In agreement with this calculation, pools 8 and 11 had significant inhibitory activity (Fig. 5b). These pools of 10 were further reduced to single clones and each individual clone was re-screened for inhibitory activity. Pool 8 contained one clone that had inhibitory activity and pool 11 contained two clones with inhibitory activity. Sequencing of these three clones revealed that these clones contained the original sequence against PGL3 (Fig 5c). In contrast, 10 clones that did not inhibit PGL3 luciferase activity had random sequences at the mutant positions (Fig. 5c).

a.

PGL3Luc siRNA
AAGGCTCCTCAGAAACAG**CTC**
CCGAGGAGTCTTTGTC**GAG**AA

PGL3Luc siRNA Library
AAGGCTCCTCAGAAACAG**NNN**
CCGAGGAGTCTTTGTC**NNN**AA

b.

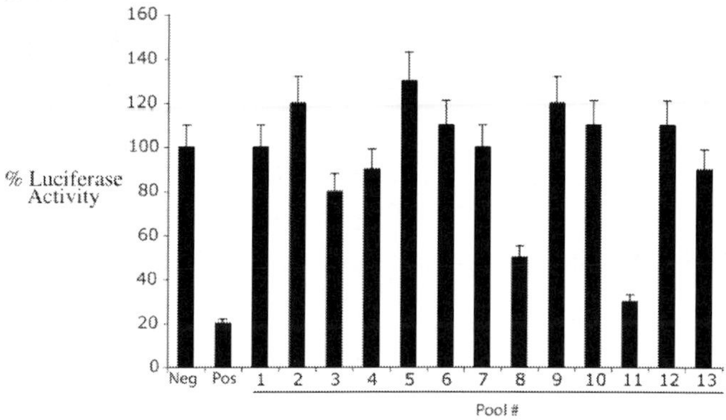

c.

Sequence from single rescued clones after deconvolution

Positive clones	Negative clones	
NNN	NNN	
1. CTC	1. ACG	2.AAG
2. CTC	3. TCG	4. GGG
3. CTC	5. GCC	6. CCG
	7. CCG	8. TTG
	9. CCC	10. GGT

Figure 5: A siRNA screen was conducted with a small siRNA library against PGL3 luciferase. **a**, Depiction of the small siRNA library against PGL3 luciferase. The target sequence for the siRNA against PGL3 siRNA that was tested in Figure 3b is shown. The bases in bold were randomized by synthesizing an oligo with random bases in place of CTC and cloned into the pHippy vector. **b**, Screening of the siRNA library. 12 pools of 10 bacterial colonies from the random siRNA library against PGL3 luciferase were screened for their ability to inhibit PGL3 luciferase activity as described in the legend to Figure 3b. **c**, The sequences of the siRNA single clones, which did or did not inhibit PGL3 luciferase. Pools 8 and 11 were broken down into individual clones and re-screened for inhibitory activity. Pool 8 contained 1 positive clone and Pool 11 contained 2 positive clones. The sequences of all three positive clones is shown on the left and corresponds to the original PGL3 siRNA. The sequences of 10 negative clones are shown on the right.

These experiments demonstrate that the pHippy system can be used for random siRNA screens. Specifically, libraries can be generated where all of the 21 nt of the siRNA are random. This library would encompass multiple targets in every gene in the human genome and could be used for phenotypic single cell assays to identify genes required for the screened phenotype, without first knowing the siRNA sequence. For instance, a random siRNA library could be used to identify genes required for Wnt signaling by screening for siRNAs that inhibit the ability of Wnt to activate Super(8X)Topflash.

CONCLUSIONS

pHippy is an siRNA expression vector, which utilizes convergent promoters to generate functional siRNAs by driving expression from both strands of the same template DNA. In tests to date, pHippy vectors directed against PGL3 luciferase knock down the expression of luciferase better than hairpin RNAs. This is likely because pHippy produces functional siRNA that do not need to be cleaved by Dicer. In addition, pHippy vectors against any target gene are easier to generate than with hairpin vectors. Specifically, the oligos needed to generate pHippy siRNA vectors are shorter and more cost effective, and the final clones are easier to identify and sequence. pHippy expression cassettes can be generated in a high throughput manner by PCR without the need to propagate through bacteria.

We have used pHippy to knock out the expression of an endogenous Wnt pathway co-receptor, as scored by measuring effects on a Wnt pathway reporter, thereby validating this approach for endogenous targets. siRNAs synthesized from pHippy are predicted to have fewer off-target effects than some siRNA approaches since pHippy-derived siRNAs have no extraneous nucleotides, a feature one might design into vectors developed with gene therapy as a goal.

One of the biggest advantages is that pHippy can be used to generate random and cDNA siRNA expression libraries, which can be used to identify genes involved in any biological phenomena that can be screened by phenotype. As a proof of principle we screened a small random siRNA library for siRNAs that inhibit PGL3 luciferase activity, and were able to recover functional siRNAs that knocked down activity of this target. This approach can now be used to identify siRNAs that modulate diverse biological phenomena.

METHODS

Cell culture and transfection

293T cells were grown in DMEM supplemented with 10% FBS and 1% Penicillin/Streptomycin under standard conditions. All transfections were performed in 24 well plates with Lipofectamine Plus or 2000 (Invitrogen) according to the manufacturer's specifications.

Confocal microscopy

2.5. $\times$ 10^5 293T cells were seeded in 12 well plates and transfected with 10 ng of pEGFPNl (Clontech), pDSREDNl (Clontech), and 30 ng of the indicated siRNA construct. 24 hours after transfection the cells were removed from the plates with PBS and live cells were sandwiched between cover-slips and glass slides and visualized for fluorescence using the appropriate lasers and filters to visualize EGFP and DSRED.

Reporter assay

Luciferase assays were preformed according to the Dual luciferase assay specifications (Promega). In all cases 293T cells were transfected with 10 ng of CMV-PGL3luciferase and 100 pg of pRLCMV (Promega), and the cells were harvested 24 hours later and assayed for luciferase activity in 96 well plate in a Berthold 96 V luminometer. Super(8X)Topflash reporter assays were performed as described [23]. 293T cells seeded in 24 well plates were transfected with 10 ng of Super(8X)Topflash, 100 pg of pRLCMV (Promega), and the indicated amount of effecter plasmids. The concentrations of all transfections were brought up to a total of 250 ng with the vector CS2+. Assays were performed as described [23] and according to the Dual luciferase assay specifications (Promega).

Multiple primer PCR

pHippy siRNA expression cassettes were generated by a single step multiple primer PCR. In short, a 10 ng of plasmid containing the human U6 promoter was used as template for PCR in a 50 μl reaction containing 2 μl of 10 pm/μl U6 primer, 2 μl of a primer encompassing the entire H1 promoter, 2 μl of 0.01 pm/μl of the gene-specific linker primer, 10 μl of 2 mM DNTPs, 10 μl of advantage buffer, and 0.5 μl tag-advantage

(Clontech). The PCR products were generate by 30 cycles of touchdown PCR program that ramped down from 60°C to 50°C.

RNAi design and recombinant DNA constructs

All recombinant constructs were generated by standard recombinant DNA techniques. The sequences of pHippy will be provided upon request. siRNAs were designed using the web based siRNA design program from the Whitehead Institute web page http://jura.wi.mit.edu/pubint/http://iona.wi.mit.edu/siRNAext/home.php. In general, siRNAs were used from previously published siRNAs or were designed by introducing the corresponding sequence for a given gene into the Whitehead siRNA design program using the following pattern AAN19TT or AAGN17CN2. The output sequences for a given gene that were not complimentary to other genes after blasting were chosen, and the corresponding oligos were designed and ordered. Oligos were based on the Whitehead output sequences and modified for cloning into pHippy by addition of 2 or 4 adenines to the 5' end of the sense and antisense versions of the Whitehead output sequences. Oligos used; PGL3luciferase, sense (5'AAaaggctcctcagaaacagctc3'), antisepses (5'AAaagagctgtttctgaggagcc3'): EGFP, sense (5' AAaagcaagctgaccctgaagttcat3'), antisense (5' AAaaatgaacttcagggtcagcttgc3'): LRP6#1 sense (5'AAaaaggttcccttccacatcct3'), antisense (5'AAaaaggatgtggaagggaacct3'): LRP6#2 sense (5'AAaaaaggttcccttccacatcctttt3'), antisense (5'AAaaaaggatgtggaagggaacctttt3'): LRP6#3 sense (5'AAaagaagatggcagccagggct3'), antisense (5'AAaaagccctggctgccatcttc3'): LRP6#4 sense (5'AAaaggcacttacttccctgcaa3'), antisense (5'AAaattgcagggaagtaagtgcc3'): LRP6#5 sense (5'AAaaaaggcacttacttccctgcaatt3'), antisense (5'AAaaaattgcagggaagtaagtgcctt3').

ACKNOWLEDGEMENTS

We thank Bryan White and Stephane Angers for comments on the manuscript. AK is an Associate, and RTM an Investigator, of the HHMI, which supported this work.

REFERENCES

1. Bernstein E, Denli AM, Hannon GJ: The rest is silence. RNA. 2001, 7: 1509-21.

2. PubMedGoogle Scholar
3. Dykxhoorn DM, Novina CD, Sharp PA: Killing the messenger: short RNAs that silence gene expression. Nat Rev Mol Cell Biol. 2003, 4: 457-67. 10.1038/nrm1129.
4. Fire A, Xu S, Montgomery MK, Kostas SA, Driver SE, Mello CC: Potent and specific genetic interference by double-stranded RNA in Caenorhabditis elegans. Nature. 1998, 391: 806-11. 10.1038/35888.
5. Jorgensen R: Altered gene expression in plants due to trans interactions between homologous genes. Trends Biotechnol. 1990, 8: 340-4. 10.1016/0167-7799(90)90220-R.
6. Brigneti G, Voinnet O, Li WX, Ji LH, Ding SW, Baulcombe DC: Viral pathogenicity determinants are suppressors of transgene silencing in Nicotiana benthamiana. Embo J. 1998, 17: 6739-46. 10.1093/emboj/17.22.6739.
7. Hamilton AJ, Baulcombe DC: A species of small antisense RNA in posttranscriptional gene silencing in plants. Science. 1999, 286: 950-2. 10.1126/science.286.5441.950.
8. Kasschau KD, Carrington JC: A counterdefensive strategy of plant viruses: suppression of posttranscriptional gene silencing. Cell. 1998, 95: 461-70. 10.1016/S0092-8674(00)81614-1.
9. Stark GR, Kerr IM, Williams BR, Silverman RH, Schreiber RD: How cells respond to interferons. Annu Rev Biochem. 1998, 67: 227-64. 10.1146/annurev.biochem.67.1.227.
10. Elbashir SM, Harborth J, Lendeckel W, Yalcin A, Wcbcr K, Tuschl T: Duplexes of 21-nucleotide RNAs mediate RNA interference in cultured mammalian cells. Nature. 2001, 411: 494-8. 10.1038/35078107.
11. Elbashir SM, Martinez J, Patkaniowska A, Lendeckel W, Tuschl T: Functional anatomy of siRNAs for mediating efficient RNAi in Drosophila melanogaster embryo lysate. Embo J. 2001, 20: 6877-88. 10.1093/emboj/20.23.6877.
12. Elbashir SM, Lendeckel W, Tuschl T: RNA interference is mediated by 21- and 22-nucleotide RNAs. Genes Dev. 2001, 15: 188-200. 10.1101/gad.862301.
13. Brummelkamp TR, Bernards R, Agami R: A system for stable

expression of short interfering RNAs in mammalian cells. Science. 2002, 296: 550-3. 10.1126/science.1068999.

14. Paddison PJ, Gaudy AA, Bernstein E, Hannon GJ, Conklin DS: Short hairpin RNAs (shRNAs) induce sequence-specific silencing in mammalian cells. Genes Dev. 2002, 16: 948-58. 10.1101/gad.981002.
15. Paddison PJ, Caudy AA, Hannon GJ: Stable suppression of gene expression by RNAi in mammalian cells. Proc Natl Acad Sci USA. 2002, 99: 1443-8. 10.1073/pnas.032652399.
16. Paul CP, Good PD, Winer I, Engelke DR: Effective expression of small interfering RNA in human cells. Nat Biotechnol. 2002, 20: 505-8. 10.1038/nbt0502-505.
17. Esposito D, Gillette WK, Hartley JL: Blocking oligonucleotides improve sequencing through inverted repeats. Biotechniques. 2003, 35: 914-920.
18. Kawasaki H, Taira K: Short hairpin type of dsRNAs that are controlled by tRNA(Val) promoter significantly induce RNAi-mediated gene silencing in the cytoplasm of human cells. Nucleic Acids Res. 2003, 31: 700-7. 10.1093/nar/gkg158.
19. Shirane D, Sugao K, Namiki S, Tanabe M, lino M, Hirose K: Enzymatic production of RNAi libraries from cDNAs. Nat Genet. 2004, 36: 190-6. 10.1038/ng1290.
20. Sen G, Wehrman TS, Myers JW, Blau HM: Restriction enzyme-generated siRNA (REGS) vectors and libraries. Nat Genet. 2004, 36: 183-9. 10.1038/ng1288.
21. Zamore PD, Tuschl T, Sharp PA, Bartel DP: RNAi: double-stranded RNA directs the ATP-dependent cleavage of mRNA at 21 to 23 nucleotide intervals. Cell. 2000, 101: 25-33. 10.1016/S0092-8674(00)80620-0.
22. Tran N, Cairns MJ, Dawes IW, Arndt GM: Expressing functional siRNAs in mammalian cells using convergent transcription. BMC Biotechnol. 2003, 3: 21-10.1186/1472-6750-3-21.
23. Schweizer L, Varmus H: Wnt/Wingless signaling through beta-catenin requires the function of both LRP/Arrow and frizzled classes of receptors. BMC Cell Biol. 2003, 4: 4-10.1186/1471-2121-4-4.

24. Kaykas A, Yang-Snyder J, Heroux M, Shah KV, Bouvier M, Moon RT: Mutant Frizzled 4 associated with vitreoretinopathy traps wild-type Frizzled in the endoplasmic reticulum by oligomerization. Nat Cell Biol. 2004, 6: 52-8. 10.1038/ncb1081.

SECTION VIII
PLASMID CURING SYSTEMS

CHAPTER

12

CURING OF PLASMID PXO1 FROM BACILLUS ANTHRACIS USING PLASMID INCOMPATIBILITY

Xiankai Liu., Dongshu Wang., Huagui Wang, Erling Feng, Li Zhu, Hengliang Wang

State Key Laboratory of Pathogen and Biosecurity, Beijing Institute of Biotechnology, Beijing, China

ABSTRACT

The large plasmid pXO1 encoding the anthrax toxin is important for the virulence of *Bacillus anthracis*. It is essential to cure pXO1 from *B. anthracis* to evaluate its role in the pathogenesis of anthrax infection. Because conventional methods for curing plasmids (e.g., curing agents or growth at elevated temperatures) can induce mutations in the host chromosomal DNA, we developed a specific and reliable method to eliminate pXO1 from *B. anthracis* using plasmid incompatibility. Three putative replication origins of pXO1 were inserted into a temperature-sensitive plasmid to generate three incompatible plasmids. One of the three plasmids successfully eliminated the large plasmid pXO1 from *B. anthracis* vaccine strain A16R and wild type strain A16. These findings provided additional information about the replication/partitioning of pXO1 and demonstrated that introducing a small incompatible plasmid can generate plasmid-cured strains of *B. anthracis* without inducing spontaneous mutations in the host chromosome.

INTRODUCTION

Bacillus anthracis, a gram-positive, rod-shaped, spore-forming bacterium, is the etiologic agent of anthrax, an acute and often fatal mammalian disease [1], [2], [3]. Virulent forms of *B. anthracis* harbor two large pathogenicity-related plasmids: pXO1 (181.6 kb), which encodes the anthrax toxin genes *pagA*, *lef*, and *cya* [2], [4]; the toxin regulatory elements *atxA* and *pagR*; the plasmid-borne germination genes *gerXC*, *-A*, and *-B* [2], [4], [5], [6]; and pXO2 (93.5 kb), which carries the genes responsible for capsule synthesis and degradation, *capA*, *-B*, *-C* and *-D*[7], [8], [9], [10], [11]. The two large plasmids of *B. anthracis* are essential for full pathogenicity; elimination of either dramatically attenuates the virulence of *B. anthracis*.

Numerous studies have attempted to characterize the role of these plasmids in virulence. Generating a plasmid-free strain is important to elucidate the crosstalk between the plasmids and host chromosome. Conventional curing strategies involve chemical agents (e.g., acridine orange, novobiocin, ethidium bromide), growth at elevated temperature, or treatment with ultraviolet (UV) light [12], [13], [14], [15], [16]. These strategies have been used successfully in many bacteria to eliminate resident plasmids. However, these methods are not specific to a particular plasmid but eliminate all resident plasmids. Furthermore, elimination of the resident plasmids is spontaneous; therefore, screening colonies for plasmid-cured strains is a tedious process. Finally, most known plasmid-curing chemical agents are toxic or mutagenic (e.g., acridine orange, ethidium bromide, and sodium dodecyl sulfate) and can therefore induce mutations in the host chromosome [17]. Mutations also occur during harsh physical curing treatments (e.g., high temperature or UV light). The phenotypic changes caused by mutation may interfere with the functional analysis of the plasmid [18]. For these reasons, there is a need to develop a curing method that is safe, reliable, and specific to the plasmid of interest.

Plasmid incompatibility is the inability of two different but related plasmids to coexist stably in the same host cell in the absence of continued selective pressure. Plasmids that share the same mechanisms for replication or partitioning are placed in the same incompatibility group. Plasmid incompatibility may be due to competition for the same replication or segregation sites, or caused by the repression of replication initiation [19]. Thus, introducing a smaller high-copy-number plasmid from the same incompatibility group may specifically eliminate a resident plasmid. This strategy has been demonstrated to be a specific, efficient, and safe technique to generate plasmid-free bacterial strains [20], [21].

Tinsley et al. isolated and characterized a minireplicon of plasmid pXO2 [22]. These researchers cloned a 2429-bp region of pXO2 into an *Escherichia coli* vector that is normally unable to replicate in *B. anthracis*, and demonstrated that this plasmid replicated when introduced into *B. anthracis*. The pXO2 replicon contains an open reading frame (ORF) encoding the putative replication initiation protein RepS and the putative origin of replication *ori*. Another ORF (*repB*) was included in this fragment but thought to be dispensable for replication of the pXO2 minireplicon [22]. Based on this information, we successfully cured the large pXO2 plasmid from the *B. anthracis* wild type strain A16 ($pXO1^+pXO2^+$) using a smaller plasmid in which a 3836-bp region of pXO2 containing *repS*, *repB*, and the putative pXO2 *ori* was inserted [23]. These results initiated our interest in eliminating the plasmid pXO1 from *B. anthracis* using plasmid incompatibility, which has not yet been reported in the literature. Little is known about the replication and partition properties of pXO1. The origin of replication and genes involved in replication initiation, plasmid partitioning, and plasmid stability have not been identified. Researchers have long sought to characterize the structure of the pXO1 replicon. Robertson et al. attempted to localize the replication origin by cloning pXO1 DNA fragments into an *E. coli* vector and transforming the plasmid into *B. subtilis*. This group identified several clones that mapped to an 11-kb region (coordinates 86249–97209) of the pXO1 sequence (GenBank accession no. AF065404; ORF72-81) (coordinate numbers and ORFs annotated by Okinaka et al. will be used throughout this report) [24]. The products encoded by the 10 ORFs did not show any sequence similarity to *rep* proteins associated with theta-replicating plasmids. In the past 10 years, data from numerous *B. anthracis* genome-sequencing projects have considerably increased our understanding of the biological roles of plasmids in this species. Plasmid pXO1 of the *B. anthracis* Sterne strain was sequenced by Okinaka et al. using a random whole-genome shotgun strategy, which enabled analysis of its replicon. The annotation predicted 143 ORFs in pXO1, but only 35 have putative functions assigned to them based on similarity to genes in open databases. Unfortunately, no sequences have been found that show significant similarity with known replication initiator proteins encoded by other plasmids [24]. Tinsley and Khan cloned a 5-kb region (ORF43–46, nt 54863–60166) of the plasmid pXO1 into an *E. coli* vector and demonstrated that this plasmid could replicate in *B. anthracis*. Mutational analysis showed that ORF45 was required for replication of the minireplicon, and this ORF was designated *repX*[25]. In a more recent effort to understand pXO1 replication, Andrei et al. isolated and characterized a new minireplicon of *B. anthracis* plasmid pXO1 using the Cre-loxP recombination system to delete large DNA segments; this region (ORF14–16, nt 19032–24981) differed from those previously described. Deletion analysis showed that only ORF14 and ORF16 were essential for replication of the minireplicon, and ORF16 was thought to be the replication initiator protein with a putative theta-replicating origin of replication [26].

In this report, we cured the virulence-associated plasmid pXO1 from *B. anthracis* vaccine strain A16R and wild type strain A16, using plasmid incompatibility to generate isogenic plasmid-cured and plasmid-containing strains. Our results will lead to a better understanding of the pXO1 plasmid and its interaction with the *B. anthracis* chromosome.

MATERIALS AND METHODS

Bacterial Strains, Plasmids, and Culture

The bacterial strains and plasmids used in this study are listed in Table 1. *B. anthracis* A16 strain was isolated from the carcass of a mule that had died of anthrax in Langfang city in the Hebei province of China in 1953. The lethal dose of A16 is 100 spores for mice and 150 spores for rabbits administered by subcutaneous injection. *B. anthracis* vaccine strain A16R was derived from A16 by exposure to UV radiation. A culture dish containing the spore suspension of A16 was placed 10 cm from the center of a UV lamp. After a 3-hour UV treatment, the irradiated suspension was spread on agar plates containing 20% bovine serum and incubated in 20% CO_2 at 37°C for 24 hours. A rough colony was chosen to streak on an agar plate containing 20% bovine serum, which was incubated in 20% CO_2 at 37°C for 24 hours. This procedure was repeated until all descendants produced rough colonies. One of these clones was carried continuously in nutrient medium for 141 passages and in mice for 30 generations. With no descendants producing smooth colonies, this clone was considered stable and designated A16R. Its virulence was greatly attenuated, primarily through loss of the bacterial capsule, and A16R is now used for the human anthrax vaccine in China. For safety reasons and ease of handling in the laboratory, an attenuated *B. anthracis* vaccine strain A16R (pXO1+ pXO2−) was used for the initial experiments in the present study. The experiments were then repeated with the wild type strain A16. The temperature-sensitive shuttle vector pKSV7 (permissive temperature, 30°C; restrictive temperature, 37°C) contains a chloramphenicol resistance gene for selection in *B. anthracis*, an ampicillin resistance gene for selection in *E. coli*, and a multiple cloning site [27]. Strains containing recombinant plasmids were cultured at 30°C, except when curing the plasmid from the host bacteria, when the strains were cultured at 37°C. Before transformation into *B. anthracis*, all recombinant plasmids were passaged through *E. coli* SCS110 (*dam dcm* mutant) to demethylate the plasmid DNA, thereby facilitating its transformation into bacteria. All strains were cultured in Luria-Bertani (LB) broth (10 g/L tryptone [Oxoid, UK], 5 g/L yeast extract [Oxoid], 10 g/L NaCl) or on LB agar plates. Transformants were selected for ampicillin resistance (100 μg/mL) in *E. coli* or for chloramphenicol resistance (10 μg/mL) in *B. anthracis*.

Table 1: Plasmids and bacterial strains used in this study.

Plasmids and strains	Relevant genotype and characteristics	Source
pKSV7	Shuttle vector, temperature-sensitive, Amp^r (gram-negative) in *E. coli* and Cm^r (gram-positive) in *B. anthracis*	[27]
pKS4K	Incompatible plasmid, constructed by inserting 4-kb fragment of pXO1 into pKSV7	This work
pKS5K	Incompatible plasmid, constructed by inserting 5-kb fragment of pXO1 into pKSV7	This work
pKS11K	Incompatible plasmid, constructed by inserting 11-kb fragment of pXO1 into pKSV7	This work
E. coli DH5α	F^-, φ80d/*lacZ*ΔM15, Δ(*lacZYA*-argF)U169, *deoR*, *recA*1, *endA*1, *hsdR*17(r_k^-,m_k^+), *phoA*, *supE*44λ^-, *thi*-1, *gyrA*96, *relA*1	This lab
E. coli SCS110	*rpsL*(StrR), *thr*, *leu*, *endA*, *thi*-1, *lacy*, *galK*, *galT*, *ara*, *tonA*, *tsx*, *dam*-, *dcm*-, *supE*44(lac-proAB), F- [traD36, proAB, laclqlacZΔM15]	This lab
B. anthracis A16R	Vaccine strain, $pXO1^+$, $pXO2^-$	This lab
B. anthracis A16	Wild type A16R; $pXO1^+$, $pXO2^+$	This lab
B. anthracis A16R5K	A16R was cured of pXO1, harbors recombinant plasmid pKS5K; $pXO1^-$, $pXO2^-$, $pKS5K^+$	This work
B. anthracis A16RO	pXO1 plasmid-cured derivative of vaccine strain A16R; $pXO1^-$, $pXO2^-$	This work
B. anthracis A16Q1	pXO1 plasmid-cured derivative of wild type A16; $pXO1^-$, $pXO2^-$	This work

doi:10.1371/journal.pone.0029875.t001

Construction of Incompatible Plasmids

Based on the literature, we evaluated three DNA fragments (11-kb, 4-kb, and 5-kb) containing putative replication origins of pXO1 (Figure 1). The vector pKSV7 was linearized at the *SmaI*restriction site with restriction enzyme (TaKaRa Biotechnology, Dalian, China) and purified after 0.6% agarose gel electrophoresis using an agarose gel DNA purification kit. The 4-kb and 5-kb fragments were amplified with Pyrobest DNA polymerase (TaKaRa Biotechnology,Dalian, China) by polymerase chain reaction (PCR) from the total genomic DNA of strain A16R with the primer pairs 4K_F/R and 5K_F/R, respectively. The resulting fragments were cloned directly into the linearized pKSV7 vector by homologous recombination using the CloneEZ kit (GenScript, Piscataway, NJ, USA), generating the plasmids pKS4K and pKS5K, respectively. The 11-kb fragment was amplified by PCR from total DNA of strain A16R with the three primer pairs 11K_1F/R, 11K_2F/R, and 11K_3F/R. These PCR products were purified with the QIAquick PCR purification kit (Qiagen, Germany) and cloned in tandem into the linearized pKSV7 vector to generate pKS11K. PCR was performed in a 100-μL reaction containing 1 μL Pyrobest DNA polymerase, 10 μL 10× polymerase buffer, 8 μL dNTP mixture, 2 μL template DNA, 2 μL each primer, and 75 μL double-distilled H_2O (ddH_2O). PCR amplification was carried out in a GeneAmp PCR System 9700 (Applied Biosystems, Carlsbad, CA, USA) with an initial denaturation step at 94°C for 5 min, followed by 30 cycles of 94°C for 40 s, 55°C to 58°C for 40 s, 72°C for 4 to 6 min, and a final extension step at 72°C for 5 min. Clones were generated by homologous recombination with the CloneEZ kit (GenScript) according to the manufacturer's instructions.

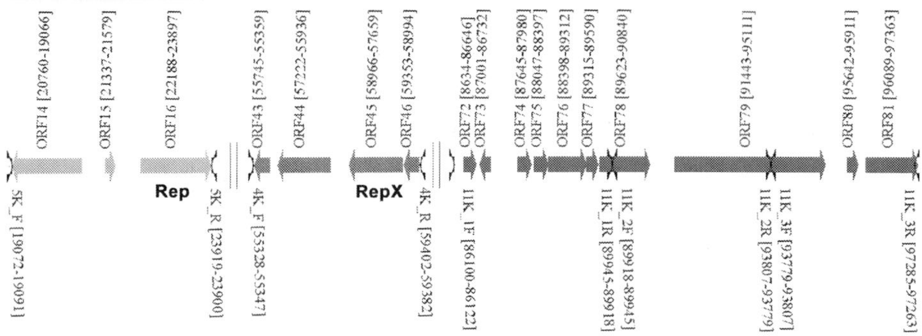

Figure 1: Three putative replicons of plasmid pXO1 (GenBank Accession no. AF065404).

Each putative replicon of pXO1 is indicated in a different color, along with primer pairs used for amplification.

The plasmids pKS4K, pKS5K, and pKS11K were introduced into chemically competent *E. coli*DH5α cells. The samples were grown at 30°C for 3 h in fresh LB broth without antibiotics on a shaker (225 rpm) and then spread onto selective agar plates containing ampicillin. After incubation at 30°C for 12 to 16 h, transformants were picked and analyzed by colony PCR. Briefly, a single colony was suspended in 100 μL ddH_2O, heated to 95°C for 2 min, and then cooled to room temperature. Cellular debris was removed by centrifugation at 15,000×*g* for 5 min, and 1 μL of the lysate was used as template for PCR amplification. PCR was performed in a 20-μL reaction containing 10 μL 2× Taq PCR Master Mix, 1 μL template, 7 μL ddH_2O, and 1 μL each primer. The primer pair pKSV7-F/R, which flanks the multiple cloning site, was used to determine whether the 4-kb or 5-kb DNA fragment were inserted into pKSV7, and three primer pairs (11K_1F/R, 11K_2F/R and 11K_3F/R) were used to determine whether the 11-kb DNA fragment was inserted. The thermal cycling conditions used to analyze transformants were same as those used to amplify fragments from the A16R total genomic DNA. Colonies that were positive for transformation, as assessed by colony PCR analysis, were analyzed by restriction digestion and DNA sequencing in an ABI Prism Model 3730XL DNA analyzer (Applied Biosystems). The recombinant plasmids pKS4K, pKS5K, and pKS11K were confirmed by sequencing and introduced into the *dam*$^-$ *dcm*$^-$ *E. coli* host strain SCS110 by electroporation (200 Ω, 25 μF, 1.8 kV) (Bio-Rad Gene Pulser II electroporator, Hercules, CA, USA) to obtain demethylated plasmid DNA for transformation into the *B. anthracis* A16R strain.

Curing of Plasmid pXO1 from *B. anthracis* A16R Strain

The electroporation-competent *B. anthracis* A16R cells were prepared as previously described [28]. The plasmids pKS4K, pKS5K, and pKS11K were isolated (Axygen Scientific, Union City, CA, USA) from *E. coli* SCS110 and introduced into *B. anthracis* A16R by electroporation (500 Ω, 25 μF, 0.6 kV). The cells were grown at 30°C for 3 h in LB broth without antibiotics on a shaker (200 rpm) and then spread onto chloramphenicol-containing agar plates. After incubation at 30°C for 18 to 22 h, transformants were analyzed by colony PCR. Positive clones were selected and designated pKS4K/A16R (pXO1$^+$pXO2$^-$pKS4K$^+$), pKS5K/A16R (pXO1$^+$pXO2$^-$pKS5K$^+$), and pKS11K/A16R (pXO1$^+$pXO2$^-$pKS11K$^+$), respectively. The three clones were passaged separately at least five times in 5 mL LB broth containing chloramphenicol (30°C, 225 rpm) for 12 h. At each passage, aliquots of each culture were diluted and spread on chloramphenicol-containing agar plates. Single clones were analyzed by colony PCR to determine the elimination of pXO1 from *B. anthracis* A16R using three pairs of primers (pag-F/R, lef-F/R, and cya-F/R). Putative pXO1-cured clones were selected, and genomic DNA was purified with TIANamp Bacteria DNA kit (Tiangen Biotech, Beijing, China) according to the manufacturer's instructions for isolating genomic DNA from gram-positive bacteria. To ensure the elimination of pXO1 from the A16R strain, 14 different pXO1 genes were analyzed by PCR using the total genomic DNA from strain A16R and the putative pXO1-cured clones as templates.

Elimination of Extraneous Plasmid pKS5K from Strain A16R5K (pXO1$^-$pXO2$^-$pKS5K)

To obtain a plasmid-cured strain without exogenous DNA, we needed to eliminate the recombinant plasmid from the host A16R strain. The pXO1-cured A16R strain containing the incompatible plasmids was passaged more than six times in 5 mL LB broth without antibiotics at 37°C for 12 h on a shaker (225 rpm). At each passage, aliquots of the cultures were diluted and spread on agar plates without antibiotics and incubated at 37°C for 12 h. Single clones were streaked onto two agar plates with or without chloramphenicol. The agar plates were incubated at 30°C overnight, and chloramphenicol-sensitive clones were considered positive for loss of the incompatible plasmid. The DNA of these clones was extracted and used as template for PCR with two primer pairs specific to plasmid pKSV7 (pKSV7P3_F/R and pKSV7P6_F/R,) to verify elimination of the incompatible plasmid.

Detection of Anthrax Toxin Protective Antigen by Western Blot Analysis

Western blot analysis was performed to verify the curing of pXO1 from *B. anthracis* A16R. The A16R strain ($pXO1^+$ $pXO2^-$) and A16R derivative strain ($pXO1^-$ $pXO2^-$) were cultured on LB agar supplemented with 5% horse serum and 0.9% sodium bicarbonate at 37°C in 5% CO_2 for 12 h. The bacteria were collected, and the cell suspension was separated by 12% bis-polyacrylamide gel electrophoresis. Proteins were transferred with Hoefer TE 77 semi-dry transfer unit (Amersham Bioscience, USA) from the gel to a polyvinylidene fluoride membrane with a constant current of 35 mA for 1 h 50 min. The transferred membrane was blocked with Tris buffered saline with 0.05% Tween 20 (TBST) containing 5% skim milk and then incubated with the mouse monoclonal antibody against protective antigen (C3; Santa Cruz, CA, USA) diluted in TBST (1:500) for 60 min at room temperature. After washing three times with TBST for 7 min, the membrane was incubated in 40 ml TBST with goat anti-mouse IgG conjugated to horseradish peroxidase (HRP) (1:10000) (Santa Cruz) for 60 min, and then washed four times with TBST for 7 min. The membrane was immersed in ECL substrate solution (Thermo Scientific, USA) for 5 min, and then exposed to x-ray film (Kodak RP X-OMAT, USA).

Use of Plasmid Incompatibility to Cure pXO1 from Wild Type Strain A16

The plasmid used to cure pXO1 from *B. anthracis* vaccine strain A16 was introduced into *B. anthracis* wild type strain A16 to cure the plasmid pXO1 using the same methods described above. Bacterial capsules were stained with India ink (Gibco, USA) and observed with a phase-contrast microscope (Eclipse TE300, Nikon, Tokyo, Japan).

RESULTS

Identification of Incompatible Plasmids

Colony PCR was performed to screen for transformants of the incompatible plasmids, and plasmids were isolated from the positive clones. Restriction analysis confirmed insertion of the 4-kb, 5-kb, and 11-kb DNA fragments into the vector pKSV7. The recombinant plasmids were confirmed by DNA sequencing and designated pKS4K, pKS5K, and pKS11K, respectively. These plasmids were introduced into *B. anthracis* to cure plasmid pXO1 from A16R.

Elimination of pXO1 from A16R by Plasmid Incompatibility

The recombinant plasmids pKS4K, pKS5K, and pKS11K were introduced into *B. anthracis*vaccine strain A16R and passaged 10 times in chloramphenicol-containing LB. The cultures were plated on selective LB agar plates containing chloramphenicol, and colony PCR was performed with the three pairs of specific primers (pag_F/R, lef_F/R, and cya_F/R) to screen for pXO1-cured colonies (Figure 2). Positive colonies underwent further PCR analysis using 14 pairs of specific primers to confirm the elimination of pXO1 from A16R. None of the primers specific to plasmid pXO1 amplified DNA from the A16R colonies transformed with pKS5K (Figure 3). These results demonstrate the elimination of pXO1 from *B. anthracis* A16R by the incompatible plasmid pKS5K. The pXO1-cured A16R derivative strain was designated A16R5K ($pXO1^{-}pXO2^{-}pKS5K^{+}$). In contrast, the recombinant plasmids pKS4K and pKS11K did not eliminate pXO1 even after passaging the cells for more than 2 months in chloramphenicol-containing LB broth.

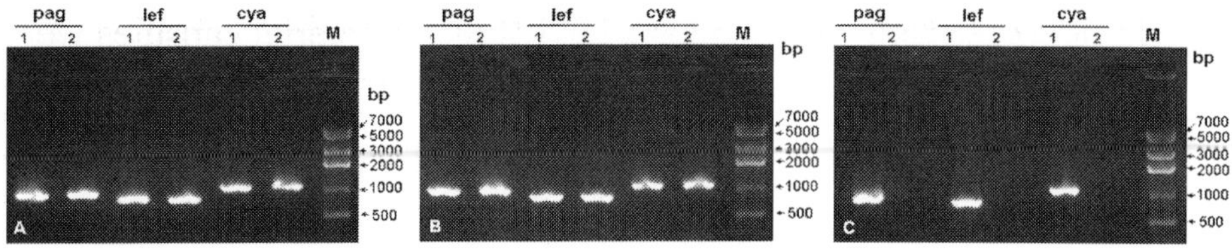

Figure 2: Colony PCR screening for pXO1 elimination with three plasmid-specific primer pairs.

Results of using recombinant plasmids pKS11K (A), pKS4K (B), and pKS5K (C) to eliminate the large plasmid pXO1 from *B. anthracis* vaccine strain A16R by plasmid incompatibility. The presence of anthrax toxin genes *pagA*, *lef*, and *cya* was determined by PCR analysis of the vaccine strain A16R (1) and the putative pXO1-cured strain A16R (2). M, DNA marker IV (Real-Times Biotechnology, Beijing, China).

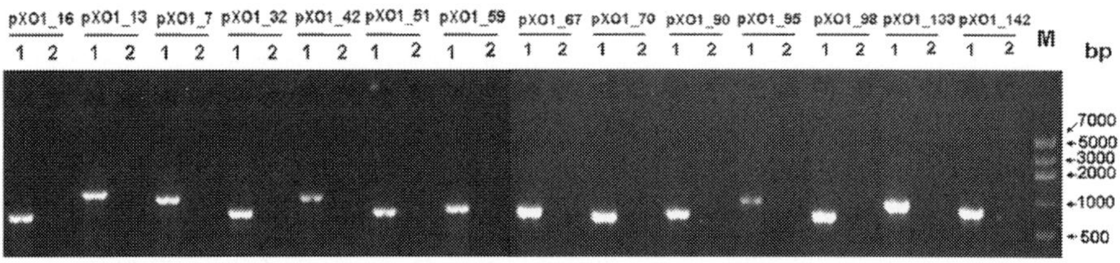

Figure 3: PCR analysis of pXO1 in *B. anthracis* vaccine strain A16R.

PCR analysis of the vaccine strain A16R (1) and the putative pXO1-cured strain A16R (2) with 14 primer pairs specific for plasmid pXO1. M, DNA marker IV (Real-Times Biotechnology).

Elimination of Exogenous Plasmid pKS5K from A16RK5 (pXO1$^-$pXO2$^-$pKS5K$^+$)

After curing pXO1 from A16R, we needed to eliminate the exogenous plasmid pKS5K from the host bacteria. To avoid damage to the chromosomal DNA of A16R, the bacteria containing pKS5K were passaged five times in LB broth without antibiotic at 37°C. The culture was plated on LB agar without antibiotics, and single clones were screened by colony PCR with two primer pairs (pKSV7P3_F/R and pKSV7P6_F/R) specific to vector pKSV7. The pXO1-cured A16R derivative strain (A16R5K) and the candidate clones were analyzed by

PCR to confirm elimination of the recombinant plasmid pKS5K. The predicted DNA fragments were amplified from DNA of A16R5K, but not from DNA of the candidate clones. These results indicate that the recombinant plasmid pKS5K was successfully eliminated from A16R5K, and this strain was named A16RO (pXO1⁻pXO2⁻) (Figure 4).

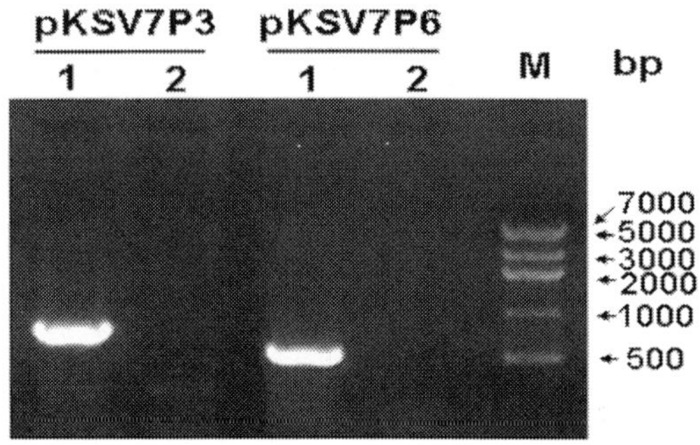

Figure 4: Elimination of exogenous plasmid pKS5K from A16R5K.

Strains A16R5K (1), and A16RO (2) were analyzed with vector-specific primer pairs pKSV7P3_F/R and pKSV7P6_F/R. M, DNA marker IV (Real-Times Biotechnology).

Phenotypic Confirmation of A16RO (pXO1⁻pXO2⁻)

To confirm elimination of pXO1 from strain A16RO, the presence or absence of the anthrax toxin protective antigen, encoded by pXO1, was determined by western blot analysis. The results revealed the presence of the protective antigen protein in the cell lysate of A16R, but this protein band was not detected in the cell lysate of A16RO, confirming that pXO1 had been eliminated (Figure 5).

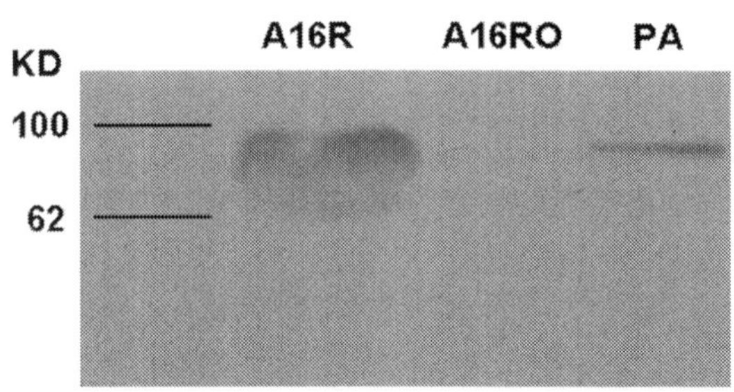

Figure 5: Western blot analysis of A16R and A16RO.

The *B. anthracis* vaccine strain A16R (pX01^{+} pX02^{-}) (left lane), A16RO (pX01^{-}pX02^{-}) (middle lane), and anthrax toxin protective antigen (PA) purified by affinity chromatography (right lane). The lines indicate the location of protein bands (Blue Plus Protein Marker, TransGen Biotechnology, Beijing, China).

Use of Plasmid pKS5K to Cure pXO1 from Wild Type Strain A16 (pXO1^{+}pXO2^{+})

The results of PCR analysis and western blot analysis of the protective antigen demonstrated that pXO1 was successfully eliminated from *B. anthracis* wild type strain A16; this pXO1-cured strain was named A16Q1 (Figure 6). Although both strains produced capsules (Figure 6 B), the plasmid-cured strain A16Q1 did not produce the anthrax toxin protective antigen (Figure 6 C), which was consistent with the results of PCR analysis (Figure 6 A).

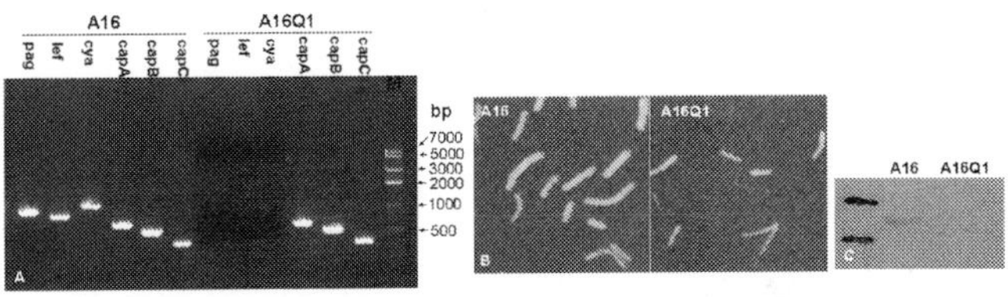

Figure 6: Use of plasmid pKS5K to cure the megaplasmid pXO1 from wild type strain A16.

(A) Results of PCR analysis show the loss of anthrax toxin genes *pagA*, *lef*, and *cya*from pXO1-cured strain A16Q1. (B) Capsule formation was observed in both wild type strain A16 and pXO1-cured strain A16Q1. (C) Results of western blot analysis show the expression of anthrax toxin protective antigen protein in pA16, but not in A16Q1.

DISCUSSION

In this study, we aimed to eliminate the large plasmid pXO1 from *B. anthracis* using plasmid incompatibility. Although this strategy has been used in some bacterial species to cure resident plasmids, it has not yet been used in *B. anthracis* other than our previous work to eliminate plasmid pXO2 from *B. anthracis* wild type strain A16. This is mainly because the replication and partitioning of the virulence-associated plasmids pXO1 and pXO2 are poorly understood. Characterizing the replication origin is important to eliminate resident plasmids from the host bacteria. However, only three studies have described the replication and partition properties of pXO1. Furthermore, the replication origins identified by the three reports differed; therefore, we tested the ability of three fragments, each containing one of the putative replication origins, to eliminate pXO1.

Three DNA fragments containing putative origin of replication of *B. anthracis* were inserted into a temperature-sensitive shuttle vector producing the curing recombinant plasmids. One fragment successfully eliminated pXO1 from vaccine strain *B. anthracis* A16R, and this result was repeated in wild type strain *B. anthracis* A16. Our results suggest that replication is not initiated at the putative pXO1 replication origins reported by Robertson et al. [24] and Tinsley et al. [25]. Robertson et al. investigated the replication origin of *B. anthracis* by transformation of a shuttle vector containing ORF72 to ORF80 of pXO1 into *B. subtilis*. However, replication of the shuttle vector in *B. subtilis* is not sufficient to infer its replication in *B. anthracis*. Tinsley et al. cloned several regions of plasmid pXO1 into the pBSCm plasmid of *E. coli* and then introduced these recombinant plasmids into *B. anthracis*. Because pBSCm is unable to replicate in *B. anthracis*, replication of the recombinant plasmids in *B. anthracis* suggested that the DNA fragments inserted into pBSCm contained the replication origin of pXO1. However, DNA regions that allow pBSCm to replicate in *B. anthracis* may not initiate replication of pXO1 in *B. anthracis*. After deleting these regions from pXO1, the plasmid should be tested in *B. anthracis*to determine whether it loses the ability to replicate.

Our results are in agreement with those of Andrei et al. [26] who characterized the pXO1 minimal replicon by DNA deletion. Andrei et al. reported that deleting a 5950-bp region of pXO1 prevented replication. We successfully used a 4848-bp DNA fragment within the 5950-bp region to cure pXO1 from *B. anthracis* using plasmid incompatibility. Our findings indicate that this fragment is the actual replication origin of plasmid pXO1.

In summary, we cured the plasmid pXO1 from *B. anthracis* A16R and A16 using plasmid incompatibility. This result has provided addition information about the replication and partitioning of pXO1. In addition, we developed a specific and reliable method to generate plasmid-cured strains of *B. anthracis* without inducing spontaneous mutations in the chromosomal DNA.

AUTHOR CONTRIBUTIONS

Conceived and designed the experiments: XKL HLW. Performed the experiments: DSW HGW ELF LZ. Analyzed the data: XKL LZ. Contributed reagents/materials/analysis tools: HGW ELF. Wrote the paper: XKL.

REFERENCES

1. Hanna PA (1998) Anthrax pathogenesis and host response. Curr Top Microbiol Immunol 225: 13–35.PA Hanna1998Anthrax pathogenesis and host response.Curr Top Microbiol Immunol2251335
2. Koehler TM (2002) Bacillus anthracis genetics and virulence gene regulation. Curr Top Microbiol Immunol 271: 143–164. TM Koehler2002Bacillus anthracis genetics and virulence gene regulation.Curr Top Microbiol Immunol271143164
3. Mock M, Fouet A (2001) Anthrax. Annu Rev Microbiol 55: 647–671.M. MockA. Fouet2001Anthrax.Annu Rev Microbiol55647671
4. Guidi-Rontani C, Pereira Y, Ruffie S, Sirard JC, Weber-Levy M, et al. (1999) Identification and characterization of a germination operon on the virulence plasmid pXO1 of Bacillus anthracis. Mol Microbiol 33: 407–414.C. Guidi-RontaniY. PereiraS. RuffieJC SirardM. Weber-Levy1999Identification and characterization of a germination operon on the virulence plasmid pXO1 of Bacillus anthracis.Mol Microbiol33407414
5. Bourgogne A, Drysdale M, Hilsenbeck SG, Peterson SN, Koehler TM (2003) Global effects of virulence gene regulators in a Bacillus anthracis strain with both virulence plasmids. Infect Immun 71: 2736–2743.A. BourgogneM. DrysdaleSG HilsenbeckSN PetersonTM Koehler2003Global effects of virulence gene regulators in a Bacillus anthracis strain with both virulence plasmids.Infect Immun7127362743
6. Pannucci J, Okinaka RT, Sabin R, Kuske CR (2002) Bacillus

anthracis pXO1 plasmid sequence conservation among closely related bacterial species. J Bacteriol 184: 134–141.J. PannucciRT OkinakaR. SabinCR Kuske2002Bacillus anthracis pXO1 plasmid sequence conservation among closely related bacterial species.J Bacteriol184134141

7. Green BD, Battisti L, Koehler TM, Thorne CB, Ivins BE (1985) Demonstration of a capsule plasmid in Bacillus anthracis. Infect Immun 49: 291–297.BD GreenL. BattistiTM KoehlerCB ThorneBE Ivins1985Demonstration of a capsule plasmid in Bacillus anthracis. Infect Immun49291297

8. Makino S, Uchida I, Terakado N, Sasakawa C, Yoshikawa M (1989) Molecular characterization and protein analysis of the cap region, which is essential for encapsulation in Bacillus anthracis. J Bacteriol 171: 722–730.S. MakinoI. UchidaN. TerakadoC. SasakawaM. Yoshikawa1989Molecular characterization and protein analysis of the cap region, which is essential for encapsulation in Bacillus anthracis.J Bacteriol171722730

9. Uchida I, Sekizaki T, Hashimoto K, Terakado N (1985) Association of the encapsulation of Bacillus anthracis with a 60 megadalton plasmid. J Gen Microbiol 131: 363–367.I. UchidaT. SekizakiK. HashimotoN. Terakado1985Association of the encapsulation of Bacillus anthracis with a 60 megadalton plasmid.J Gen Microbiol131363367

10. Uchida I, Makino S, Sasakawa C, Yoshikawa M, Sugimoto C, et al. (1993) Identification of a novel gene, dep, associated with depolymerization of the capsular polymer in Bacillus anthracis. Mol Microbiol 9: 487–496.I. UchidaS. MakinoC. SasakawaM. YoshikawaC. Sugimoto1993Identification of a novel gene, dep, associated with depolymerization of the capsular polymer in Bacillus anthracis.Mol Microbiol9487496

11. Vietri NJ, Marrero R, Hoover TA, Welkos SL (1995) Identification and characterization of a trans-activator involved in the regulation of encapsulation by Bacillus anthracis. Gene 152: 1–9.NJ VietriR. MarreroTA HooverSL Welkos1995Identification and characterization of a trans-activator involved in the regulation of encapsulation by Bacillus anthracis.Gene15219

12. Brana H, Benada O, Navratil O, Cejka K, Hubacek J (1983) Stability of the hybrid plasmid pIM138 and its curing by some eliminating agents. Folia Microbiol (Praha) 28: 441–445.H. BranaO. BenadaO. NavratilK. CejkaJ. Hubacek1983Stability of the hybrid plasmid pIM138 and its curing by some eliminating agents.Folia Microbiol (Praha)28441445
13. El-Mansi M, Anderson KJ, Inche CA, Knowles LK, Platt DJ (2000) Isolation and curing of the Klebsiella pneumoniae large indigenous plasmid using sodium dodecyl sulphate. Res Microbiol 151: 201–208.M. El-MansiKJ AndersonCA IncheLK KnowlesDJ Platt2000Isolation and curing of the Klebsiella pneumoniae large indigenous plasmid using sodium dodecyl sulphate.Res Microbiol151201208
14. Mesas JM, Rodriguez MC, Alegre MT (2004) Plasmid curing of Oenococcus oeni. Plasmid 51: 37–40.JM MesasMC RodriguezMT Alegre2004Plasmid curing of Oenococcus oeni.Plasmid513740
15. Keyhani J, Keyhani E, Attar F, Haddadi A (2006) Sensitivity to detergents and plasmid curing in Enterococcus faecalis. J Ind Microbiol Biotechnol 33: 238–242.J. KeyhaniE. KeyhaniF. AttarA. Haddadi2006Sensitivity to detergents and plasmid curing in Enterococcus faecalis.J Ind Microbiol Biotechnol33238242
16. Spengler G, Molnar A, Schelz Z, Amaral L, Sharples D, et al. (2006) The mechanism of plasmid curing in bacteria. Curr Drug Targets 7: 823–841.G. SpenglerA. MolnarZ. SchelzL. AmaralD. Sharples2006The mechanism of plasmid curing in bacteria.Curr Drug Targets7823841
17. Hovi M, Sukupolvi S, Edwards MF, Rhen M (1988) Plasmid-associated virulence of Salmonella enteritidis. Microb Pathog 4: 385–391.M. HoviS. SukupolviMF EdwardsM. Rhen1988Plasmid-associated virulence of Salmonella enteritidis.Microb Pathog4385391
18. Poppe C, Gyles CL (1988) Tagging and elimination of plasmids in Salmonella of avian origin. Vet Microbiol 18: 73–87.C. PoppeCL Gyles1988Tagging and elimination of plasmids in Salmonella of avian origin.Vet Microbiol187387
19. Novick RP (1987) Plasmid incompatibility. Microbiol Rev 51:

381–395.RP Novick1987Plasmid incompatibility.Microbiol Rev51381395

20. Ni B, Du Z, Guo Z, Zhang Y, Yang R (2008) Curing of four different plasmids in Yersinia pestis using plasmid incompatibility. Lett Appl Microbiol 47: 235–240.B. NiZ. DuZ. GuoY. ZhangR. Yang2008Curing of four different plasmids in Yersinia pestis using plasmid incompatibility.Lett Appl Microbiol47235240
21. Uraji M, Suzuki K, Yoshida K (2002) A novel plasmid curing method using incompatibility of plant pathogenic Ti plasmids in Agrobacterium tumefaciens. Genes Genet Syst 77: 1–9.M. UrajiK. SuzukiK. Yoshida2002A novel plasmid curing method using incompatibility of plant pathogenic Ti plasmids in Agrobacterium tumefaciens.Genes Genet Syst7719
22. Tinsley E, Naqvi A, Bourgogne A, Koehler TM, Khan SA (2004) Isolation of a minireplicon of the virulence plasmid pXO2 of Bacillus anthracis and characterization of the plasmid-encoded RepS replication protein. J Bacteriol 186: 2717–2723.E. TinsleyA. NaqviA. BourgogneTM KoehlerSA Khan2004Isolation of a minireplicon of the virulence plasmid pXO2 of Bacillus anthracis and characterization of the plasmid-encoded RepS replication protein.J Bacteriol18627172723
23. Wang H, Liu X, Feng E, Zhu L, Wang D, et al. (2011) Curing the Plasmid pXO2 from Bacillus anthracis A16 Using Plasmid Incompatibility. Curr Microbiol 62: 703–709.H. WangX. LiuE. FengL. ZhuD. Wang2011Curing the Plasmid pXO2 from Bacillus anthracis A16 Using Plasmid Incompatibility.Curr Microbiol62703709
24. Okinaka RT, Cloud K, Hampton O, Hoffmaster AR, Hill KK, et al. (1999) Sequence and organization of pXO1, the large Bacillus anthracis plasmid harboring the anthrax toxin genes. J Bacteriol 181: 6509–6515.RT OkinakaK. CloudO. HamptonAR HoffmasterKK Hill1999Sequence and organization of pXO1, the large Bacillus anthracis plasmid harboring the anthrax toxin genes.J Bacteriol18165096515
25. Tinsley E, Khan SA (2006) A novel FtsZ-like protein is involved in replication of the anthrax toxin-encoding pXO1 plasmid in

Bacillus anthracis. J Bacteriol 188: 2829–2835.E. TinsleySA Khan2006A novel FtsZ-like protein is involved in replication of the anthrax toxin-encoding pXO1 plasmid in Bacillus anthracis.J Bacteriol18828292835

26. Pomerantsev AP, Camp A, Leppla SH (2009) A new minimal replicon of Bacillus anthracis plasmid pXO1. J Bacteriol 191: 5134–5146. AP PomerantsevA. CampSH Leppla2009A new minimal replicon of Bacillus anthracis plasmid pXO1.J Bacteriol19151345146
27. Smith K, Youngman P (1992) Use of a new integrational vector to investigate compartment-specific expression of the Bacillus subtilis spoIIM gene. Biochimie 74: 705–711.K. SmithP. Youngman1992Use of a new integrational vector to investigate compartment-specific expression of the Bacillus subtilis spoIIM gene.Biochimie74705711
28. Shatalin KY, Neyfakh AA (2005) Efficient gene inactivation in Bacillus anthracis. FEMS Microbiol Lett 245: 315–319.KY ShatalinAA Neyfakh2005Efficient gene inactivation in Bacillus anthracis.FEMS Microbiol Lett245315319

CITATIONS

CHAPTER 1

Kuenne C, Voget S, Pischimarov J, Oehm S, Goesmann A, Daniel R, et al. (2010) Comparative Analysis of Plasmids in the Genus Listeria. PLoS ONE 5(9): e12511. https://doi.org/10.1371/journal.pone.0012511

CHAPTER 2

Constante M, Grünberg R, Isalan M (2011) A Biobrick Library for Cloning Custom Eukaryotic Plasmids. PLoS ONE 6(8): e23685. https://doi.org/10.1371/journal.pone.0023685.

CHAPTER 3

Watve MM, Dahanukar N, Watve MG (2010) Sociobiological Control of Plasmid Copy Number in Bacteria. PLoS ONE 5(2): e9328. https://doi.org/10.1371/journal.pone.0009328

CHAPTER 4

Gong CC, Klumpp S (2017) Modeling sRNA-Regulated Plasmid Maintenance. PLoS ONE 12(1): e0169703. https://doi.org/10.1371/journal.pone.0169703

CHAPTER 5

Klümper U, Droumpali A, Dechesne A and Smets BF (2014) Novel assay to measure the plasmid mobilizing potential of mixed microbial communities. Front. Microbiol. 5:730. doi: 10.3389/fmicb.2014.00730

CHAPTER 6

Funnell BE (2016) ParB Partition Proteins: Complex Formation and Spreading at Bacterial and Plasmid Centromeres. Front. Mol. Biosci. 3:44. doi: 10.3389/fmolb.2016.00044.

CHAPTER 7

Wood DO, Hines A, Tucker AM, Woodard A, Driskell LO, Burkhardt NY, et al. (2012) Establishment of a Replicating Plasmid in Rickettsia prowazekii. PLoS ONE 7(4): e34715. https://doi.org/10.1371/journal.pone.0034715

CHAPTER 8

Okazaki S, Noisangiam R, Okubo T, Kaneko T, Oshima K, Hattori M, et al. (2015) Genome Analysis of a Novel Bradyrhizobium sp. DOA9 Carrying a Symbiotic Plasmid. PLoS ONE 10(2): e0117392. https://doi.org/10.1371/journal.pone.0117392

CHAPTER 9

Nicholson BA, West AC, Mangiamele P, Barbieri N, Wannemuehler Y, Nolan LK, et al. (2016) Genetic Characterization of ExPEC-Like Virulence Plasmids among a Subset of NMEC. PLoS ONE 11(1): e0147757. https://doi.org/10.1371/journal.pone.0147757

CHAPTER 10

Shintani M, Sanchez ZK and Kimbara K (2015) Genomics of microbial plasmids: classification and identification based on replication and transfer systems and host taxonomy. Front. Microbiol. 6:242. doi: 10.3389/fmicb.2015.00242.

CHAPTER 11

Ajamete Kaykas and Randall T Moon, A plasmid-based system for expressing small interfering RNA libraries in mammalian cells, https://doi.org/10.1186/1471-2121-5-16.

CHAPTER 12

Liu X, Wang D, Wang H, Feng E, Zhu L, Wang H (2012) Curing of Plasmid pXO1 from Bacillus anthracis Using Plasmid Incompatibility. PLoS ONE 7(1): e29875. https://doi.org/10.1371/journal.pone.002987

INDEX